RELIABILITY AND OPTIMIZATION OF STRUCTURAL SYSTEMS, V

IFIP Transactions B: Applications in Technology

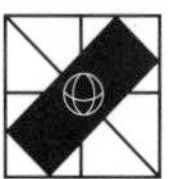

International Federation for Information Processing

Technical Committee 5:

Computer Applications in Technology

IFIP Transactions Abstracted/Indexed in:
INSPEC Information Services
Index to Scientific & Technical Proceedings®
CompuMath Citation Index®, Research Alert™, and SciSearch®

B-12

RELIABILITY AND OPTIMIZATION OF STRUCTURAL SYSTEMS, V

Proceedings of the IFIP WG7.5 Fifth Working Conference on
Reliability and Optimization of Structural Systems
Takamatsu, Kagawa, Japan, 24-26 March, 1993

Edited by

P. THOFT-CHRISTENSEN
Department of Building Technology & Structural Engineering
University of Aalborg
Aalborg, Denmark

H. ISHIKAWA
Department of Information Science
Kagawa University
Takamatsu, Kagawa, Japan

1993

NORTH-HOLLAND
AMSTERDAM • LONDON • NEW YORK • TOKYO

ELSEVIER SCIENCE PUBLISHERS B.V.
Sara Burgerhartstraat 25
P.O. Box 211, 1000 AE Amsterdam, The Netherlands

Keywords are chosen from the ACM Computing Reviews Classification System, ©1991, with permission.
Details of the full classification system are available from
ACM, 11 West 42nd St., New York, NY 10036, USA.

ISBN: 0 444 81605 4
ISSN: 0926-5481

This book is printed on acid-free paper.

Printed in The Netherlands

PREFACE

This proceedings volume contains 32 papers presented at the 5th Working Conference on "Reliability and Optimization of Structural Systems", held at the the General Education & Training Center Shikoku Electric Power Co., Takamatsu-Shi, Kagawa, Japan, on March 24 - 26, 1993. The Working Conference was organised by the IFIP (International Federation for Information Processing) Working Group 7.5 of Technical Committee 7 and was the fifth in a series, following similar conferences held at the University of Aalborg, Denmark, May 1987, at the Imperial College, London, UK, September 1988, at the University of California, Berkeley, California, USA, March 1990, and at the Technical University of Munich, Germany, September 1991. The Working Conference was attended by 126 participants from 10 countries.

The objectives of Working Group 7.5 are:

- to promote modern structural systems optimization and reliability theory,
- to advance international cooperation in the field of structural system optimization and reliability theory,
- to stimulate research, development and application of structural system optimization and reliability theory,
- to further the dissemination and exchange of information on reliability and optimization of structural systems
- to encourage education in structural system optimization and reliability theory.

At present the members of the Working Group are:

A. H.-S. Ang, U.S.A.
G. Augusti, Italy
M. J. Baker, UK
P. Bjerager, Norway
C. A. Cornell, U.S.A.
R. B. Corotis, U.S.A.
A. Der Kiureghian, U.S.A.
O. Ditlevsen, Denmark
L. Esteva, Mexico
D. M. Frangopol, U.S.A.
H. Furuta, Japan
S. Garribba, Italy
M. Grigoriu, U.S.A.
M. Grimmelt, Germany
H. Ishikawa, Japan
N. C. Lind, Canada
H. O. Madsen, Denmark
R. E. Melchers, Australia
F. Moses, U.S.A.
Y. Murotsu, Japan
A. S. Nowak, U.S.A.
R. Rackwitz, Germany
P. Sniady, Poland
C. G. Soares, Portugal
J. D. Sørensen, Denmark
P. Thoft-Christensen (chairman), Denmark
Y. K. Wen, U.S.A.

Members of the Organizing Committee were:

P. Bjerager, Norway
D. M. Frangopol, U.S.A.
H. Furuta (co-chairman), U.S.A.
Y. Murotsu (co-chairman), Japan
R. Rackwitz, Germany
P. Thoft-Christensen (co-chairman), Denmark

Members of the Local Organizing Committee were:

H. Ishikawa (chairman), Japan
H. Furuta, Japan
Y. Murotsu, Japan

The Working Conference received financial support from IFIP, Kagawa University, several Japanese organizations and the University of Aalborg.

On behalf of WG 7.5 and TC-7 the co-chairmen of the Conference would like to express their sincere thanks to the sponsors, to the members of the Organizing Committee for their valuable assistance, and to the authors for their contributions to these proceedings. Special thanks are due to Mrs. Kirsten Aakjær, University of Aalborg, for her efficient work as conference secretary.

June 1993

P. Thoft-Christensen H. Ishikawa

CONTENTS

Lectures

Short Presentations

Reliability and Optimization of Structural Systems, V (B-12)
P. Thoft-Christensen and H. Ishikawa (Editors)
Elsevier Science Publishers B.V. (North-Holland)

Supply Reliability of Electric Power Systems

Y. Minato

Shikoku Research Institute Inc.
2109 Yashimanishimachi, Takamatsu City, 761-01 Japan

1. INTRODUCTION

In every country, assurance of stable supply of energy is essential to the sound growth of the economy and to the improvement in the quality of national life.

Japan is not favored with domestic energy resources, and is dependent for more than 80 percents of primary energy supply on imports. Hence assurance of the stability of energy demand and supply is very important to Japan.

With regard to the circumstances relating to utilization of energy, the global environmental problems have been closed up in recent years in addition to the issues of increase of the world population and resource restraints. In effectively assuring energy supply, we must pay due attention to the relationships of these issues and take appropriate alleviating measures.

In such circumstances, the electric utility industry, playing a role in the energy industry, has been positively promoting, on the supply side, research and development of new power generation methods such as fuel cell for higher energy efficiency. On the side of electricity utilization, the industry has been positively developing technologies for higher energy efficiency, which effectively utilize characteristics unique to electricity, such as high-efficiency heat pumps and use of far-infrared radiation.

With regard to the operation of plants and equipment, the highest priority has been given to the assurance of safety of nuclear power generation. A variety of carefully-thought-out safety measures have been taken to free the minds of local residents from anxiety.

The power generation plants have been growing in size partly due to the difficulties in siting new power sources. As a result, if an accident happens on a power plant including the transmission facilities, the accident will have a very large impact. The industry, therefore, has been energetically developing techniques for preventing and predicting accidents on facilities, and also for preventing human errors during operation.

This paper outlines the approach to reserve supply capability which is the key to the assurance of power supply reliability, and then the records of supply reliability of the distribution systems in Japan. We wish this paper serves to any extent in increasing the interest in the electric utility industry.

2. POWER DEMAND

Electricity is essential to the daily life and to the economic and industrial activities. With the advancement of electronics, electricity has been expanding its applications, through diversification of life style, factory automation of industries, and advanced communication in the society.

First, we examine, in outline, how the power demand in Japan has developed to the present state. There are ten general electric utilities in Japan. These companies, except Okinawa Electric Power Company, interconnect their electric power systems with interconnection ties.

2.1 Electric Energy Demand (kWH)

In Japan, the demand for power grew at two-figure rates in the high economic growth period (from 1955 to 1965); the rates were greater than the economic growth rates of the period. At the time of the first oil crisis (fiscal 1974), the demand for power recorded the first negative growth rate in the postwar period. In and after 1975, the power demand showed a steady growth at rates of an order of 3 % partly due to the effects of the second oil crisis.

With the two oil crises as a momentum, energy saving was advanced in Japan. The heavy-electric-power-consuming industries disappeared, and the industrial structure shifted towards less-electric-power-consuming industries.

Since 1981 the power demand established the domestic demand-pull type growth, and marked the growth of about 5 % almost comparable to that of the economic growth.

The circumstances of the above-mentioned period are shown in Figures 1 and 2. In terms of electric energy per unit GNP, the power demand for industry showed a tendency to decrease since the oil crises. On the other hand, the power demand for non-industrial use has been running steady.

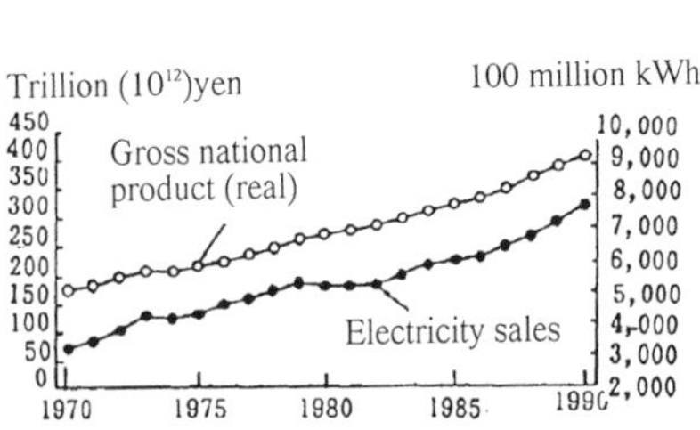

Figure 1 GNP and power demand

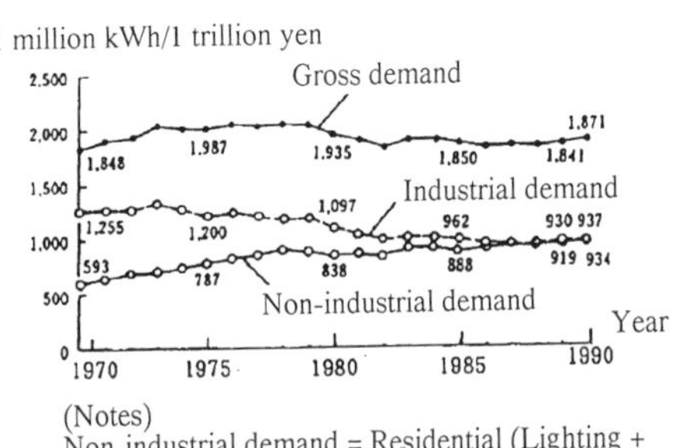

(Notes)
Non-industrial demand = Residential (Lighting + Night-only) + Commercial + Low-voltage + Railway + Public and miscellaneous

Industrial demand = Gross demand − (Non-industrial demand)

Figure 2 Electric energy per unit GNP

In recent years, however, the economy has been hovering at low levels due to the collapse of the bubbles in the securities and land markets. Hence the growth rates of power demand have been decelerated, and are expected to develop at an order of 2 percents.

2.2 Maximum Electricity Demand (kW)

In the period from 1955 to 1965, the annual maximum electricity demand occurred in the winter; in December or January.

The demand for cooling began to increase rapidly in the latter half of the 1950s. At the same time, the demands for lighting and heating also increased significantly. As a result, the annual load factor increased.

With the rapid increase in the demand for cooling, some electric utilities began to mark the annual maximum demand during the daytime in August around 1966. Since 1968 the total maximum demand of all the utilities has shifted to the daytime (14:00 to 15:00) in the summer (July or August).

As a result of the increase in the demand for cooling, the maximum power began to be influenced significantly by the weather conditions including the atmospheric temperature.

Moreover, partly due to the growth of processing and assembly industries that operate during the daytime, the maximum demand marked a mean growth rate of 10 % or over in the period from 1965 to 1975. Since 1990, load-levelling measures have been taken to suppress this tendency, such as the introduction of seasonal and time-of-use rates, and development and diffusion of heat storage type high-efficiency appliances. The maximum demand, however, is increasing steadily.

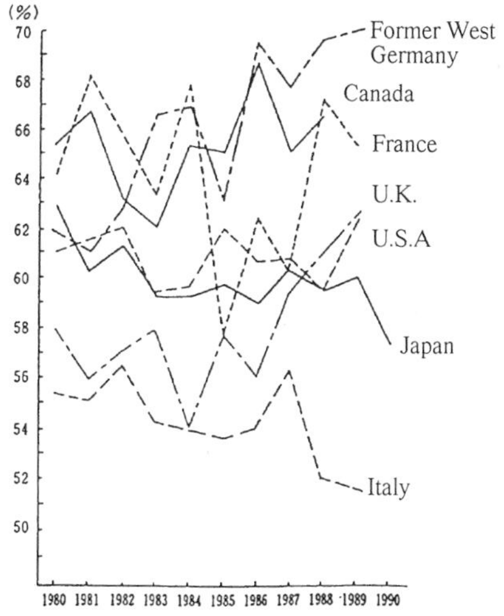

Figure 3 Transition of annual load factors of some countries

3. PLAN OF POWER DEMAND AND SUPPLY

The plan of power demand and supply is an important plan which considers the balance between power demand and supply capability and assures stable supply of power.

In a short-term plan of power demand and supply, the installed generating capacity is considered as a given condition. Hence the plan is developed to minimize the operating cost (mainly the fuel expenses) while maintaining the supply reliability. On the other hand, the long-term plan of power demand and supply is examined to define a construction plan of power sources and transmission and distribution facilities that allow the minimization of the total of the new capital expenditure and the fuel expenses over a long period while maintaining the supply reliability.

In the following, we outline our approach to the plan of power demand and supply, and the direction of power source development in Japan.

3.1 Approach to Plan of Power Demand and Supply

Figure 4 shows the hourly power demands by a daily load curve. A supply capability is planned to match this load curve. First, pumped storage hydropower, reservoir type hydropower and regulating reservoir type hydropower are combined to match the peak portion of the load curve. Then the flat supply capabilities of the runoff-river plants, etc. are subtracted.

The remaining load is called the balance load. Nuclear power and thermal power capabilities are appropriated to the base portion of the balance load by the ascending order of kWH cost to develop a power generation plan. This is the basic approach to the plan of power demand and supply. It should be noted that the plan of hydropower must meet various restraints such as the annual operation of plan of each reservoir and the minimum discharge which maintains each river.

Thermal and nuclear power units need periodic inspection. Hence a reserve power must be held to assure stable power supply when some units are in scheduled outage.

Generally speaking, when a day of the maximum demand falls on a day of drought, the reserve power will be reduced. To grasp such conditions, the reserve power to be used in the plan of power demand and supply is determined by combining the average demand of top three and the hydropower supply capability of the Vth flow point (average of the lowest five days of the series and parallel duration curves).

An example of a monthly plan of power demand and supply for one year is shown in Figure 5.

If a short-term plan of power demand and supply indicates that the reserve capability will fall below the target value, a planned interchange of power from another utility will be required. If

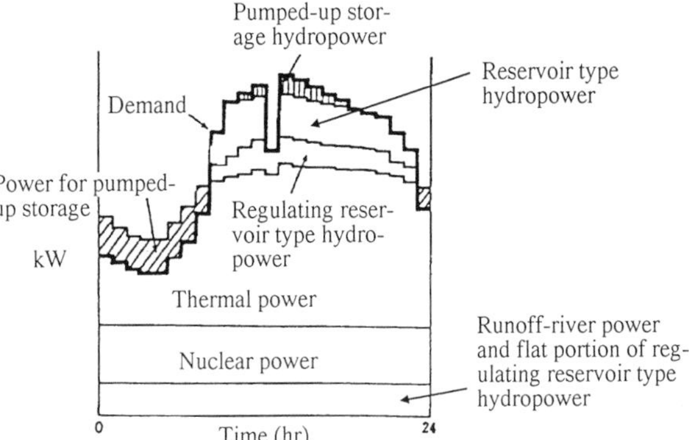

Figure 4 Daily load curve and power generation curves of various supply capabilities

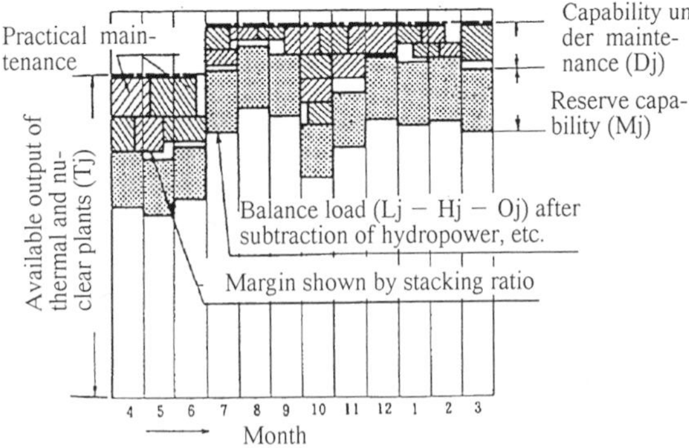

Figure 5 Reserve capability and thermal and nuclear capability under maintenance by month

a long-term plan of power demand and supply indicates a deficiency, it will be necessary to consider installation of a new power source.

3.2 Direction of Power Source Development in Japan

According to the electric power development plan of 1991, the generation mix and the configuration of generated energy up to the year 2000 of Japan will develop as shown in Figure 6.

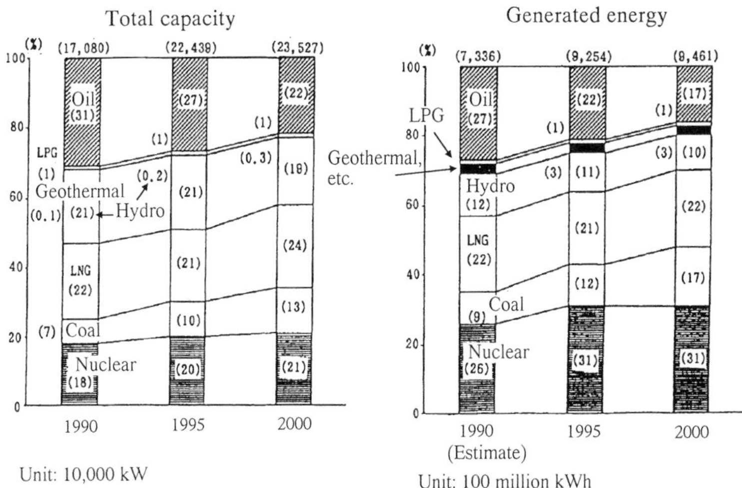

Figure 6 Generation mix and configuration of generated energy by year

As can be seen clearly from the configuration of power plants, the future power sources to be developed will be nuclear power, coal-thermal power, hydropower, and geothermal power.

Since Japan will not develop oil-thermal power plants according to an agreement among the industrialized nations of the International Energy Agency, the plants will be limited to the existing ones. Hence the weight of the oil-thermal power plants will decrease year after year.

As for the hydropower, potential sites for development have been depleted. Hence the development of pure pumped storage plants has a large weight.

With regard to the generated energy of the year 2000, the nuclear power generation has the largest share because the fuel cost of nuclear power is cheapest and it bears the base load. Nuclear power is followed by LNG-thermal power. Then thermal power of oil and coal and hydropower have almost the same shares.

4. APPROACH TO RESERVE CAPABILITY

Reserve capability is essential to the stable supply of electricity. In determining a reserve capability, it is necessary to take the following into account: the supply capability that meets the maximum demand, the demand variation, the hydropower supply capability at the time of drought, and uncertain events such as forced outages of thermal or nuclear power units.

Moreover, various power units require a periodic inspection to maintain a certain level of reliability. Every unit must be shut down for a certain period to make repair works and works for restoring degraded functions.

The difference between the installed capability minus this scheduled outage power and the average demand of top three is normally called the "reserve capability." If this reserve capability is excessive, the generating cost will rise, and it is not economical although the supply reliability will be improved.

On the other hand, if the reserve capability is deficient, the generating cost will be lower, but the customers will be forced to tolerate service interruption and disturbance of productive activities in case of a contingent accident.

These points must be considered in establishing an appropriate reserve capability and the corresponding reliability. This reliability has been expressed up to now in a unit of "expected number of days with deficient capability." The reliability has been set at about 0.3 day with deficient capability per year over many years.

The reserve capability corresponding to this reliability or 0.3 expected day with deficient capability per year is the target reserve capability. The ratio of this reserve capability to the average demand of top three is called the target reserve capability ratio.

This target reserve capability ratio must be adjusted whenever a significant change occurs in the power system; such as a big change in the pattern of demand, a change in the generation mix, a change in the system interconnection or its capacity.

4.1 Supply Reliability of an Isolated System

When an electric utility operates its power system without any interconnection tie with another utility's system, the former system is called an isolated system.

The main factors for the calculation of the supply reliability of an isolated system include the demand variation probability, the trouble probabilities of thermal and nuclear power units, and the flow variation probability of hydropower plants. Such probabilities are treated as follows.

It is assumed that all transmission lines and distribution lines are completely interconnected within the isolated system, and there will be no difficulties in supply at the time of a trouble or troubles.

1) Demand Variation Error

In operating the power system, the electric utility makes a power generation plan on the previous day by estimating the maximum demand of the next day.

Based on the relationship between this estimated maximum power and the actual maximum power, the demand variation error is defined by

Demand variation error = [(Actual maximum power − Estimated maximum power)/Estimated maximum power] × 100 (%).

This variation error depends on some factors including the change in atmospheric temperature, and Table 1 shows the results of some surveys. The distribution of demand variation errors is shown in Figure 7.

2) Flow Variation Probabilities

The available power of hydropower generation is computed from the river flow records from 1942 to the present. The flow variation probabilities are included in the records.

Table 1
Standard deviation and cumulative probability

	Standard deviation (σ)	Cumulative probability of error rate of 6 %
1 year	2.7	98.8
8 months	2.8	98.3
12 months	2.4	99.3

In balancing the capability against the maximum power, the utility uses the hydropower supply capability at the Vth flow point.

The hydropower supply capability is based on an inflow duration curve obtained by halving the addition of the series and parallel duration curves. The representative points are defined as shown in Table 2.

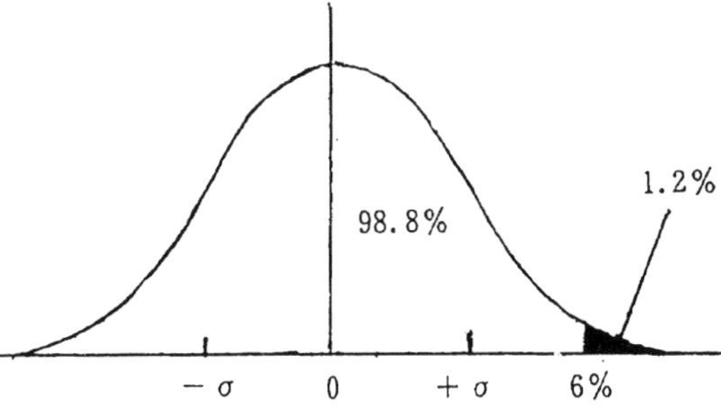

Figure 7 Distribution of demand variation errors

Table 2
Definitions of flow points

Flow point	Available power	Remarks
Ist flow point (the highest flow day)	$\frac{P_{SH5}+P_{PH5}}{2}$	
IInd flow point (plentiful flow day)	P_{PH5}	
IIIrd flow point (ordinary flow day)	Monthly average potential output	Used as the standard of average flow, such as the standard for computing the flow rate.
IVth flow point (drought flow day)	P_{PL5}	
Vth flow point (the lowest flow day)	$\frac{P_{SL5}+P_{PL5}}{2}$	Normally used in examining the maximum power balance. Usually called L5.

(Notes)
P_{SH5} = Average of top five days of the series river inflow duration curve (potential output)
P_{PH5} = Average of top five days of the parallel river inflow duration curve (potential output)
P_{SL5} = Average of bottom five days of the series river inflow duration curve (potential output)
P_{PL5} = Average of bottom five days of the parallel river inflow duration curve (potential output)

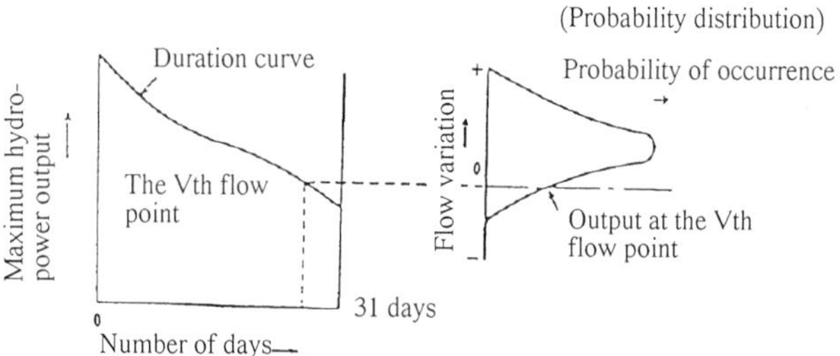

Figure 8 Distribution of flow variation probabilities

3) Forced Outage Probabilities of Hydro, Thermal and Nuclear Power Units

According to a survey of the actual records of all the electric utilities of Japan, the forced outage rates of the hydro, thermal and nuclear power units are as follows:

Hydro power (including pumped storage) .... 0.5 %

Thermal power (also applies to nuclear power)

Unit of 325 MW or over ..... 2.5 %

Unit of 264 MW or less ..... 2.0 %

The rate at the initial start of operation is 5.0 %.

The method of calculation of the forced outage power distribution is identical for hydropower, thermal power and nuclear power. The probability of r units in forced outage out of n units grouped by capacity, P(r), is defined by the following formula using binomial distribution:

$$P(r) = {}_{n}C_{r}\, P^{r} \cdot (1 - P)^{n-r}$$

where P: Forced outage probability; r = 0, 1, 2, 3, ...

In addition to grouping by capacity, the respective forced outage probabilities of hydro, thermal and nuclear power unit groups are determined. Then the forced outage probability distribution for the entire power units is determined by accumulation of the respective probabilities. The results are as shown in Figure 9.

4) Probability Distribution of Overall Output Variation

The overall output variation probability distribution, which integrates the demand variation, flow variation, and forced outage probabilities of various types of power units, may be expressed as shown in Figure 10. At the origin of the overall output variation, the forced outage power is zero and the hydropower output is at the Vth flow point.

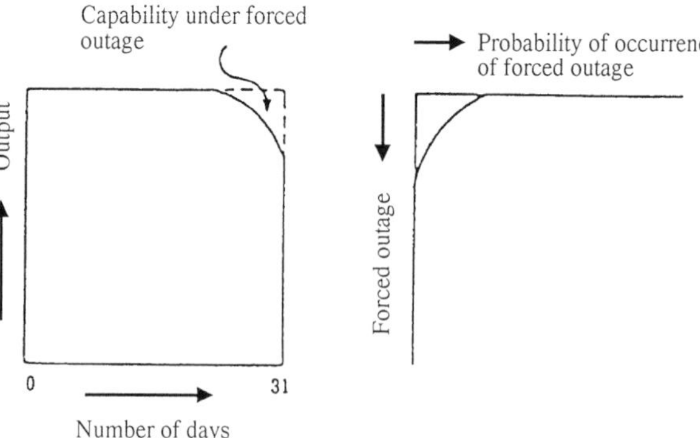

Figure 9 Forced outage capability distribution

5) Relationship Between the Expected Number of Days with Deficient Capability (Reliability) and the Reserve Capability

The expected number of days with deficient capability is as shown in Figure 11 when the average demand of the top three is H_3 and the reserve capability is R. When the overall power capability variation probability P (A_i) is combined, the expected number of days with deficient capability is determined from

Expected number of days with deficient capability = $\Sigma\ E_i \cdot P(A_i)$

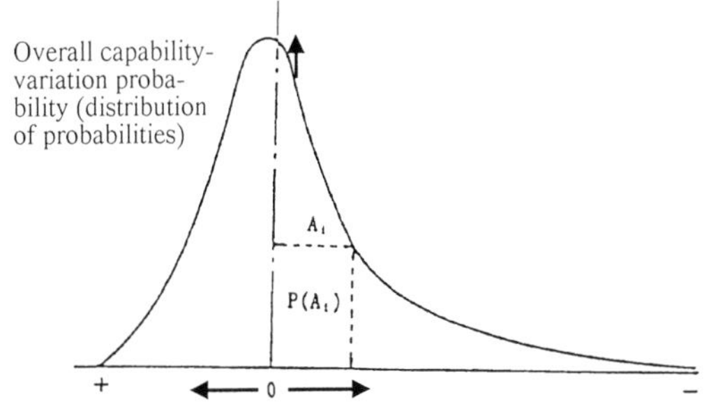

Figure 10 Probability distribution of overall capability variation

When the reserve capability is used as a parameter to compute the expected number of days with deficient capability, the relationship between the expected number of days with deficient capability and the reserve capability is as shown in Fig. 12.

As clearly seen in the diagram, when the expected number of days of deficient capability is reduced, the required reserve capability will increase exponentially.

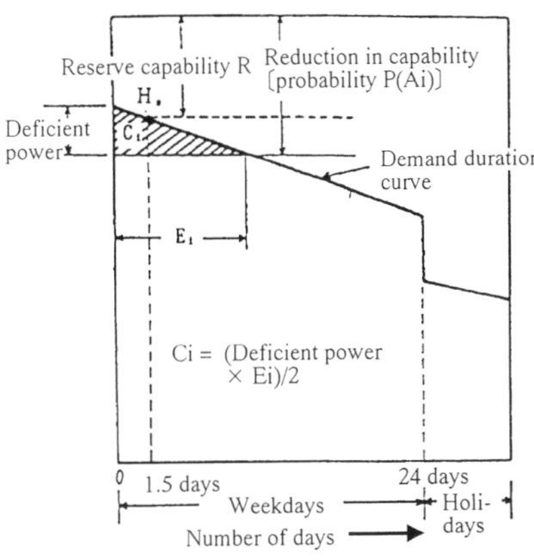

Figure 11 Relation of reserve capability with other factors

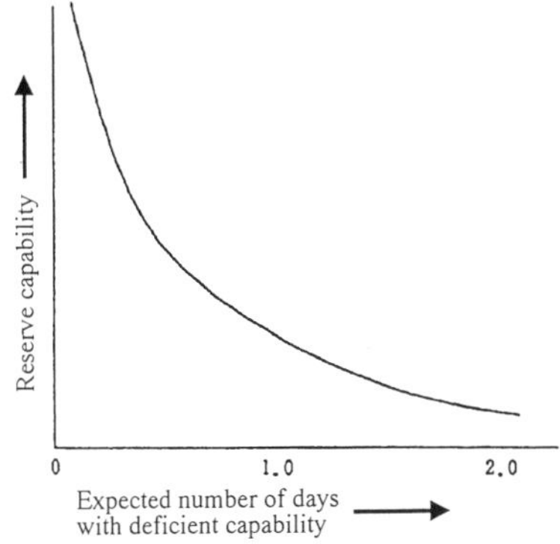

Figure 12 Reserve capability and expected number of days with deficient capability

4.2 Reserve Capability for Interconnected Systems

The reliability or the expected number of days with deficient capability of each power system has been set at 0.3 day/year. We have already outlined the method for computing the reserve capability corresponding to 0.3 day/year for an isolated system. When some power systems are interconnected with each other, non-isochronism of demand, flow and occurrence of trouble are introduced. As a result, the required reserve capability is saved.

This saving partly depends on the capacities of interconnection ties. The reduction is largest when the capacity of each interconnection tie is infinite.

1) Reserve Capability When Interconnection Tie Capacities are Infinite

Suppose three power companies A, B and C are interconnected to each other with infinite-capacity interconnection ties. In this case, to compute the expected number of days with deficient capability for the three companies as a whole, it is not sufficient to simply sum up the average demands of top three and the supply capabilities at the Vth flow point of the respective companies. These factors must be synthesized for each time point.

The relationship between the expected number of days with deficient capability for the three companies and the overall reserve capability thus determined is as shown in Figure 13.

Compare the two cases, one with each company operating its own isolated system and the other with three companies operating interconnected systems, in terms of the expected number of days with deficient capability and the reserve capability. A saving in the reserve capability is $(A_i + B_i + C_i) - T_i$.

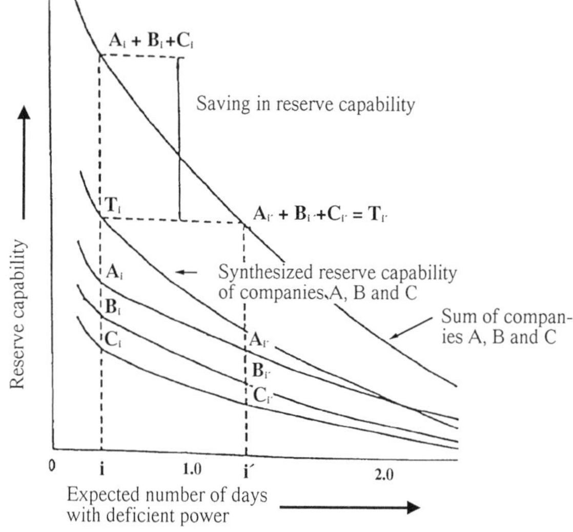

Figure 13 Allocation of reserve capability for an identical expected number of days with deficient capability

This saving results from the non-isochronism of demand and flow. The respective reserve capabilities of the three companies are determined by the point of $T_{i'} = A_{i'} + B_{i'} + C_{i'}$ on the curve. The marginal capabilities are allocated to the three companies in such a way that when their systems are isolated from each other all systems have the same level of reliability.

2) Reserve Capabilities When Interconnection Tie Capacities are Taken into Consideration

Suppose the power systems of three companies A, B and C are interconnected as shown in Figure 14. We will examine the relationship between the reliability and the reserve capability of this case.

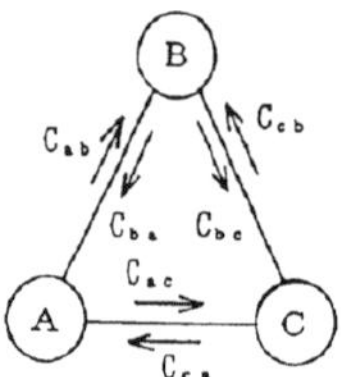

Figure 14 Interconnection ties of three companies

C is an interconnection tie capacity. When power is interchanged with another company, Cab and Cba are not equal to each other.

Suppose the demand, trouble and flow variation cumulative probability distributions of the three companies A, B and C are as shown in Figures 15 and 16. The respective reserve capabilities are assumed. Events for the respective companies are determined by the Monte-Carlo method, and the expected number of days with deficient capability are computed.

In this case, if no company experiences any deficit in the reserve capability, there will be no problem. If one company has a deficit, the other two companies will send powers in proportion to the ratio of their reserve capabilities. If two companies have deficits, the other company will send powers to the two companies in proportion to the ratio of their deficient powers and within the limits of the respective interconnection tie capacities.

The calculation according to the Monte-Carlo method is repeated till the expected number of days with deficient capability converges.

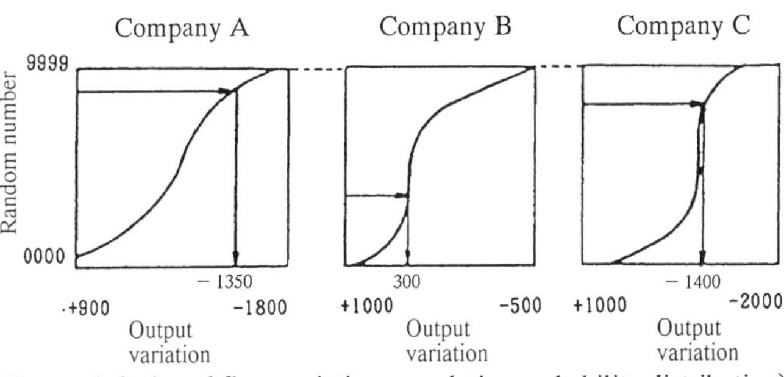

(Demand, fault and flow variation cumulative probability distribution)

Figure 15 Capability distributions of three companies

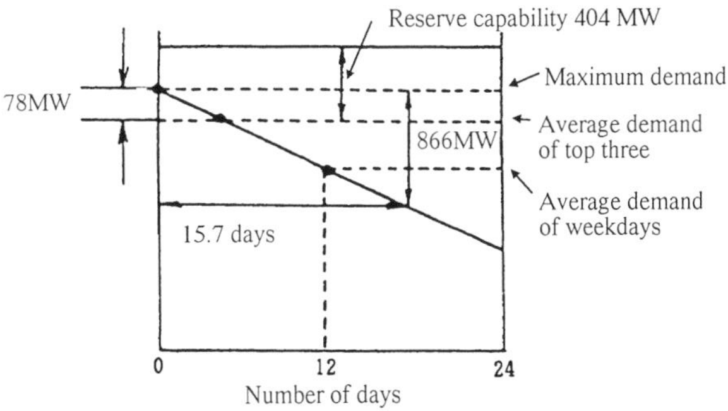

Figure 16 A case of demand duration curve and number of days with deficient capability of the company A

For such interconnected power systems it will take a very long time to use the reserve capability as a parameter and determine a reserve capability corresponding to a reliability or expected number of days with deficient capability of 0.3 day/year. Hence it is a normal practice to use the reserve capability as a parameter, determine the expected number of days with deficient capability corresponding to it, and estimate a reserve capability by developing an empirical formula from the results.

For the above-mentioned interconnected systems the reserve capability corresponding to an expected number of days with deficient capability of 0.3 day/year is from 8 to 10 %. The breakdown of this figure is 7 % for the demand estimation error, flow and contingencies such as forced outages, and 1-3 % for business fluctuation or the like which is difficult to handle by demand estimation.

5. RECORDS OF SUPPLY RELIABILITY OF DISTRIBUTION SYSTEMS

The target supply reliability of power sources is, as explained above, to hold a reserve capability which meets the expected number of days with deficient capability of 0.3 day/year . It is difficult, however, to extend such an approach to the transportation facilities and establish a supply reliability in a unified manner. Analysis and treatment pose difficulties. There are several reasons. Transmission, transformation and distribution differ in function and weight of role. The facilities are diverse, complicated, yet related with each other. Moreover, they are deployed over a wide area and are exposed to the severe natural environment.

Efforts have been made, in the designs of transmission facilities, to enhance their reliability. For instance, transmission lines are provided with overhead ground wires, arrestors, differential insulation, etc. so that, in the case of a double circuit transmission line, power can be transmitted through one circuit even if the other circuit is faulted by lightning. Moreover, assistance can be made by looping operation of transmission lines.

Distribution lines are arranged for quicker detection of a faulted section when a fault occurs on a line due to lightning, etc. Power can be transmitted to the affected distribution line from a sound distribution line.

With such measures, the Japanese electric utilities have been attempting to avoid service interruption and reduce the service interruption areas.

The number of faults on transportation facilities of the Japanese electric utilities in fiscal 1989 is as shown in Table 3.

Table 3
Number of faults on transportation facilities in fiscal 1989

Facility / Cause	Transmission		Transformation		Distribution	
	Number of incident	%	Number of incident	%	Number of incident	%
Defective facility	2	1	18	31	307	5
Defective maintenance	2	1	15	26	686	12
Natural phenomenon	(185) 227	(52) 64	(6) 9	(10) 15	(1,667) 2,800	(28) 46
By accident and by design	49	14	1	1	688	12
Accidental contact	46	13	7	12	853	14
Other faults	8	2	1	1	20	0
Miscellaneous	20	5	8	14	684	11
Total	354	100	59	100	6,038	100

Notes: Data of the ten electric power companies and Electric Power Development Co., Ltd.
Figures in parentheses indicate faults caused by lightning.

Of these troubles on the transmission facilities, about one half of those occurred on transmission and distribution facilities were caused by natural phenomena. In particular, the most frequent cause of the troubles on the transmission facilities is lightning.

As mentioned before, such troubles do not directly result in service interruption. Service interruption can be avoided by a variety of system assistance measures.

In view of such trouble conditions, the utilities define, for the distribution system, the supply reliability with the interruption frequency per customer and the interruption duration.

Forced interruption frequency = Σ (Number of affected customers × Interruption frequency)/Total number of customers.

Forced interruption duration = Σ (Number of affected customers × Interruption duration)/Total number of customers.

Interruption for work such as distribution work is also managed by defining the work interruption frequency per customer and the work interruption duration as follows:

Work interruption frequency = Σ (Number of affected customers × Interruption frequency)/Total number of customers.

Work interruption duration = Σ (Number of affected customers × Interruption duration)/Total number of customers.

The actual interruption frequencies and interruption durations are shown in Figures 17 and 18.

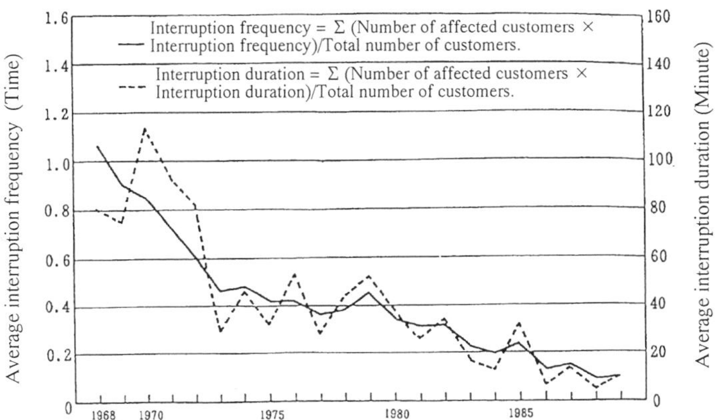

Figure 17 Annual forced interruption frequency and duration per customer (interruptions caused by troubles on the power sources are excluded)

Interruptions caused by the upstream facilities of the system including power source, transmission and transformation are not included in the records of interruption frequencies and durations. The number of such interruptions is very small relative to those relating to distribution.

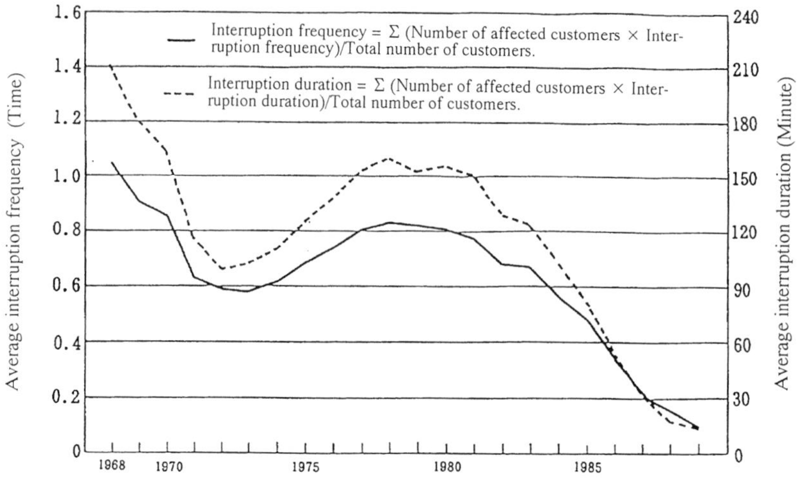

Figure 18 Annual work interruption frequency and duration per customer (interruptions caused by troubles on the power sources are excluded)

Forced interruptions and work interruptions of some countries using indicators similar to those of the distribution system reliability of Japan are as shown in Table 4. It can be seen that the reliability in Japan is relatively high.

Table 4
Present state of reliability in some nations (fiscal 1986)

	Country	Japan	U.S.A 1	U.S.A 2	U.S.A 3	U.S.A 4	U.S.A 5	U.S.A 6	Canada 1	U.K.	France
Forced interruption	Average interruption frequency (Time)	0.29	1.91	1.62	1.03	0.92	1.12	0.95	2.2	0.71*1	—
	Average interruption duration	10	—	80	72.7	108	119	79	4.0*2	70.7	234
Work interruption	Average interruption frequency (Time)	0.36	α	α	α	0.03	0.14	α	2.0	0.08	—
	Average interruption duration (Minute)	59	—		α	3	50	α	1.2*2	23.4*1	90

(Notes) *1: Records of fiscal 1984. *2: It is not certain whether unit is "minute" or "hour."
(Remarks)
U.S.A. 1: Florida Power & Light Co. U.S.A. 2: Florida Power Corp.
U.S.A. 3: Public Service Co. of Colorado. U.S.A. 4: Pennsylvania Power & Light Co.
U.S.A. 5: Boston Edison Co. U.S.A. 6: Union Electric Co.
Canada 1: Hydro – Quebec Co. (Excluding interruptions caused by power source troubles and works.)
U.K. : Up to 1984, 12 distribution boards of England and Wales + 2 distribution boards of Scotland. In and after 1985, only those of England and Wales.
France: Electricite de France (EDF).
Sources: Report Nos. 274, 249, 236, 200 of Japan Electric Power Information Center, Inc.
Report No. 91 of Japan Electric Power Information Center, Inc.
EDF statistical annual of 1988.

6. CONCLUDING REMARKS

At present the Japanese economy is in the doldrums due to the collapse of the bubbles. When the government take effective economic measures, and the structural distortions of the securities market, financial market and the like are corrected, the reliability of the economy will be improved to launch an economic recovery. In a longer perspective, the size of the Japanese economy is expected to make a steady growth.

With the improvement in living and culture and with the advancement of automation of production processes, the demand for electricity is expected to grow at a rate higher than the economic growth rate. In assuring stable demand and supply of electricity, the most important issue is the development of new power sources.

Power source siting, however, tends to be difficult because of the following reasons:

- o Power source development does not create employment for many people of the local communities relative to other industries.
- o The image of a power station as a "source of public nuisance" is deeply established in the general public.
- o The general public still hold a deep-rooted sense of unease about the safety of nuclear power generation.
- o Farmers and fishermen have deep attachment to the land and the sea, the bases of their subsistence.

Before siting a power source such issues must be handled properly to find solutions and moderation measures. Since the length of pregnancy is quite long, power source development should not be influenced by temporary economic ups and downs.

Now the turn of the century is drawing on. Aiming to solve the environmental problems, the Japanese electric utilities have been positively promoting research and development of the following advanced power generation systems:

[Fuel Cells]

Fuel cell plants are static and are expected to serve as on-site power sources. The plants have a high thermal efficiency. When waste heat utilization is included, the total efficiency will be about 70 %. The utilities are conducting demonstration tests on phosphoric acid fuel cell plants.

Research programs are also in progress to raise the capacity and extend the life of molten carbonate fuel cell plants which can be used in combination with coal gasification.

[Coal Gasification Combined Cycle Power Generation]

To make coal clean, positive efforts are being made to promote coal gasification. The gas generated by coal gasification is combusted in a gas turbine to generate electricity. Then the waste heat is utilized to generate steam for conventional power generation. Development of feature technologies is in progress to achieve the sending end efficiency of 43 % or over for 250,000 kW class power plants.

[Advanced Battery Power Storage Systems]

1000-kW sodium-sulfur and zinc-bromine advanced battery power storage plants were constructed. Tests were made on the plants to demonstrate their load levelling functions as on-site power generation plants alternative to pumped storage power systems. A good prospect of practical utilization has been obtained by the demonstration tests.

In addition to the promotion of power source siting, research and development of advanced power generation systems of the future, and development of technologies for higher reliability of all power system facilities, it is important to raise the electric energy efficiency in various fields of electricity utilization. Hence the utilities have been developing more efficient equipment.

REFERENCES

1. Outline of Power Source Development, MITI, 1991.
2. Explanation of the Power Demand Estimation and the Power Supply Plan Computation System, Japan Electric Power Survey Committee, 1993.
3. Comprehensive Manual of Distribution Technology, Yasutsugu Sekine, Ohm-sha, 1991.

Reliability and Optimization of Structural Systems, V (B-12)
P. Thoft-Christensen and H. Ishikawa (Editors)
Elsevier Science Publishers B.V. (North-Holland)

Load Model for Highway Bridges

A. S. Nowak

Department of Civil and Environmental Engineering
University of Michigan, Ann Arbor, MI 48109, USA

The paper deals with the development of load models for LRFD bridge design code. The considered load components include dead load, static live load and dynamic live load. The statistical data is based on the available truck surveys. Distribution functions are derived for moments and shears. The distribution functions are extrapolated to calculate maximum moments and shears for longer time periods, up to 75 year life time for a newly designed bridge. Live load model is derived for one lane and multi-lane bridges. It was observed that for the spans up to about 30-40m lane loading is governed by one truck. For longer spans, the largest effect is produced by two trucks following behind each other. The correlation between truck weights is considered. For two lane bridges, the maximum load effect is caused by two fully correlated trucks, side-by-side in adjacent lanes. The maximum 75 year moment in a grider is caused by two maximum 2 month trucks. Each such a truck is about 85% of the maximum 75 year truck. Dynamic portion of live load is modeled by simulations. Three parameters are considered: bridge dynamics, vehicle dynamics and road roughness. Extensive simulations indicate that dynamic load, as a fraction of live load, decreases with increasing truck weight. It was also found that dynamic load caused by two trucks side-by-side is lower (as a fraction of live load) than the effect of a single truck.

Keyword Codes: J.2, J.6

Keywords: Physical Sciences and Engineering, Computer-aided Engineering

1. INTRODUCTION

The paper deals with the derivation of statistical models for dead load, static and dynamic components of live load and load combinations. The models are developed using the available statistical data, surveys and observations. Load components are treated as random variables. Their variation is described by the cumulative distribution function (CDF), mean value and coefficient of variation. The relationship among various load parameters is described in terms of the coefficient of correlation.

The presented results can serve as a basis for the calculation of load and resistance factors in the LRFD design code.

2. DEAD LOAD

Dead load, D, is the gravity load due to the self weight of the structural and nonstructural elements permanently connected to the bridge. Because of different degrees of variation, it is convenient to consider the following components of D:

D_1 = weight of factory made elements (steel, precast concrete members),
D_2 = weight of cast-in-place concrete members,
D_3 = weight of the wearing surface (asphalt).
D_4 = miscellaneous weight (e.g. railing, luminaries).

All components of D can be treated as normal random variables. The statistical parameters, ratio of mean to nominal (bias factor) and coefficient of variation (V) are listed in Table 1 [1]. For miscellaneous items (weight or railings, curbs, luminaries, signs, conduits, pipes, cables, etc.), the statistical parameters (means and coefficients of variation) are similar to those of D_1, if the considered item is factory-made with the high quality control measures, and D_2, if the item is cast-in-place, with less strict quality control.

Table 1. Statistical Parameters of Dead Load

Component	Bias	V
Factory-made members	1.03	0.08
Cast-in-place members	1.05	0.10
Asphalt	75 mm*	0.25
Miscellaneous	1.03-1.05	0.08-0.10

*mean thickness

3. LIVE LOAD

Live load, L, covers a range of forces produced by vehicles moving on the bridge. Traditionally, the static and dynamic effects are considered separately. Therefore, in this study, L covers only the static component. The dynamic component is denoted by I.

The effect of live load depends on many parameters including the span length, truck weight, axle loads, axle configuration, position of the vehicle on the bridge (transverse and longitudinal), number of vehicles on the bridge (multiple presence), girder spacing, and stiffness of structural members (slab and girders).

The live load model is based on the Ontario truck survey [2]. The study covered about 10,000 selected trucks (only trucks which appeared to be heavily loaded were measured and included in the data base). The uncertainties involved in the analysis are due to limitations and biases in the survey data. Even though 10,000 trucks is a large number, it is very small compared to the actual number of heavy vehicles in a bridge life time. It is also reasonable to expect that some extremely heavy trucks purposefully avoided the weighing stations. A considerable degree of uncertainty is caused by unpredictability of the future trends with regard to configuration of axles and weights.

For each truck in the survey, bending moments and shear forces were calculated for a wide range of simple spans. Negative moments were calculated for continuous two-equal span bridges. The results served as a basis for the development of cumulative distribution functions (CDF) of moments and shears. The upper portions of the CDF's are plotted on normal probability paper in Fig. 1 for moments, Fig. 2 for shears, and Fig. 3 for negative moments, for spans from 9 to 60m [1, 3]. The horizontal scale is in terms of the HS-20 live load moment (truck or lane load, whichever governs), as shown in Fig. 4 [4]. The vertical scale, z, is the inverse standard normal distribution function.

The maximum moments and shears for various time periods are determined by extrapolation of the distributions shown in Fig. 1, 2 and 3. Let N be the total number of trucks in time period of T. It is assumed that the surveyed trucks represent about two week traffic. Therefore, in T = 75 years, the number of trucks, N, will be about 2,000 times larger than in the survey. This will result in N = 20 million trucks. The probability level corresponding to N is 1/N, and for N = 20 million, it is 1/20,000,000 = 5_{10}-8, which corresponds to z = 5.33 on the vertical scale in Fig. 1, 2 and 3.

The mean moments and shears calculated for time periods from 1 day to 75 years are presented in Fig. 5, 6 and 7, for positive moments, shears and negative moments, respectively. For comparison, the means are also plotted for an average truck. The coefficients of variation for the maximum truck moments and shears can be calculated by transformation of the distribution functions. Each function can be raised to a certain power, so that the calculated earlier mean maximum moment (or shear) becomes the mean value after the transformation. The slope of the transformed CDF determines the coefficient of variation. The resulting coefficients of variation vary from 0.11 for 75 years to 0.2 for 1 day.

The maximum moment or shear can be caused by more than one truck. Multiple presence (more than one truck on the bridge simultaneously) can occur in one lane or in adjacent lanes. An important part of the statistical model is information about frequency of such occurrence and correlation between the weights of trucks. Correlation expresses the similarity in weight. Highly correlated trucks carry a similar load, loaded by the same crew, owned by the same company and so on. The statistical parameters for multiple presence were determined by simulations.

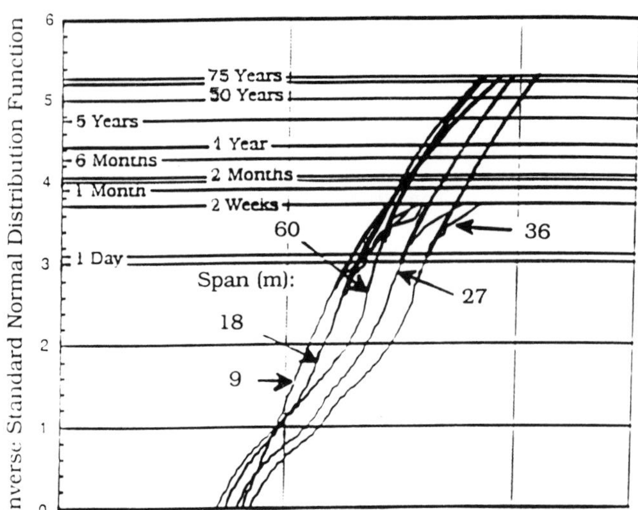

Fig. 1. Extrapolated CDF's of Moments for Simple Spans.

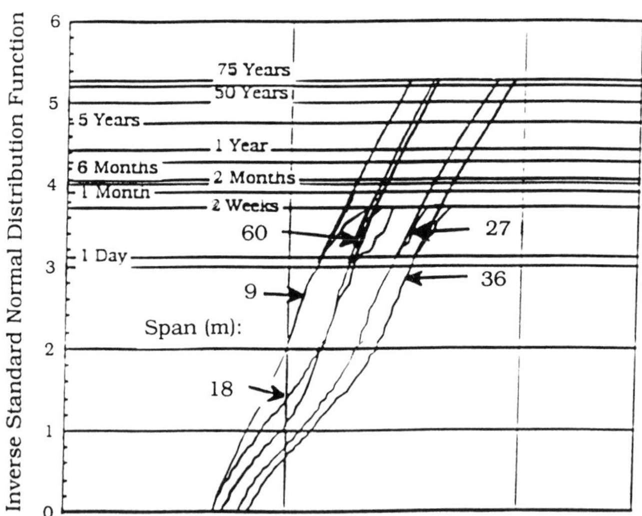

Fig. 2. Extrapolated CDF's of Shears.

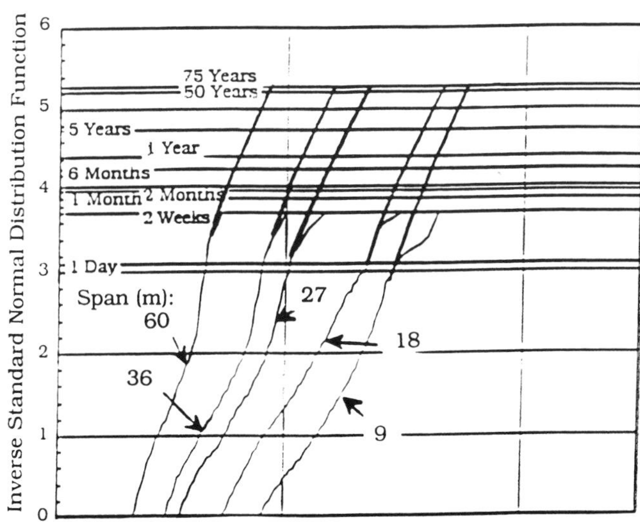

Fig. 3. Extrapolated CDF's of Negative Moments.

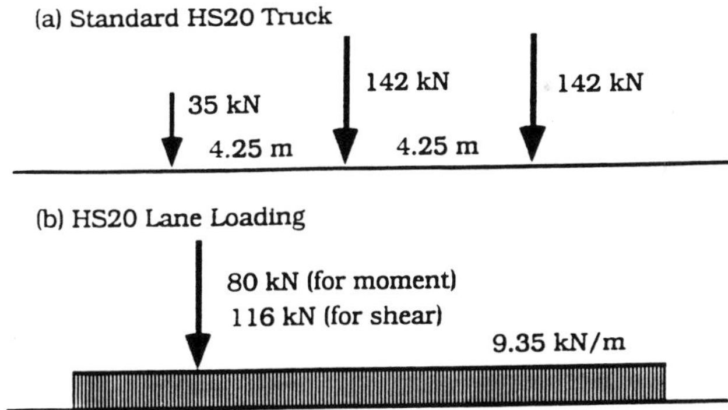

Fig. 4. HS-20 Truck and Lane Laoding (AASHTO 1992).

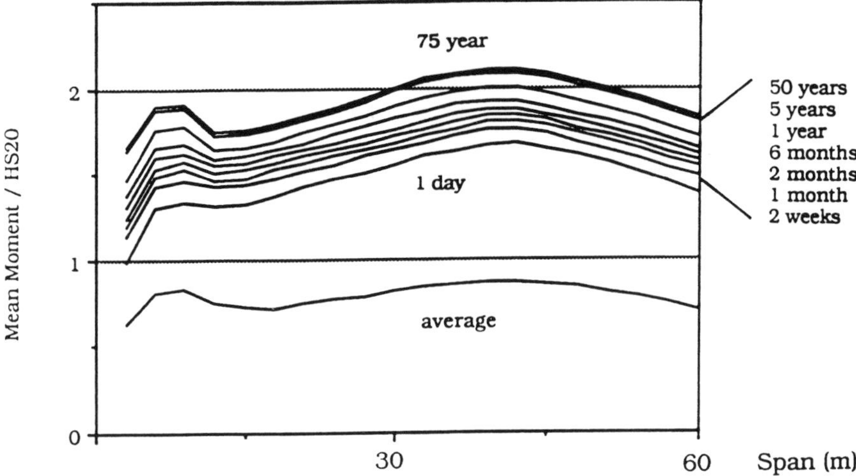

Fig. 5. Mean Maximum Moments for Simple Spans.

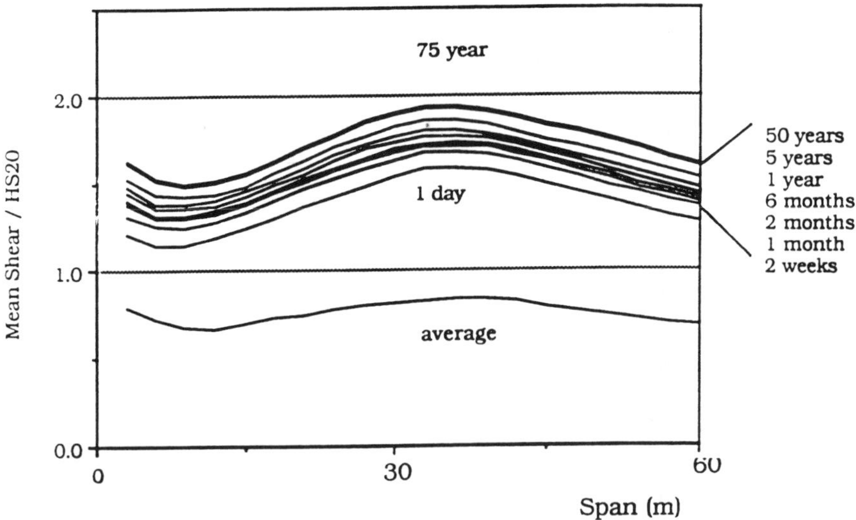

Fig. 6. Mean Maximum Shears.

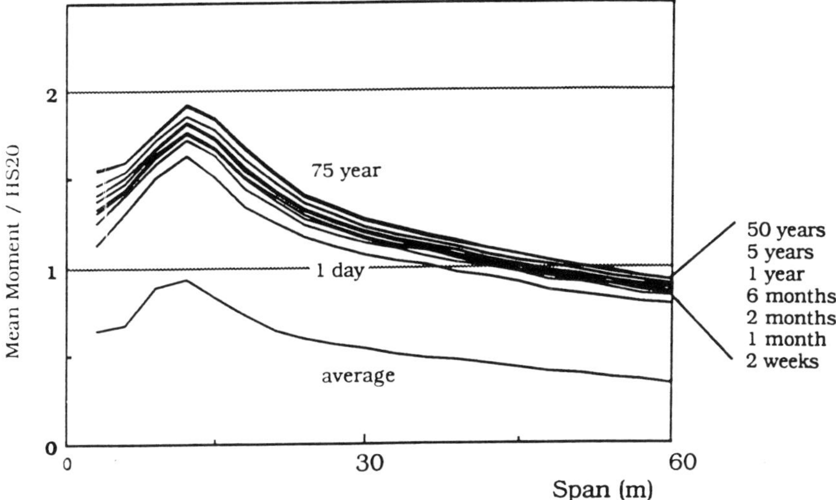

Fig. 7. Mean Maximum Negative Moments.

The effect of two trucks travelling in the same lane depends on the headway distance. It was assumed, conservatively, that the headway distance is 5 m. The simulations showed that a single truck governs up to about 40 m span for positive moments, 35 m for shears and 15 m for negative moments. For longer spans, two fully correlated trucks produce the maximum moment. It was observed that, on average, about every 600th truck is followed by a fully correlated truck. By selecting every 600th truck the sample space is reduced to 30,000 trucks. The maximum of 30,000 is the maximum 1 month truck and it corresponds to z = 3.99 on the vertical scale on normal probability paper.

For two lane bridges, the calculations were performed for one truck or two trucks with various degrees of correlation. In most cases, the governing case is the effect of two side-by-side trucks with full correlation. It is assumed that such an event occurs for every 400th truck. The sample space is reduced to 50,000 vehicles, the maximum of 50,000 is the maximum 2 month truck and it corresponds to z = 4.11. The mean maximum 2 month truck moment is about 85% of the mean maximum 75 year moment for all the spans.

The truck traffic volume is expressed by the average daily truck traffic (ADTT). The presented results correspond to ADTT = 1,000 (in one direction). For other ADTT's and number of lanes from one to four, the mean maximum value of the live load moment can be calculated by applying multi-lane factors listed in Table 2.

Table 2. Multiple Lane Factors.

ADTT (in one direction)	Number of Lanes (in one direction) 1	2	3	4
100	0.95	0.80	0.55	0.45
1,000	1.00	0.85	0.70	0.50
10,000	1.05	0.90	0.75	0.55

4. DYNAMIC LOAD

The dynamic load is defined as the maximum dynamic deflection, D_{dyn}, divided by the maximum static deflection, D_{sta}, as shown in Fig. 8.

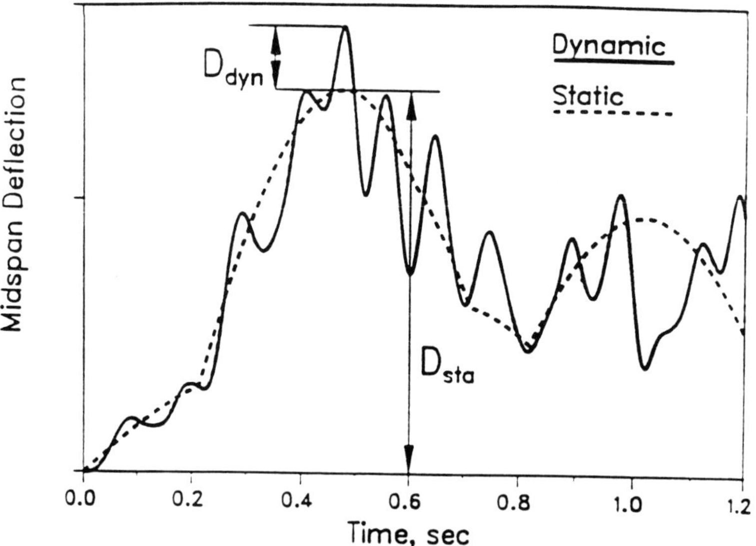

Fig. 8. Static and Dynamic Deflection.

Dynamic load is basically determined by three factors: road roughness, bridge dynamics (natural vibration properties) and vehicle dynamics (suspension system). There is little field data available, in particular this concerns heavy trucks. Therefore, the derivation of the dynamic load model is based on the numerical simulations [5]. Dynamic load effect, I, is considered as an equivalent static load effect added to the live load, L. The objective of this analysis is to determine the parameters (mean and coefficient of variation) of the dynamic load to be added to the maximum 75 year live load.

Dynamic bridge tests were carried out by Billing [6]. The results are available for 22 bridges and 30 spans, including prestressed concrete girders and slabs, steel girders (hot-rolled sections, plate girders, box girders), steel trusses and rigid frames. The measurements were taken using test vehicles (weights from 241 to 580 kN), and a normal traffic. The means and standard deviations, as a fraction of the static live load, are given in Table 3.

Table 3. Dynamic Load Factors from Test Results

Type of Structure	Mean		Standard deviation	
	Range	Average	Range	Average
P/C AASHTO girders	0.05-0.10	0.09	0.03-0.07	0.05
P/C box & slabs	0.10-0.15	0.14	0.08-0.40	0.30
Steel girders	0.08-0.20	0.14	0.05-0.20	0.10
Rigid frame, truss	0.10-0.25	0.17	0.12-0.30	0.26

Considerable differences between the distribution functions for very similar structures indicate the importance of other factors mentioned above (e.g. surface condition). Interpretation of these results is difficult because the dynamic loads are separated from the static live loads. It has been observed that the dynamic load, as a fraction of live load, decreases for heavier trucks. It is expected, that the largest dynamic load fractions in the survey correspond to light-weight trucks.

To verify these observations, a computer procedure was developed for simulation of the dynamic bridge behavior [5]. The procedure includes the effect of road surface roughness, bridge dynamics (frequency of vibration) and vehicle dynamics (suspension system).

Road surface roughness is one of the major parameters. The quantification of the degree of roughness is very difficult. In this study the International Roughness Index (IRI) was used. Simulation of the dynamic load requires the generation of a road profile, which is done by using a Fourier transform of the power spectral density (PSD) function. The PSD function of the road profile has, in general, an exponential form. For the worst condition of older pavements, IRI = 6. It is also close to the mean value of the survey data collected on highways in Michigan. Therefore, IRI = 6 is proposed for the development of design criteria.

The bridge is modeled as a prismatic beam. Modal equations of motion are formulated. Three fundamental modes of vibration are considered. It isassumed that the load is a mixture of 3 axle single trucks and 5 axle tractor-trailers. Each truck is composed of a body, suspension system and tires. The body is subjected to a rigid body motion including the vertical displacement and pitching rotation. Suspensions are assumed to be of multi-leaf type springs. In the simulations a nonlinear force-deflection equation was used.

Static and dynamic deflections are calculated for typical girder bridges. It was observed that dynamic deflections are almost constant while static deflections increase linearly with truck weight. In the result, dynamic load decreases for heavier trucks. Typical relationships between dynamic load and truck weight are shown in Fig. 9, 10 and 11, for a 5 axle truck on a steel girder bridge, 3 axle truck on steel girder bridge and 5 axle truck on prestressed concrete girder bridge. Examples of the simulated static and dynamic deflections are presented on the normal probability paper in Fig. 12 for a steel girder bridge and Fig. 13 for a prestressed concrete girder bridge, both for the span of 30 m.

In most cases, the maximum live load is governed by two trucks side-by-side. The corresponding dynamic loads are calculated for two trucks by superposition of single truck effects. The obtained average dynamic load for one truck is less than 17% of live load, and for two trucks it is less than 12%. The coefficient of variation is 0.80.

5. JOINT EFFECT OF DEAD LOAD, LIVE LOAD AND DYNAMIC LOAD

The maximum 75 year combination of dead load, D, live load, L, and dynamic load, I, is modeled using the statistical parameters derived for D, L and I.

It is assumed that the live load is a product of two parameters, LP, where L is the static live load and P is the live load analysis factor (influence factor). The mean value of P is 1.0 and the coefficient of variation is 0.12.

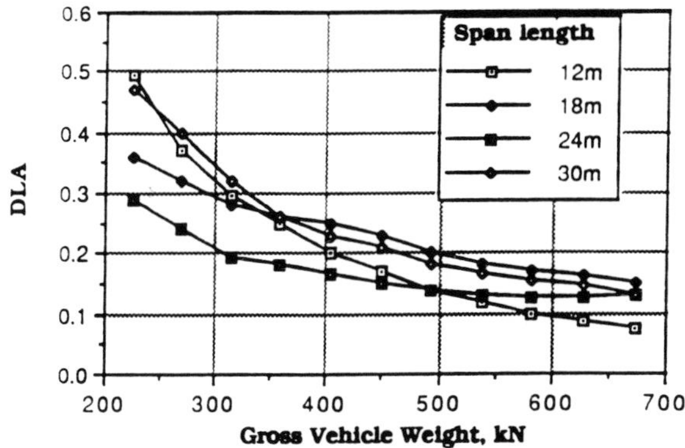

Fig. 9. Dynamic Load vs. Truck Weight for 5 Axle Truck on Steel Girder Bridge.

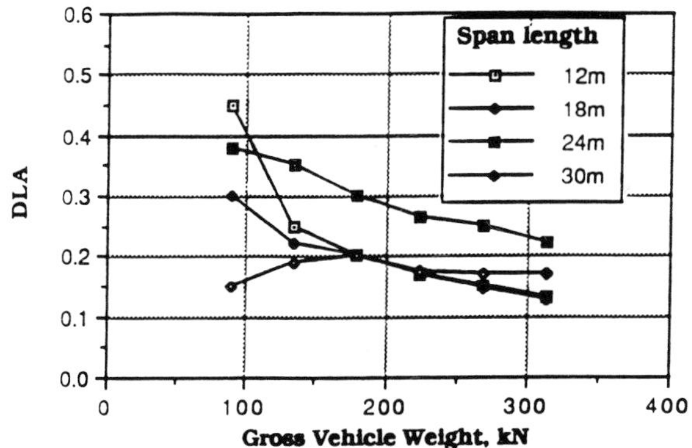

Fig. 10. Dynamic Load vs. Truck Weight for 3 Axle Truck on Steel Girder Bridge.

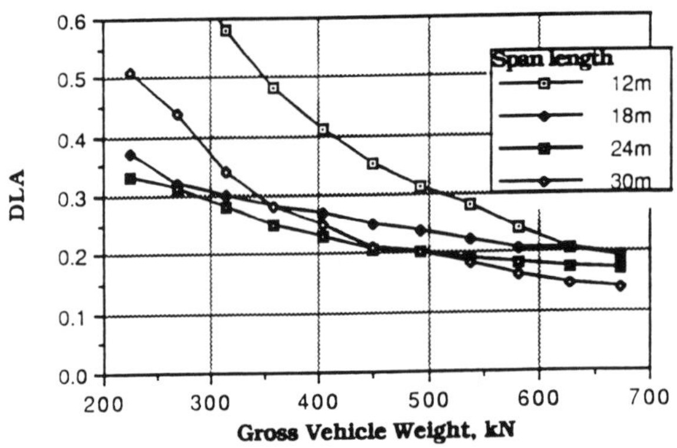

Fig. 11. Dynamic Load vs. Truck Weight for 5 Axle Truck on Prestressed Concrete Girder Bridge.

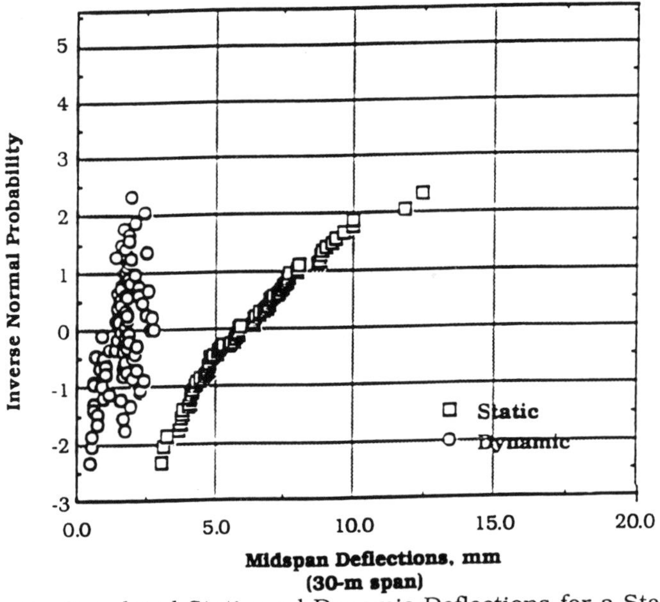

Fig. 12. Simulated Static and Dynamic Deflections for a Steel Girder Bridge

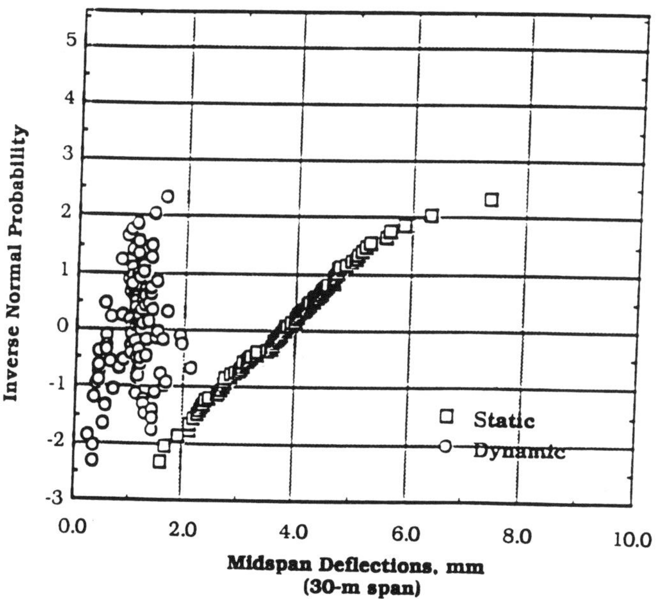

Fig. 13. Simulated Static and Dynamic Deflections for a Prestressed Concrete Girder Bridge.

The statistical parameters of (L + I) depend on span length and they are different for a single lane and two lanes. For a single lane, the coefficient of variation is 0.19 for most spans, and 0.205 for very short spans. For two lane bridges, it is 0.18 for most spans, and 0.19 for very short spans.

6. CONCLUSIONS

The bridge load models are reviewed including dead load, live load and dynamic load.

Because of a difference in variation it is convenient to differentiate between factory-made components (lower variation) and cast-in-place components (higher variation). Asphalt surface is subject to a considerable variation.

Live load is considered for single lanes and multilane bridges. For one lane, a single truck governs up to 30-40 m span. For longer spans, two fully correlated trucks following behind each other produce the largest effect. For two lanes, the maximum moments and shears are produced by two side-by-side trucks with fully correlated weights. Each truck is about 85% of the mean maximum 75 year truck. Multi-lane factors are provided for various ADTT's and number of lanes.

A procedure is summarized for simulation of the dynamic load including the effect of road roughness, bridge dynamics and vehicle dynamics. The results indicate that the dynamic load, as a fraction of live load, decreases for heavier trucks. It is also lower for two trucks compared to a single truck. Average values of dynamic load are under 17% of live load for a single truck and under 12% for two trucks.

The presented load models can be used for the development of load and resistance factors in the bridge design code.

ACKNOWLEDGEMENTS

The presented research was conducted in conjunction with NCHRP Project 12-33, with Dr. John M. Kulicki (Principal Investigator) which is gratefully acknowledged. Doctoral students at the University of Michigan played an important role in computations and the development of load models. Thanks are due to Young-Kyun Hong, Eui-Seung Hwang and Hani Nassif.

REFERENCES

1. Nowak, A.S., 1993, "Live Load Model for Highway Bridges", Journal of Structural Safety, submitted.
2. Agarwal, A.C. and Wolkowicz, M., 1976, "Interim Report on 1975 Commercial Vehicle Survey", Research and Development Division, Ministry of Transportation, Downsview, Ontario, Canada.
3. Nowak, A.S. and Hong, Y-K., 1991, "Bridge Live Load Models", ASCE Journal of Structural Engineering, Vol. 117, No. 9.
4. AASHTO, 1992, "Standard Specifications for Highway Bridges", American Association of State Highway and Transportation Officials, Washington, DC.
5. Hwang, E-S. and Nowak, A.S., 1991, "Simulation of Dynamic Load for Bridges", ASCE Journal of Structural Engineering, Vol. 117, No. 5, pp. 1413-1434.
6. Billing, J.R., 1984, "Dynamic Loading and Testing of Bridges in Ontario", Canadian Journal of Civil Engineering, Vol. 11, No. 4, pp. 833-843.

Reliability and Optimization of Structural Systems, V (B-12)
P. Thoft-Christensen and H. Ishikawa (Editors)
Elsevier Science Publishers B.V. (North-Holland)

Reliability-Based Inspection Planning for Structural Systems

J. D. Sørensen

University of Aalborg, Sohngaardsholmsvej 57
DK-9000 Aalborg, Denmark

ABSTRACT

A general model for reliability-based optimal inspection and repair strategies for structural systems is described. The total expected costs in the design lifetime is minimized with the number of inspections, the inspection times and efforts as decision variables. The equivalence of this model with a preposterior analysis from statistical decision theory is discussed. It is described how information obtained by an inspection can be used in a repair decision. Stochastic models for inspection, measurement and repair actions are presented. The general model is applied for inspection and repair planning for concrete bridges and offshore steel platforms.

Keywords: Reliability, Optimization, Inspection planning, Repair planning.

1. INTRODUCTION

During the last few decades significant achievements have been obtained in development of efficient techniques to evaluate the reliability of components and systems. Especially the development of FORM, SORM and simulation techniques has been important, see e.g. Madsen et al. [1] and Engelund & Rackwitz [2].

In the same period efficient methods to solve non-linear optimization problems have also been developed, e.g. the sequential quadratic optimization algorithms NLPQL [3] and VMCWD [4]. These developments have made it possible to solve problems formulated in a decision theoretical framework, for example reliability-based inspection and repair planning, formulated as a preposterior decision problem, see e.g. Skjong [5], Thoft-Christensen and Sørensen [6], Fujita et al. [7], Madsen et al. [8], Madsen and Sørensen [9], Fujita et al. [10] and Sørensen & Thoft-Christensen [11].

Methods which can be used to determine cost optimal inspection and repair strategies for structural systems are of great interest in a number of application areas. In this paper two areas are considered, namely concrete bridges and offshore steel platforms.

In concrete bridges degradation of the reinforcement bars due to corrosion may imply that expensive repair work is needed. For offshore steel platforms fatigue cracks can be very important for the reliability. Further inspection and repair of cracks are expensive. Therefore cost optimal planning of inspections and repair is of great practical interest. In section 2 the decision theoretical background is described. Sections 3 and 4 contain detailed descriptions of the two application areas.

2. OPTIMAL INSPECTION AND MAINTENACE PLANNING

Development of cost optimal plans for inspection and maintenance of structures can be based on the classical decision theory, see e.g. Raiffa and Schlaifer [12] and Benjamin and Cornell [13]. A very important aspect of the methodology is that it provides a consistent basis within the framework of Bayesian statistics for updating the uncertainties involved in the problem, such as loading (significant wave heigth, the marine growth) or damage (the crack growth) when new information is obtained.

In the general case the parameters defining the inspection plan are the number of inspections N, the time intervals between inspections $\overline{\Delta t} = (\Delta t_1, \cdots, \Delta t_N)$ and the inspection qualities $\overline{e} = (e_1, \cdots, e_N)$. These inspection parameters are written as $\overline{i} = (N, \overline{\Delta t}, \overline{e})$. The outcome of an inspection is assumed to be modelled by a random variable S. A decision rule d is then applied to the outcome of the inspection to decide whether or not repair should be performed. The different uncertain parameters (stochastic variables) modelling the state of nature are collected in a vector $\overline{X}$. The design of the structure is represented through the design variable z.

A rational inspection and repair plan can then be obtained using Bayesian decision analysis, see Raiffa & Schlaifer [12]. The decision tree is shown in figure 1.

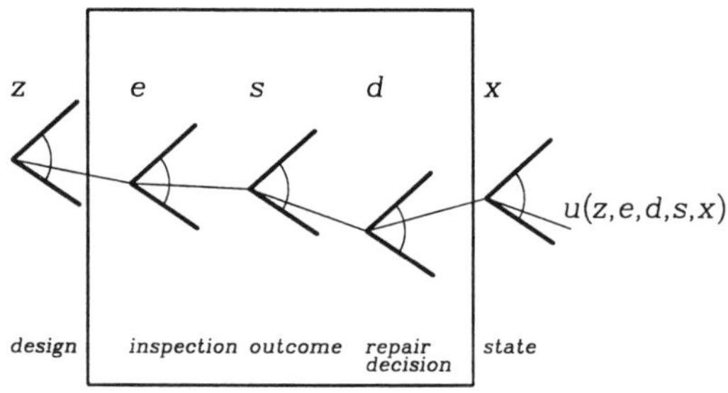

Figure 1. Decision tree.

It is assumed that the decision maker chooses the strategy $z, \overline{i}, d$ which maximizes the expected utility

$$u^* = \max_{z} \max_{\overline{i}} \max_{d} \quad u(z, \overline{i}, d) \tag{1}$$

where the expected utility u is

$$u(z, \overline{i}, d) = E_{\overline{X}, S|z,\overline{i}}[u(z, \overline{i}, S, d(S), \overline{X})] \tag{2}$$

$E_{\overline{X}, S|z,\overline{i}}[\cdot]$ is the expectation with respect to the stochastic variables $\overline{X}$ and S. This corresponds to the normal form of decision analysis, see Raiffa & Schlaifer [12]. The result of the preposterior analysis is the cost optimal inspection and repair plan.

If the utility function is related to the total costs and the benefits are neglected, (1)-(2) can equivalently be formulated as an optimization problem where the total expected cost is minimized:

$$\min_{z}\min_{\bar{i}}\min_{d} \quad C_T(z,\bar{i},d) = E_{\overline{X},S|z,\bar{i}}[C_{TOT}(z,\bar{i},S,d(S),\overline{X})] \tag{3}$$

where $C_{TOT}(z,\bar{i},S,d(S),\overline{X})$ is the total cost.
If the total expected costs are divided into initial, inspection, repair and failure costs and a constraint related to a minimum level of reliability $\beta_{\min}$ is added then the optimization problem can be written

$$\min_{z,\bar{i},d} \quad C_T(z,\bar{i},d) = C_I(z) + C_{IN}(z,\bar{i},d) + C_R(z,\bar{i},d) + C_f(z,\bar{i},d) \tag{4}$$

$$s.t. \quad \beta(T_L,z,\bar{i},d) \geq \beta_{\min} \tag{5}$$

$C_T(z,\bar{i},d)$ is the total expected cost in the design lifetime T_L. C_I is the expected initial cost, C_{IN} is the expected inspection cost, C_R is the expected cost of repair and C_f is the expected failure cost. $\beta(T)$ is the generalized reliability index defined by

$$\beta(T) = -\Phi^{-1}(P_F(T)) \tag{6}$$

where Φ is the standardized normal distribution function and $P_F(T)$ is the probability of failure in the time interval $[0,T]$.
The constraint on the minimum reliability (5) is not absolutely necessary since the reliability is already incorporated in the objective function through the expected cost of the failure term, but it is included to take into account code demands. Other constraints, e.g. on the maximum of the individual costs or simple bounds on the optimization variables can be included in the problem, see section 4.
In figure 1 a frame is used to show that inspection and repair decisions are taken in an adaptive way. Each time an inspection or repair is performed the inspection and repair plans for the remaining lifetime are updated.
In the following, cost optimal inspection and repair planning for concrete bridges and offshore steel platforms are considered. However, it should be noted that the above techniques can also be used for experiment planning, see Sørensen et al. [14], or for design of new structural systems where the expected costs due to construction and possible failure are minimized with the design parameters (e.g. geometrical quantities) as decision variables.

3. OPTIMAL INSPECTION PLANNING FOR BRIDGES

Inspection regulations from a number of countries show that inspections are typically performed after a scheme as shown in figure 2, see Sørensen & Thoft-Christensen [11]:

- routine inspections performed with a fixed short-term interval, e.g. 1 year,
- detailed inspections performed with a fixed long-term interval, e.g. 5 years and

- structural assessments (special inspections) performed when the routine or detailed inspections show a need for it.

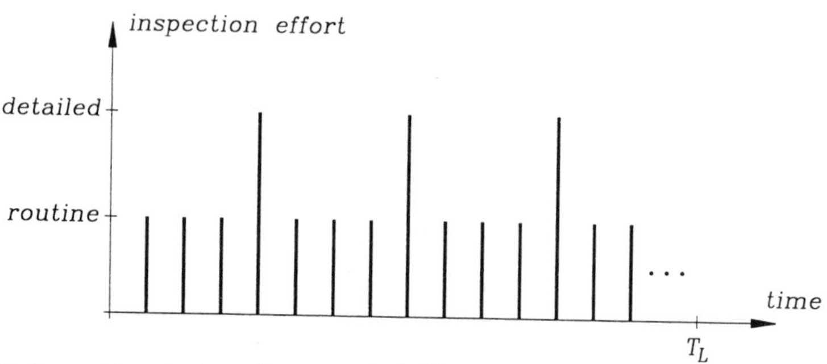

Figure 2. Inspection strategy for concrete bridges.

The decision on structural assessments (special inspections) can e.g be made on the basis of the updated reliability index $\beta^U(T_L)$ corresponding to the design lifetime T_L of the bridge, see Thoft-Christensen [15]. If $\beta^{\min}$ is some minimum acceptable reliability index then a structural assessment is performed if

$$\beta^U(T_L) \leq \beta^{\min} \tag{7}$$

It is noted that the inspection planning is not based on cost considerations. However, expert knowlege should be included in the decision process, if possible.

3.1 Decision Model for Repair Planning

When an inspection has been performed then based on the inspection result a decision on repair actions has to be taken. Assume that n_R different repair types are available and that the repair types are denoted $I_R = 0, 1, 2, ..., n_R$ where $I_R = 0$ corresponds to no repair.

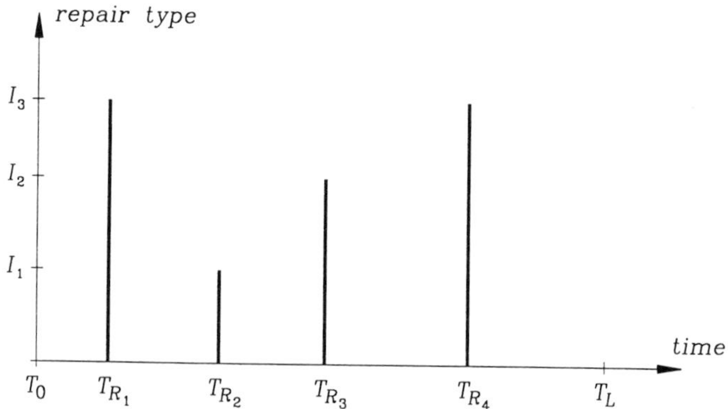

Figure 3. Repair plan.

The repair decision variables could be:

- The total number of repairs N_R in the remaining lifetime.
- Times $T_{R_i}, i = 1, ..., N_R$ of repairs.
- Types of repair $I_{R_i}, i = 1, ..., N_R$ to be used.

The repair plan is illustrated in figure 3 where T_0 is the time when the repair decisions are made. From the following optimization problem a cost-benefit optimal decision can be determined, see (1)-(5):

$$\max_{N_R,\overline{I}_R,\overline{T}_R} \quad B_T(N_R,\overline{I}_R,\overline{T}_R) = B(N_R,\overline{I}_R,\overline{T}_R) - C_R(N_R,\overline{I}_R,\overline{T}_R) - C_F(N_R,\overline{I}_R,\overline{T}_R) \tag{8}$$

$$s.t. \quad \beta^U(T_L, N_R,\overline{I}_R,\overline{T}_R) \geq \beta^{\min} \tag{9}$$

where

B_T : Total expected cost benefits
B : Expected benefits in the remaining lifetime of the bridge.
C_R : Expected repair cost in the remaining lifetime of the bridge.
C_F : Expected failure cost in the remaining lifetime of the bridge.
T_L : Expected lifetime of the bridge.
β^U : Updated reliability index.
$\beta^{\min}$: Minimum reliability index for the bridge (related to critical element or to the total system).

The decision model is used in an adaptive way where the stochastic model is updated after each structural assessment or repair and a new optimal repair decision is made. Therefore, it is mainly the time and type of the first repair after a structural assessment which is of practical interest.

3.2 Cost Modelling

Inspection costs are not included in the optimization problem. They can be assumed not to influence the optimal choice of repair action.

The benefits capitilized to year T_0 are modelled by:

$$B(N_R,\overline{I}_R,\overline{T}_R) = \sum_{i=[T_0]+1}^{[T_L]} B_i \frac{1}{(1+r)^{T_i - T_0}} \tag{10}$$

where

$[T]$ is the integer part of T measured in years
B_i is the benefits in year i. B_i is dependent on the traffic volume in year T_i and the average benefits for one vehicle passing the bridge.
r is the real rate of interest.

The expected capitalized repair costs are modelled by:

$$C_R(N_R,\overline{I}_R,\overline{T}_R) = \sum_{i=1}^{N_R}(1 - P_F^U(T_{R_i}))C_{R_0}(I_{R_i}, T_{R_i})\frac{1}{(1+r)^{T_{R_i} - T_0}} \tag{11}$$

where $C_R = 0$ if $I_R = 0$ (no repair), $P_F^U(T_R)$ is the updated probability of failure in the time interval $[T_0, T_R]$ and $C_{R_0}(I_{R_i}, T_{R_i})$ is the cost of a type I_R repair at the time T_{R_i}. The capitalized expected costs due to failure are approximated by:

$$C_F(N_R, \overline{I}_R, \overline{T}_R) \approx \sum_{i=1}^{N_R+1} C_F(T_{R_i})(P_F^U(T_{R_i}) - P_F^U(T_{i-1}))\frac{1}{(1+r)^{T_{R_i}-T_0}} \tag{12}$$

where $C_F(T)$ is the cost of failure at the time T (dependent on the direct failure costs and the functional failure costs (loss of benefit)).

3.3 Reliability Estimations

It is assumed that N_{INSP} inspections have been performed. The inspection events are modelled by the events I_i , $i = 1, ..., N_{INSP}$.

For a **single failure mode** (component) the updated probability of failure in the time interval $]T_0, T]$ for $T_0 < T \leq T_{R_1}$ is determined by

$$P_F^U(T) = P(M_F(T) \leq 0 | M_F(T_0) > 0 \cap I_1 \cap ... \cap I_{N_{INSP}}) \tag{13}$$

where $M_F(T)$ is the failure safety margin at the time T. The safety margin is time-dependent due to degrations caused by e.g. chloride attack or carbonation. The safety margin can e.g. model shear failure of a cross-section of the bridge.

The updated probability of failure at the time T using repair type I^R for $T_{R_i} < T \leq T_{R_{i+1}}$ is estimated by:

$$\begin{aligned} P_F^U(T) =& P_F^U(T_{R_i}) + \\ & P(M_F^{I_R, i-1}(T_{R_i}) > 0 \cap M_F^{I_R, i}(T) \leq 0 | M_F(T_0) > 0 \cap I_1 \cap ... \cap I_{N_{INSP}}) \end{aligned} \tag{14}$$

where $M_F^{I_R,i}(T)$ is the failure safety margin at the time T assuming repair type I_R at the time T_{R_i}. In this way the probability of failure can be determined for all times in $[T_0, T_L]$.

Next a **series system modelling** is considered. It is assumed that m failure modes are considered. The updated probability of failure in the time interval $]T_0, T]$ for $T_0 < T \leq T_{R_1}$ is:

$$P_F^U(T) = P(\cup_{j=1}^m M_{F_j}(T) \leq 0 | \cap_{j=1}^m M_{F_j}(T_0) > 0 \cap I_1 \cap ... \cap I_{N_{INSP}}) \tag{15}$$

where $M_{F_j}(T)$ is the failure safety margin at the time T for failure mode j.

The updated probability of failure at the time T for $T_{R_i} < T \leq T_{R_{i+1}}$ using repair type I^R is:

$$\begin{aligned} P_F^U(T) =& P_F^U(T_{R_i}) + \\ & P(\cap_{j \notin J} M_{F_j}(T_{R_i}) > 0 \ \cap_{j \in J} M_{F_j}^{I_R, i-1}(T_{R_i}) > 0 \\ & \bigcap \cup_{j \notin J} M_{F_j}(T) \leq 0 \ \cup_{j \in J} M_{F_j}^{I_R, i}(T) \leq 0 | \\ & \cap_{j=1}^m M_{F_j}(T_0) > 0 \cap I_1 \cap ... \cap I_{NINSP}) \end{aligned} \tag{16}$$

$M_{F_j}^{I_R,i}(T)$ is the failure safety margin for failure mode j at the time T assuming repair type I_R at the time T_{R_i}. J is the set of failure modes in the cross-sections which are repaired. The updated reliability index $\beta^U(T_L)$ at the time T_L is $\beta^U(T_L) = -\Phi^{-1}(P_F^U(T_L))$.

4. Optimal Reliability-Based Inspection Planning for Offshore Platforms

In this section the methodologies described in section 2 are applied to offshore steel platforms with fatigue cracks. Let N be the number of inspections in the design lifetime T_L. Inspections are assumed to be performed at times $0 \leq T_1 \leq T_2 \leq ... \leq T_N < T_L$ with inspection efforts (qualities) $e_1, e_2, ..., e_N$.

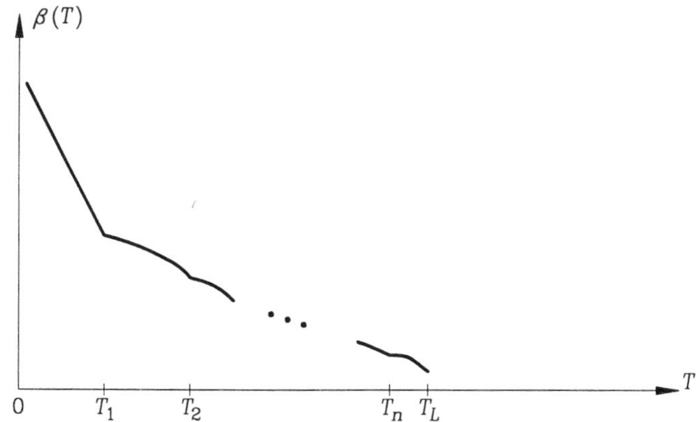

Figure 4. Forecasted reliability index as function of time with inspections at times $T_1, T_2, ..., T_N$.

For a fatigue sensitive structural detail the reliability is measured by the generalized reliability index β. In figure 4 the reliability index as a function of the time T is sketched. Generally the reliability index is decreasing with time due to deterioration. At each inspection there is a possibility of repair. This possibility implies that the slope of the reliability index curve is changed such that the slope is smaller after an inspection than before.

If repair is assumed to be performed when a crack is detected and has been measured to have a crack size a larger than a critical level a_r then the total number of repair realizations (branches) is 3^n, see figure 5, where 0, 1 and 2 signify no crack detected (no repair), crack detected, but too small to be repaired and repair, respectively.

The basis for determination of the optimal inspection and repair strategy is the preposterior analyses, see section 2. The parameters defining the inspection plan are $\bar{i} = (N, t_1, ..., t_N, e_1, ..., e_N)$ where $t_i = T_i - T_{i-1}$ $(T_0 = 0), i = 1, ..., N$ are the time intervals between inspections. The outcome of an inspection is a measured crack length, modelled by a stochastic variable a_d. After inspection it is decided if repair should be performed. This decision is based on the parameter a_r, the repair crack length limit. The uncertain quantities modelling e.g. significant wave height and the coefficients in Morison's equation are denoted $\overline{X}$. Finally the design parameter z can be the thickness of a tubular element.

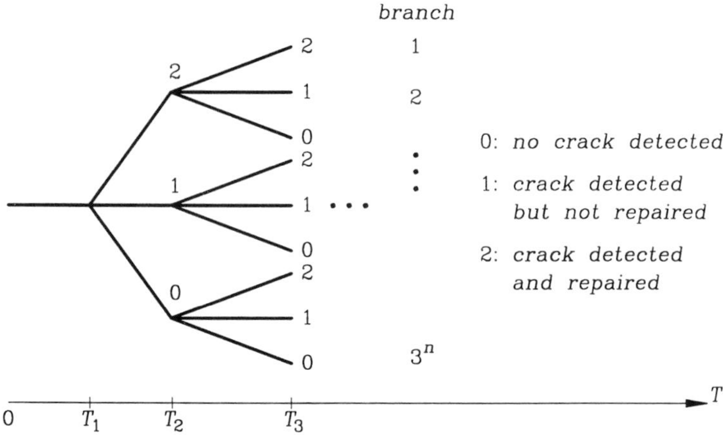

Figure 5. Repair realizations.

Using the total expected costs $C_T(N,\bar{t},\bar{e},a_r,z)$ in the design lifetime ($T_{n+1} = T_L$) as objective function the cost optimal inspection strategy can be determined from the following optimization problem, see (4)-(5)

$$\min_{N,\bar{t},\bar{e},a_r,z} \quad C_T(N,\bar{t},\bar{e},a_r,z) = C_0(z) + C_{IN}(N,\bar{t},\bar{e},a_r,z) + C_R(N,\bar{t},\bar{e},a_r,z) + C_F(N,\bar{t},\bar{e},a_r,z) \tag{17}$$

$$s.t. \quad \beta(T_L,N,\bar{t},\bar{e},a_r,z) \geq \beta^{min} \tag{18}$$

$$C_0(z) \leq C_0^{\max} \tag{19}$$

$$C_{IN}(N,\bar{t},\bar{e},a_r,z) \leq C_{IN}^{\max} \tag{20}$$

$$C_R(N,\bar{t},\bar{e},a_r,z) \leq C_R^{\max} \tag{21}$$

$$C_F(N,\bar{t},\bar{e},a_r,z) \leq C_F^{\max} \tag{22}$$

$$t^{\min} \leq T_L - T_N \leq t^{\max} \tag{23}$$

$$t^{\min} \leq t_i \leq t^{\max} \quad ,i = 1,...,N \tag{24}$$

$$e^{\min} \leq e_i \leq e^{\max} \quad ,i = 1,...,N \tag{25}$$

$$z^{\min} \leq z \leq z^{\max} \tag{26}$$

$$a_r^{\min} \leq a_r \leq a_r^{\max} \tag{27}$$

where

$\beta^{\min}$	is the minimum acceptable reliability index
$t^{\min}, t^{\max}$	are the minimum and maximum time intervals between inspections,
$e^{\min}, e^{\max}$	are the minimum and maximum inspection efforts,
$a_r^{\min}, a_r^{\max}$	are the minimum and maximum bounds on the repair level parameter, and
$z^{\min}, z^{\max}$	are the minimum and maximum bounds on the design parameter

$$C_{IN}(N,\bar{t},\bar{e},a_r,z) = \sum_{i=1}^{N} C_{IN_i}(\bar{e})(1-P_F(T_i))\frac{1}{(1+r)^{T_i}} \tag{28}$$

is the total capitalized expected inspection costs. The ith term represents the capitalized inspection costs at the ith inspection when failure has not occurred earlier. $C_{IN_i}(\bar{e})$ is the inspection cost of the ith inspection, $P_F(T_i)$ is the probability of failure in the time interval $[0, T_i]$ and r is the real rate of interest.

$$C_R(N,\bar{t},\bar{e},a_r,z) = \sum_{i=1}^{N} C_{R_i} P_{R_i} \frac{1}{(1+r)^{T_i}} \tag{29}$$

is the total capitalized expected repair costs. C_{R_i} is the cost of a repair at the ith inspection and P_{R_i} is the probability of performing a repair after the ith inspection when failure has not occurred earlier.

The total capitalized expected costs due to failure are estimated from

$$C_F(n,\bar{t},\bar{e},a_r,z) \approx \sum_{i=1}^{n+1} C_F(T_i)(P_F(T_i)-P_F(T_{i-1}))\frac{1}{(1+r)^{T_i}} \tag{30}$$

where $C_F(T)$ is the cost of failure at the time T. $\beta(T)$ is the generalized reliability index given by (6). The probabilities P_{R_i} and $P_F(T_i)$ are calculated taking into account all branches in figure 5, see Madsen & Sørensen [9].

Just after construction the design variable is fixed and some information of the stochastic variables may be available. The design parameter is then fixed in the above optimization problem and the probabilities can be updated (e.g. using Bayesian techniques) based on the new information. The optimization problem is solved and an updated inspection plan is determined.

When an inspection is performed the obtained information (e.g. measured crack size) is used to update the stochastic model. Based on the above optimization problem it can be determined whether it is optimal to perform repair or not at this inspection time. The optimization problem is modified by fixing the design variable, the first inspection time and effort. Using the updated probabilities two optimization problems are solved, namely one where repair is assumed and one where no repair at the inspection time is assumed. The difference between the total expected costs of the two cases is compared with the cost of a repair and it can then be decided if repair should be performed.

Finally when a repair decision is taken an updated optimal inspection plan for the remaining lifetime is determined. In this way an adaptive inspection model is obtained.

4.1 Modelling of Inspection Effort

The effectiveness of an inspection depends on a number of factors such as the inspection method and the people performing the inspection. Usually the reliability of the inspection is modelled by a *pod* (probability of detection) curve defined by

$$p(a) = P(\text{Detection of crack}|a) \tag{31}$$

where a is the crack length. The *pod* curve is also equal to the distribution function of the smallest detectable crack size, here denoted a_d

$$F_{a_d}(a) = p(a) \tag{32}$$

A commonly used form of the *pod* curve is an exponential function

$$p(a) = \begin{cases} 0 & \text{for } 0 < a < a_0 \\ \Delta(1 - \exp(-\lambda(e)(a - a_0))) & \text{for } a_0 \leq a \end{cases} \tag{33}$$

where Δ is a parameter ($0 \leq \Delta \leq 1$) which gives the probability of detecting a very large crack. a_0 is the minimum crack size below which a crack cannot be detected. $\lambda(e)$ determines the inspection effectiveness.

If a crack (or defect) is detected then a measurement of the actual length can be performed. Errors in connection with this measurement depend on the measurement technique used. An often used probabilistic model of measurement errors is a simple additive model where the measured crack length a_m is modelled by

$$a_m = a + \epsilon \tag{34}$$

a is the crack length and ϵ is a normally distributed random variable modelling the mesurement error. ϵ is assumed to have zero mean and a standard deviation which depends on the measurement technique.

4.2 Modelling of Repair

An important aspect in the above reliability-based model for inspection and repair planning is the choice of repair strategy when specific inspection results are obtained. In Madsen & Sørensen [9] the following repair strategies are considered:

Strategy 1 All detected cracks are repaired by welding.

Strategy 2 Only detected cracks larger than a certain size are repaired (by welding).

Strategy 3 All detected cracks are repaired. Cracks smaller than a certain size are repaired by grinding, while cracks larger than this size are repaired by welding.

Strategy 4 All detected cracks are repaired by replacement of the element.

After repair using one of these strategies the stochastic model has to be modified according to the repair performed.

4.3 Updating of Inspection Plans

When new information from measurements and inspections is obtained the reliability estimates of the structure can be updated using Bayesian techniques, see e.g. Lindley [16] and Madsen [17]. During the lifetime of the structure two types of information are likely to be collected. The first type of information is information about the functional relationship between uncertain variables. Such a functional relationship can be the crack length. The second type of information is observations of one or more of the stochastic variables, e.g. measurements of the significant wave height, the wave period, the thickness of marine growth, etc. This type of information is actual samples of the uncertain basic variables.

Both types of information are taken into account by the use of Bayes' theorem, see e.g. Lindley [16]. The information gathered during an inspection can be expressed in terms of event margins, and updating with respect to this information can be regarded as general event updating. The general information is assumed to be modelled by inequality and equality events. Updating of the probability of failure can be performed using Bayesian methods, see Madsen [17] and Rackwitz & Schrupp [18].

4.4 Simplified Inspection Planning

The numerical effort in determination of an optimal inspection plan by solving the above optimization problems can be rather large. However, some simplifications described below can be used.

The probabilities for failure and repair can be approximated by using only branches in the repair events corresponding to no-repair. This is a reasonable approximation if the reliability level is high and only few repairs are expected. Further only the next inspection time and quality can be used as optimization variables.

The inspection planning can be based on reliability requirements only, i.e. cost considerations are neglected. A minimum reliability level is chosen, for example based on code requirements. The generalized reliability index $\beta(t)$ is determined as a function of time for the component. When $\beta(t)$ decreases to $\beta_{\min}$ an inspection has to be performed. The next inspection time is estimated assuming that no defects are found by the inspection. In this way the total inspection plan can be determined, see Pedersen et al. [19].

Finally some reduction in computation time can be obtained by discretizing the decision variables, see Faber et al. [20].

4.5 Evaluation of Inspection and Maintenance Plans

When an optimal inspection and repair plan has been determined by solving the above optimization problems the decision makers will usually be very interested in knowing the sensitivities of the expected total cost and of the decision variables with respect to changes in the cost parameters and some of the statistical parameters.

The following optimization problem which corresponds to the general optimization problem defined by (17)-(27) is considered.

$$\begin{aligned} \min_{\bar{i}} \quad & C(\bar{i},\bar{p},\bar{q}) = C_0(\bar{i},\bar{p}) + \sum_j C_j(\bar{i},\bar{p})P_j(\bar{i},\bar{q}) \\ \text{s.t.} \quad & P_f(\bar{i},\bar{q}) \le P_f^{\max} \end{aligned} \tag{35}$$

where $\bar{i}$ are decision/design variables, $\bar{p}$ are quantities defining the costs and $\bar{q}$ are quantities defining the stochastic model. P_j denotes a probability (failure or repair), P_f denotes a failure probability and $P_f^{\max}$ is the maximum accepted failure probability.

The summation in equation (35) is performed over the number of terms in(28)-(30). The sensitivity of the total expected costs C with respect to the elements in $\overline{p}$ and $\overline{q}$ is obtained from, see Haftka & Kamat [21] and Enevoldsen [22]

$$\frac{dC}{dq_i} = \sum_j C_j \frac{\partial P_j}{\partial q_i} + \lambda \frac{\partial P_f}{\partial q_i} \tag{36}$$

$$\frac{dC}{dp_i} = \sum_j P_j \frac{\partial C_j}{\partial p_i} + \frac{\partial C_0}{\partial p_i} \tag{37}$$

where λ is the Lagrangian multiplier associated with the constraint in (35).
The sensitivity of the decision variables $\overline{i}$ with respect to p_i and q_i can be calculated using the formulas given below which are obtained from a sensitivity analysis of the Kuhn Tucker conditions related to the optimization problem defined in (35). $\frac{\partial \overline{i}}{\partial p_i}$ is obtained from

$$\begin{bmatrix} \overline{\overline{A}} & \overline{B} \\ \overline{B}^T & 0 \end{bmatrix} \begin{bmatrix} \frac{\partial \overline{i}}{\partial p_i} \\ \frac{\partial \lambda}{\partial p_i} \end{bmatrix} = \begin{bmatrix} \overline{C} \\ 0 \end{bmatrix} \tag{38}$$

The elements in the matrix $\overline{\overline{A}}$ and the vectors $\overline{B}$ and $\overline{C}$ are

$$A_{rs} = \frac{\partial^2 C_0}{\partial x_r \partial x_s} + \sum_j (P_j \frac{\partial^2 C_j}{\partial x_r \partial x_s} + 2\frac{\partial P_j}{\partial x_r}\frac{\partial C_j}{\partial x_s} + C_j \frac{\partial^2 P_j}{\partial x_r \partial x_s}) + \lambda \frac{\partial^2 P_f}{\partial x_r \partial x_s} \tag{39}$$

$$B_r = \frac{\partial P_f}{\partial x_r} \qquad C_r = -\frac{\partial^2 C_0}{\partial x_r \partial p_i} - \sum_j (\frac{\partial^2 C_j}{\partial x_r \partial p_i} + \frac{\partial P_j}{\partial x_r}\frac{\partial C_j}{\partial p_i}) \tag{40}$$

$\frac{\partial \overline{i}}{\partial q_i}$ is obtained from similar formulas. It is seen that the sensitivity of the objective function (the total expected cost) with respect to some parameters can be determined on the basis of the first order sensitivity coefficients of the probabilites and of the cost functions, see (36)-(37). However, calculation of the sensitivities of the decision parameters is much more complicated because it involves estimation of the second order sensitivity coefficients of the probabilites, see e.g. Enevoldsen [22].

5. EXAMPLE

The following example is taken from Sørensen et al. [23]. A fatigue sensitive structural component in an offshore structure is considered. The crack is described by the crack size a. Using Paris' equation to describe the crack growth and assuming constant geometry function then a safety margin $M_F(T)$ corresponding to failure defined by having a crack larger than a_c is

$$M_F(T) = a_c - a_T \tag{41}$$

where

$$a_T = \left(a_0^{\frac{2-m}{2}} + \frac{2-m}{2} C \pi^{\frac{m}{2}} \nu T E\left[\left((z_0 + \frac{z_1}{z})\Delta\sigma\right)^m\right]\right)^{\frac{2}{2-m}} \tag{42}$$

a_0 is the initial crack size, m and C are parameters in Paris law, $\nu(H_S, T_Z)$ is the expected number of stress ranges per year, H_S is the significant wave height, T_Z is the zero crossing period, $E[\Delta\sigma(H_S, T_Z)^m]$ is the expected value of stress ranges $\Delta\sigma$ to the power m, z is a design parameter and z_0 and z_1 model the influence of the design parameter.

An event margin $M_D(T)$ modelling detection of a crack is written

$$M_D(T) = a_d - a_T \tag{43}$$

An event margin $M_R(T)$ used to model the repair decision is written

$$M_R(T) = (a_r + \epsilon) - a_T \tag{44}$$

where a_r is the repair level parameter and ϵ is the measurement uncertainty. ν and $E[\Delta\sigma^m]$ are modelled as functions of H_S and T_Z. This relationship is described by response surfaces. The tripod offshore structure in Karadeniz [24] is considered and the response surfaces generated in Bryla et al. [25] are used.

The following constants are used : m=3, a_c=40 mm and T_L= 30 years. H_S and T_Z are modelled by dependent Weibull distributions, see Bryla et al. [25]. a_0 is exponentially distributed with expected value 0.1 mm. $\ln C$ is normally distributed with expected value -33 and standard deviation 0.47. a_d is modelled by (31)-(33) with Δ=1, a_0=0 and $\lambda(e) = e$. ϵ is normally distributed with standard deviation 0.1 mm. All lengths and forces are in mm and N.

At each inspection new stochastic variables for a_d and ϵ are generated and at each repair new stochastic variables for a_0 and C are generated.

The initial cost $C_0(z)$ is modelled by $C_0(z) = 0 + 0.00125z$. The inspection cost factor $C_{IN_i}(\bar{e})$ is modelled by $C_{IN_i}(\bar{e}) = 0.0125 + 0.1e_i^2$. The C_{R_i} factor in the repair costs is taken as $C_{R_i} = 1.25$ and C_F is assumed to be constant equal to 250. All costs are in mio. D-mark.

The lower limit $\beta^{\min}$ of the reliability index is chosen as 3.29. The real rate of interest r is 0.02. Upper and lower limits of the design parameter z are 60 and 40 mm, respectively.

Inspection planning:

The optimization problem (2)-(12) is solved using NLPQL [3]. The results with n=1 and n=2 are (C is the total expected costs in the lifetime in mio. D-mark and a_r and z are in mm) :

n	t_1	t_2	e_1	e_2	C
1	9.4		0.65		0.191
2	7.9	8.6	0.37	0.33	0.189

z = 60mm and a_r = 5mm both for n = 1 and for n = 2. The results show that it is a little cheaper to perform two inspections than one inspection.

Repair planning:

After the first inspection is performed the decision whether repair should be performed or not is based on the following calculations : If the n=2 result is used the first inspection is performed after 7.9 years. Assuming that the crack measurement corresponds to the expected value of a after 7.9 years the solution of the updated optimization problem is

n	t_1	t_2	e_1	e_2	C
1		7.1		0.38	0.093

If this cost for the remaining lifetime is compared with a repair cost (C_R=1.25) it is seen that no repair should be performed. The next inspection time is thus after 7.1 years. Adding the initial costs and the inspection costs at the first inspection gives the total expected cost in the whole lifetime as $C = 0.190$.

6. CONCLUSIONS

A model for reliability-based optimal inspection and repair planning is described in general and for application with concrete bridges and offshore steel platforms. For concrete bridges a decision model for optimal repair planning is described. The total expected cost-benefits related to the repair and possible failure are maximized with respect to the parameters in the repair plan. For single components in offshore steel platforms the total expected costs in the lifetime are minimized with the number of inspections, the inspection times and efforts as decision variables. It is described how information obtained by an inspection can be used in the repair decision process and to update the inspection plan.

Stochastic models for inspection, measurement and repair actions are presented. Further it is shown how sensitivities of the expected total costs associated with the optimal inspection and repair plan and of the inspection parameters can be estimated. These sensitivities are determined with respect to critical parameters in the models which the decision maker assumes are subject to uncertainty but which are not included in the stochastic model. Examples of such parameters are cost coefficients and statistical parameters.

A simple example is presented which illustrates the inspection and repair planning process.

7. ACKNOWLEDGEMENTS

Part of this paper is supported by the research project "Risk Analysis and Economic Decision Theory for Structural Systems" supported by the Danish Technical Research Council.

8. REFERENCES

[1] Madsen, H.O., Krenk, S. and Lind, N.C.: *Methods of Structural Safety*. Prentice-Hall, 1986.

[2] Engelund, S. & R. Rackwitz: A Benchmark Study on Importance Sampling Techniques in Structural Reliability. To appear in Structural Safety, 1993.

[3] Schittkowski, K.: NLPQL : A FORTRAN Subroutine Solving Non-Linear Programming Problems. Annals of Operations Research, 1986.

[4] Powell, M.J.D.: VMCWD: A FORTRAN Subroutine for Constrained Optimization. Report DAMTP 1982/NA4, Cambridge University, England, 1982.

[5] Skjong, R.: Reliability-Based Optimization of Inspection Strategies. Proc. ICOSSAR'85 Vol. III. pp. 614-618, 1985.

[6] Thoft-Christensen, P. and Sørensen, J.D.: Optimal Strategies for Inspection and Repair of Structural Systems. Civil Engineering Systems, Vol. 4, June 1987, pp. 94-100.

[7] Fujimoto, Y., Itagaki, H., Itoh, S., Asada, H. and Shinozuka, M.: Bayesian Reliability Analysis of Structures with Multiple Components. Proceedings ICOSSAR 89, San Francisco 1989, pp. 2143-2146.

[8] Madsen, H.O., Sørensen, J.D. and Olesen, R.: Optimal Inspection Planning for Fatigue Damage of Offshore Structures. Proceedings ICOSSAR 89, San Francisco 1989, pp. 2099-2106.

[9] Madsen, H.O. and Sørensen, J.D.: Probability-Based Optimization of Fatigue Design Inspection and Maintenance. Presented at Int. Symp. on Offshore Structures, July 1990, University of Glasgow.

[10] Fujita, M., Schall, G. and Rackwitz, R.: Adaptive Reliability Based Inspection Strategies for Structures Subject to Fatigue. Proceedings ICOSSAR 89, San Francisco 1989, pp. 1619-1626.

[11] Sørensen, J.D. & Thoft-Christensen, P.: Inspection Strategies for Concrete Bridges. Proc. IFIP WG 7.5 1988, Vol. 48, pp. 325-335, Springer-Verlag.

[12] Raiffa, H. and Schlaifer, R.: Applied Statistical Decision Theory. Harward University Press, Cambridge, Mass., 1961.

[13] Benjamin, J.R. and Cornell, C.A.: Probability, Statistics and Decision for Civil Engineers. McGraw-Hill, 1970.

[14] Sørensen, J.D., Faber, M.H. and Kroon, I.B.: Risk Based Optimal Fatigue Testing. Probabilistic Mechanics and Structural and Geotechnical Reliability, Proceedings of the sixth speciality conference, Denver, Colorado, 1992, pp. .

[15] Thoft-Christensen, P.: Reliability-Based Expert Systems for Optimal Maintenance of Concrete Bridges. Proc. of the Structures Congress 93, Irvine, California, ASCE, pp. 1053-1058, 1993.

[16] Lindley, D.V.: Introduction to Probability and Statistics from a Bayesian Viewpoint, Vol 1+2. Cambridge University Press, Cambridge 1976.

[17] Madsen, H.O.: Model Updating in Reliability Theory. Proceedings of ICASP5, pp. 564-577, 1987.

[18] Rackwitz, R. & Schrupp, K.: Quality Control, Proof Testing and Structural Reliability. Structural Safety, Vol. 2, 1985, pp. 239-244.

[19] Pedersen, C., Nielsen, J.A., Riber, J.P., Madsen, H.O., Krenk, S.: Reliability Based Inspection Planning for the Tyra Field. Proceedings of OMAE 1992, Calgary, Canada, pp. 255-263.

[20] Faber, M.H., I.B. Kroon & J.D. Sørensen: Sensitivities in Structural Maintenance Planning. Submitted to Reliability Engineering and Systems Safety, 1993.

[21] Haftka, R.T & M.P. Kamat : Elements of Structural Optimization. Martinus Nijhoff, The Hague, 1985.

[22] Enevoldsen, I. : Sensitivity Analysis of a Reliability-Based Optimal Solution. Strucutral Reliability Theory, Paper no. 101, The University of Aalborg, 1992 - accepted for publication in ASCE, Journal of Engineering Mechanics.

[23] Sørensen, J.D., Faber, M.H., Thoft-Christensen, P. and Rackwitz, R.: Modeling in Optimal Inspection and Repair. Proceedings of OMAE 1991, Vol. II, Stavanger, Norway, pp. 281-288.

[24] Karadeniz, H.: Advanced Stochastic Analysis Program for Offshore Structures. User's manual for SAPOS. TU Delft, August 1989.

[25] Bryla, Ph., Faber, M.H. & Rackwitz, R.: Second Order Methods in Time Variant Reliability Problems. Proceedings of OMAE 1991, Vol. II, Stavanger, Norway, pp. 143-150.

Reliability and Optimization of Structural Systems, V (B-12)
P. Thoft-Christensen and H. Ishikawa (Editors)
Elsevier Science Publishers B.V. (North-Holland)

Series System Second Order Bounds on Vector Process Outcrossing Rates

O. Ditlevsen

Department of Structural Engineering, Technical University of Denmark
Building 118, DK-2800 Lyngby, Denmark

Abstract The reasoning that leads to the well–known second order bounds on the failure probability of a random variable series system reliability problem is applied to obtain similar formulae for upper and lower bounds on the mean outcrossing rate of a general vector process out of the safe set of a series system given that the stream of outcrossings satisfies a general regularity condition. It is demonstrated that a direct application of the lower failure probability bound on outcrossing rates can be erroneous.

History of second order bounds

In the structural reliability theory a series system is characterized by having a failure event $\mathcal{F}$ which is the union of several failure events $\mathcal{F}_1,\ldots,\mathcal{F}_m$. For a time invariant problem the system failure probability $P(\mathcal{F})$ is bounded in terms of the probabilities $P(\mathcal{F}_i)$, $i = 1,\ldots,m$, and $P(\mathcal{F}_i \cap \mathcal{F}_j)$, $i,j = 1,\ldots,m$, by the inequalities

$$P(\mathcal{F}) \le \sum_{i=1}^{m} P(\mathcal{F}_i) - \sum_{i=2}^{m} \max_{j<i}\{P(\mathcal{F}_i \cap \mathcal{F}_j)\} \tag{1}$$

$$P(\mathcal{F}) \ge P(\mathcal{F}_1) + \sum_{i=2}^{m} \max\{P(\mathcal{F}_i) - \sum_{j=1}^{i-1} P(\mathcal{F}_i \cap \mathcal{F}_j), 0\} \tag{2}$$

In the domain of small failure probabilities these inequalities are essential improvements of the simple inequalities

$$\max_i\{P(\mathcal{F}_i)\} \le P(\mathcal{F}) \le \sum_{i=1}^{m} P(\mathcal{F}_i) \tag{3}$$

used in early papers on reliability investigations of structural systems, Cornell (1967) Moses and Kinser (1967), Ang and Amin (1968). The upper bound given by eq. (1) was first formulated within the field of structural systems reliability by Vanmarcke (1973) on the basis of a particular failure set representation formulated in Moses and Kinser (1967).

The fact that the upper bound in eq. (1) often is a surprisingly good approximation to the failure probability $P(\mathcal{F})$, given that $P(\mathcal{F})$ is small, became clear after the formulation of the lower bound in eq. (2), Ditlevsen (1979).

After the completion of this development it was recognized among structural reliability theoreticians that both bounds (1) and (2) were formulated independently within the field of mathematical statistics, Kounias (1968), Hunter (1977).

Shortcoming of the intuition in applications on mean outcrossing rates

For time variant problems the failure probability $P(\mathcal{F})$ corresponding to a given time period is interpreted as the probability that a vector process $\mathbf{X}(t)$ crosses out of the safe set $\mathcal{S}$ and into the failure set $\mathcal{F}$ within the considered time period. Assume that the vector process $\mathbf{X}(t)$ obeys the following quite general regularity condition for the stream of outcrossings out of $\mathcal{S}$: There is a function $\nu(t)$ of time t such that the probability of having an outcrossing in the time interval from t to $t+\Delta t$ is $\nu(t)\Delta t+o(\Delta t)$ where $o(\Delta t)/\Delta t \to 0$ for $\Delta t \to 0$, while the probability of having $i \geq 2$ outcrossings in the interval is $o(\Delta t)p_i$ where $2p_2+3p_3+...ip_i+... < \infty$. Then it is seen by direct calculation that $\nu(t)$ is the mean outcrossing rate at t (i.e. the mean number of outcrossings per time unit at time t). For example, Gaussian vector processes with differentiable sample paths satisfy this regularity condition given that the boundary $\partial\mathcal{S}$ is sufficiently regular to ensure the existence of well defined outcrossing points.

Focusing on the event of failure in the time interval $[t,t+\Delta t[$ and relying on intuition the inequalities (1) and (2) should be applicable on

$$P(\mathcal{F}) \simeq \nu_{\partial\mathcal{S}}\Delta t \tag{4}$$

with

$$P(\mathcal{F}_i) \simeq \nu_{\partial\mathcal{S}_i}\Delta t \tag{5}$$

$$P(\mathcal{F}_i \cap \mathcal{F}_j) \simeq (\nu_{\partial\mathcal{S}_i \cap \mathcal{F}_j} + \nu_{\partial\mathcal{S}_j \cap \mathcal{F}_i})\Delta t \tag{6}$$

in which $\nu_{\partial\mathcal{S}} = \nu(t)$, $\nu_{\partial\mathcal{S}_i}$ is the mean outcrossing rate from the ith safe set $\mathcal{S}_i$ (complement to $\mathcal{F}_i$) at time t , and $\nu_{\partial\mathcal{S}_i \cap \mathcal{F}_j}$ is the mean crossing rate from $\mathcal{S}_i \cap \mathcal{F}_j$ into $\mathcal{F}_i \cap \mathcal{F}_j$ at time t . Thus the inequalities (1) and (2) become

$$\nu_{\partial\mathcal{S}} \leq \sum_{i=1}^{m} \nu_{\partial\mathcal{S}_i} - \sum_{i=2}^{m} \max_{j<i}\{\nu_{\partial\mathcal{S}_i \cap \mathcal{F}_j} + \nu_{\partial\mathcal{S}_j \cap \mathcal{F}_i}\} \tag{7}$$

$$\nu_{\partial\mathcal{S}} \geq \nu_{\partial\mathcal{S}_1} + \sum_{i=2}^{m} \max\{\nu_{\partial\mathcal{S}_i} - \sum_{j=1}^{i-1} (\nu_{\partial\mathcal{S}_i \cap \mathcal{F}_j} + \nu_{\partial\mathcal{S}_j \cap \mathcal{F}_i}), 0\} \tag{8}$$

$$\geq \sum_{i=1}^{m} \nu_{\partial\mathcal{S}_i} - \sum_{k=1}^{m} \sum_{\substack{i=1 \\ i \neq k}}^{m} \nu_{\partial\mathcal{S}_k \cap \mathcal{S}_i} \tag{9}$$

It turns out that the intuition fails with respect to the derivation of the inequality (8). Consider a three–dimensional vector process $\mathbf{X}(t)$ which in its probabilistic structure is rotational symmetric with respect to the origin. Let $\mathcal{S}$ be the first octant of an orthogonal coordinate system. Then $\partial\mathcal{S}_1$, $\partial\mathcal{S}_2$, $\partial\mathcal{S}_3$ are the three coordinate planes. Assume that $\nu_{\partial\mathcal{S}_i} = 4$. Then $\nu_{\partial\mathcal{S}_i \cap \mathcal{F}_j} = 2$ for all $i \neq j$ and eq. (7) gives

$$\nu_{\partial\mathcal{S}} \leq (4+4+4) - (2+2) - \max\{2+2,2+2\} = 4 \tag{10}$$

while eq. (8) gives

$$\nu_{\partial\mathcal{S}} \leq 4 + \max\{4-(2+2),0\} + \max\{4-(2+2) - (2+2),0\} = 4 \tag{11}$$

Together eqs. (10) and (11) imply that $\nu_{\partial\mathcal{S}} = 4$. But this is obviously wrong because it is directly seen that $\nu_{\partial\mathcal{S}} = 1+1+1 = 3$. However, the inequality (9) gives $\nu_{\partial\mathcal{S}} \geq 0$ and it turns out that both the inequalities (7) and (9) are generally valid. In fact, the next section gives a valid derivation that even implies a sharpening of both inequalities.

Outcrossing rate upper bound for general series systems

The first part of the derivation follows closely the same line of reasoning as in Ditlevsen (1979). For completeness this reasoning is repeated here.

Let $\mathbf{X}(t)$ be the actual vector process and consider a series system reliability problem with the single mode safe sets $\mathcal{S}_1,\dots,\mathcal{S}_m \subset \mathbb{R}^n$ and the corresponding boundaries (mode limit states) $\partial\mathcal{S}_1 \subset \mathcal{S}_1,\dots, \partial\mathcal{S}_m \subset \mathcal{S}_m$. The problem is to bound the outcrossing rate $\nu(t)$ of $\mathbf{X}(t)$ out of the intersection $\mathcal{S}_1 \cap \dots \cap \mathcal{S}_m$. Focus on the time interval from t to $t+\Delta t$ and let

$$1_{\mathcal{S}_i} = \begin{cases} 1 \text{ if } \mathbf{X}(\tau) \in \mathcal{S}_i \text{ for some } \tau \in [t,t+\Delta t[\\ 0 \text{ otherwise} \end{cases} \tag{12}$$

$$1_{\partial\mathcal{S}_i} = \begin{cases} 1 \text{ if } \mathbf{X}(\tau_1) \in \mathcal{S}_i \text{ and } \mathbf{X}(\tau_2) \in \mathcal{F}_i \\ \quad \text{for some } \tau_1,\tau_2 \text{ such that } t \leq \tau_1 < \tau_2 < t+\Delta t \\ 0 \text{ otherwise} \end{cases} \tag{13}$$

where $\mathcal{F}_i$ is the ith failure set, that is, the complementary set to $\mathcal{S}_i$. We obviously have that

$$\sum_{i=1}^{m} 1_{\mathcal{S}_1}\cdot\ldots\cdot 1_{\mathcal{S}_{i-1}} 1_{\partial\mathcal{S}_i} 1_{\mathcal{S}_{i+1}}\cdot\ldots\cdot 1_{\mathcal{S}_m} \tag{14}$$

is zero if and only if there is no crossing out of $\mathcal{S}_1 \cap \dots \cap \mathcal{S}_m$ in the time interval.

Assuming that an intersection between $\mathbf{X}(t)$ and $\partial\mathcal{S}_i \cap \partial\mathcal{S}_j$, $i \neq j$, occurs with probability $o(\Delta t)$ in the considered time interval it is clear that the expectation of the random variable (14) under the previously formulated regularity condition is $\nu(t)\Delta t+o(\Delta t)$.

With sufficient generality we can concentrate on finding the expectation $\nu_1\Delta t+o(\Delta t)$ of the first term

$$1_{\partial\mathcal{S}_1} 1_{\mathcal{S}_2}\cdot\ldots\cdot 1_{\mathcal{S}_m} = 1_{\mathcal{S}_2\cap\partial\mathcal{S}_1}\cdot\ldots\cdot 1_{\mathcal{S}_m\cap\partial\mathcal{S}_1} \tag{15}$$

in eq. (14). Writing $1_{\mathcal{S}_i\cap\partial\mathcal{S}_1}$ as 1_i, $i = 2,\dots,m$, and setting $1_1=1$ we have

$$1-1_2\cdot\ldots\cdot 1_m = \sum_{i=2}^{m} 1_1\cdot\ldots\cdot 1_{i-1}(1-1_i) = 1-1_2+\sum_{i=3}^{m}\max\left\{\left[1-\sum_{j=2}^{i-1}(1-1_j)\right](1-1_i),0\right\} \tag{16}$$

Taking expectation and using the general expectation property $E[\max\{X,Y\}] \geq \max\{E[X],E[Y]\}$ we get with obvious notation that

$$\begin{aligned}
&1-\nu_1\Delta t + o(\Delta t) \geq 1-\nu_{\mathcal{S}_2\cap\partial\mathcal{S}_1}\Delta t + \\
&\sum_{i=3}^{m}\max\left\{1-\nu_{\mathcal{S}_i\cap\partial\mathcal{S}_1}\Delta t - \sum_{j=2}^{i-1}\left[1-(\nu_{\mathcal{S}_i\cap\partial\mathcal{S}_1}+\nu_{\mathcal{S}_j\cap\partial\mathcal{S}_1}-\nu_{\mathcal{S}_i\cap\mathcal{S}_j\cap\partial\mathcal{S}_1})\Delta t\right],0\right\} \\
&= 1-\nu_{\mathcal{S}_2\cap\partial\mathcal{S}_1}\Delta t + \\
&\sum_{i=3}^{m}\max\left\{3-i+\Delta t\left[-\nu_{\mathcal{S}_i\cap\partial\mathcal{S}_1}+\sum_{j=2}^{i-1}(\nu_{\mathcal{S}_i\cap\partial\mathcal{S}_1}+\nu_{\mathcal{S}_j\cap\partial\mathcal{S}_1}-\nu_{\mathcal{S}_i\cap\mathcal{S}_j\cap\partial\mathcal{S}_1})\right],0\right\} \\
&= 1-\nu_{\mathcal{S}_3\cap\mathcal{S}_2\cap\partial\mathcal{S}_1}\Delta t
\end{aligned} \tag{17}$$

for Δt sufficiently small. Thus it follows that

$$\nu_1 \leq \nu_{\mathcal{S}_3\cap\mathcal{S}_2\cap\partial\mathcal{S}_1} \tag{18}$$

and the best upper bound on $\nu(t)$ we can achieve from the second order bounds is therefore

$$\nu(t) \leq \sum_{k=1}^{m}\min_{\substack{i\neq j\\ i\neq k\\ j\neq k}}\left\{\nu_{\mathcal{S}_i\cap\mathcal{S}_j\cap\partial\mathcal{S}_k}(t)\right\} \tag{19}$$

Since (Morgan's rule)

$$1_{\mathcal{S}_i\cap\mathcal{S}_j\cap\partial\mathcal{S}_k} = 1_{\partial\mathcal{S}_k} - 1_{\mathcal{F}_i\cap\partial\mathcal{S}_k} - 1_{\mathcal{F}_j\cap\partial\mathcal{S}_k} + 1_{\mathcal{F}_i\cap\mathcal{F}_j\cap\partial\mathcal{S}_k} \tag{20}$$

the upper bound may alternatively be written as

$$\nu(t) \le \sum_{k=1}^{m} \nu_{\partial\mathcal{S}_k} - \sum_{k=1}^{m} \max_{\substack{i\neq j\\ i\neq k\\ j\neq k}} \left\{ \nu_{\mathcal{F}_i\cap\partial\mathcal{S}_k} + \nu_{\mathcal{F}_j\cap\partial\mathcal{S}_k} - \nu_{\mathcal{F}_i\cap\mathcal{F}_j\cap\partial\mathcal{S}_k} \right\} \tag{21}$$

The equality sign is valid for $m \le 3$. It is seen that the inequality (7) is consistent with this result but also that the inequality (21) is a sharpening of the inequality (7).

Outcrossing rate lower bound for general series system

Going back to the first part of eq. (16) we have

$$1 - 1_2 \cdot \ldots \cdot 1_m \le 1 - 1_2 + \sum_{i=3}^{m} \min_{1<j<i} \{1_j(1-1_i)\} \tag{22}$$

which by taking expectation gives

$$1 - \nu_1 \Delta t + o(\Delta t) \le 1 - \nu_{\mathcal{S}_2\cap\partial\mathcal{S}_1} \Delta t + \sum_{i=3}^{m} \min_{1<j<i} \{\nu_{\mathcal{S}_j\cap\partial\mathcal{S}_1} - \nu_{\mathcal{S}_i\cap\mathcal{S}_j\cap\partial\mathcal{S}_1}\}\Delta t \tag{23}$$

or

$$\nu_1 \ge \nu_{\mathcal{S}_2\cap\partial\mathcal{S}_1} - \sum_{i=3}^{m} \min_{1<j<i} \{\nu_{\mathcal{F}_i\cap\mathcal{S}_j\cap\partial\mathcal{S}_1}\}$$

$$= \nu_{\partial\mathcal{S}_1} - \nu_{\mathcal{F}_2\cap\partial\mathcal{S}_1} - \sum_{i=3}^{m} \min_{1<j<i} \{\nu_{\mathcal{F}_i\cap\partial\mathcal{S}_1} - \nu_{\mathcal{F}_i\cap\mathcal{F}_j\cap\partial\mathcal{S}_1}\}$$

$$= \nu_{\partial\mathcal{S}_1} - \sum_{i=2}^{m} \nu_{\mathcal{F}_i\cap\partial\mathcal{S}_1} + \sum_{i=3}^{m} \max_{1<j<i} \{\nu_{\mathcal{F}_i\cap\mathcal{F}_j\cap\partial\mathcal{S}_1}\} \tag{24}$$

Thus we get the lower bound

$$\nu(t) \ge \sum_{k=1}^{m} \nu_{\partial\mathcal{S}_k} - \sum_{k=1}^{m} \sum_{i\neq k}^{m} \nu_{\mathcal{F}_i\cap\partial\mathcal{S}_k} + \sum_{k=1}^{m} \sum_{i\neq k}^{m} \max_{\substack{j<i\\ j\neq k}} \{\nu_{\mathcal{F}_i\cap\mathcal{F}_j\cap\partial\mathcal{S}_k}\} \tag{25}$$

where the equality sign is valid for $m \leq 3$. It is seen that the inequality (9) is consistent with this result but also that the inequality (25) is a sharpening of the inequality (9).

The bound (25) may possibly be improved by using an optimal numbering of the failure modes for each of the terms in $\nu(t) = \nu_1 + ... + \nu_m$ where ν_i corresponds to the ith term in eq. (14).

Acknowledgement

This work has been financially supported by the Danish Technical Research Council

References

Ang, A.H.–S. and Amin, M. (1968): Reliability of structures and structural systems. *J. Eng. Mech.*, ASCE, 94(2), 671–691.

Cornell, C.A. (1967): Bounds on the reliability of structural systems. *Journal of Struct. Div.*, ASCE, 93(1), 171–200.

Ditlevsen, O. (1979). Narrow reliability bounds for structural systems. *J. Struct. Mech.*, 7(4), 435–451.

Hunter, D. (1977): Approximating percentage points of statistics expressible as maxima. *TIMS Studies in the Management Sciences*, 7, North Holland Publ. Comp.

Kounias, E.G. (1968): Bounds for the probability of a union, with applications. *Ann. Math. Stat.*, 39(6), 2154–2158.

Moses, F. and Kinser, D.E. (1967): Analysis of structural reliability, *J. Struct. Div.*, ASCE, 93(5), 147–164.

Vanmarcke, E.H. (1973): Matrix formulation of reliability analysis and reliability–based design. *Computers & Structures*, 3, 757–770.

Reliability and Optimization of Structural Systems, V (B-12)
P. Thoft-Christensen and H. Ishikawa (Editors)
Elsevier Science Publishers B.V. (North-Holland)
1993 IFIP

Optimal Bayesian Designs for Fatigue Tests

S. Engelund, V. Bouyssy, R. Rackwitz

Technical University of Munich
Arcisstrasse 21, D-8000 Munich 2, Germany

ABSTRACT
Cost optimal plans for the estimation of parameters describing the fatigue crack initiation lifetime of a structural element are determined on the basis of Bayesian decision analysis. Bayesian statistics allow prior information to be taken into account. Thereby already existing test results can be used in the analysis. A cost optimal test plan is determined for a given set of known distribution parameters. The plan, however, can be very sensitive towards changes of the parameters and, therefore, depends strongly on the quality of the initial guesses for the parameters. If a number of prior experimental results are available and if the purpose of the experiments is to validate the distribution parameters determined from these experiments it is possible to find a robust experimental plan.

1. INTRODUCTION

Whenever critical fatigue sensitive structural elements are identified in important structures special full size tests are frequently performed. Such tests are almost always very costly and careful planning of tests is mandatory. Thereby, it can be useful to consider prior information from tests previously made with similar structural elements, materials and manufacturing methods. By the tests and possibly the prior information the parameters of the lifetime distribution need to be estimated. The designer of an experimental plan faces the question : how to choose the experimental conditions in such a way that the experiment contains a maximum of information? The decision will among other things have to be based on the available apparatus, economic considerations and the prior information about the experimental outcomes. The scope of this paper is to review and apply a methodology based on classical decision theory which enables the experimenter to determine an optimal experimental plan.

2. EXPERIMENTAL PLANNING

In figure 1 the main steps in experimental planning are shown. On the basis of some prior knowledge a probabilistic model for the experimental outcomes must be formulated. Then, an experimental plan is determined such that the amount of information from the additional tests is maximized or, by economical considerations, the total cost are minimized. The experimental outcomes can then be used to verify the selected model as well as to plan additional experiments if the information obtained so far still shows to large scatter. Experimentation is terminated when a given accuracy criterion for the parameters is met and the adopted model is verified by a statistical test. In the following it is assumed that there is no doubt about the model.

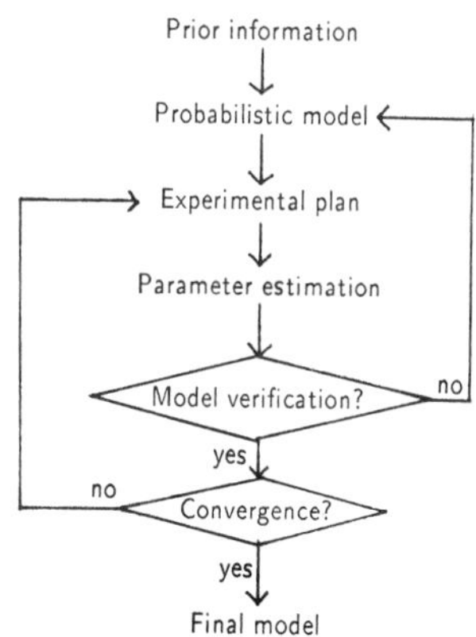

Figure 1: Experimental planning.

The planning of fatigue experiments or, more generally, for accelerated testing has had an interesting historical development. Already Chernoff [3] formulated almost all basic concepts. He introduced the concept of locally optimal designs, that is, designs which are optimal if the parameters were known to be at given values. Then, the outcome of a future experiment can be determined in terms of a probability distribution function accounting for the inherent variability in fatigue lives. It follows that in this context, any experimental plan depends strongly on the choice of the next observations. In the presence of statistical uncertainties the assumption of known parameters appears somewhat self contradictory and suggests by itself Bayesian reasoning. But Bayesian reasoning, of course, does not solve the question where to place the uncertain future outcomes. At most, a probability distribution can be assigned to the unknown parameters. Chernoff [3] performs a maximum likelihood estimation for all parameters in a polynomial for the mean of an exponential distribution. The optimal plan is the plan which minimizes the sum of the variances of the parameters (so–called A–optimality). This optimality criterion fails to take into account the influence of the parameters on the reliability of a structural element. Later a number of similar optimality criteria have been proposed. They are all based on some functional of the accuracy of the parameter estimates as expressed by the parameter covariance matrix, e.g. the trace (A–optimality), the determinant (D–optimality), the maximum eigenvalue (E–optimality) or some quadratic form (C–optimality) (see Pilz [13]). Chernoff argues in a classical context that, if a sensitivity analysis shows that the design is nearly optimal for a sufficiently large range of parameters around the given values such a plan would be satisfying in practical applications.

Chernoff's findings have been generalized in various directions. Mann [9] determined optimal designs for the Weibull distribution with constant shape parameter and with a scale parameter which is a polynomial function of the reciprocal of the applied stress. The test plan is designed such that the least squares curve has minimum variance for a given stress level. Escobar and Meeker [4] investigated the more general case where the lifetime follows a location–scale distribution and the location parameter is a function of the stress level. The optimal plan is determined by minimizing the variance of the best linear unbiased estimator of a percentile of the lifetime distribution at working stress. Clearly, this leads to simple reliability oriented experimental plans. Nelson and Kielpinski [11] used a result of Chernoff [3] as basis for the planning of life tests for normal and lognormal distributions. Chernoff [3] showed that if the parameter vector $\boldsymbol{\theta} = f(\theta_1, \theta_2, \dots \theta_r)$ is to be estimated there is an asymptotically optimal design which involves repeating at most r of the available experiments. For a simple linear model this implies that it is only necessary to perform experiments at two different stress levels. The high stress level should be as large as possible in order to minimize the variance. The low stress level and the proportion of tests to be performed at the low stress level are determined by minimizing the large–sample variance of the maximum likelihood estimate of the mean at a specified working stress. Further results on these lines may be found in Kielpinski and Nelson [7], Meeker and Nelson [10], Nelson and Meeker [12]. It is clear that these general findings carry over also to a Bayesian analysis. General Bayesian considerations for experimental design can be found in Pilz [13] and Chaloner [2]. Chaloner [2] determines optimal designs for linear models where the observations are normally distributed. The designs are optimal for estimating a linear combination of the regression parameters by minimizing its variance (e.g. the variance of lifetime at working stress).

Minimizing the variance of certain lifetime estimators, at mentioned, does not yet account for the skewness in the loss function in the original space. Sørensen et al. [15] suggested a criterion involving directly the experimental cost and the monetary consequences when using the experimental results for predictions of the performance of the object under consideration. Moreover, they suggested to determine optimal plans by using Bayesian decision theory (see Raiffa and Schlaifer [14]) thus enabling incorporation of prior information. This is also the approach to be discussed in more detail in this paper. The optimal plan is defined as the plan which minimizes the expected cost

$$E_{\mathbf{Y}}[C_T] = E_{\mathbf{Y}}[C_F\ P_f] + E_{\mathbf{Y}}[E_c] \tag{1}$$

where C_F denotes the cost of failure, C_T the total cost, P_f the failure probability and E_c the experimental cost. Index $\mathbf{Y}$ indicates that the expectation is taken with regard to the experimental outcomes $\mathbf{y}$. Assume that the unknown parameters $\mathbf{p}$ in the prediction model can be regarded as stochastic variables. On the basis of prior information (objective or subjective) a prior density $f'(\mathbf{p})$ of the parameters is determined or selected. If no prior information is available a noninformative prior can be used. From observations $\mathbf{y}$ of the stochastic variable $\mathbf{Y}$, the a posterior density of the parameters can be determined

$$f''(\mathbf{p} \mid \mathbf{x},\mathbf{y}) = \frac{1}{c} f'(\mathbf{p})\ L(\mathbf{p} \mid \mathbf{x},\mathbf{y}) \tag{2}$$

where $L(\mathbf{p} \mid \mathbf{x},\mathbf{y})$ is the likelihood function and $f''(\mathbf{p} \mid \mathbf{x},\mathbf{y})$ the posterior density of the parameters $\mathbf{p}$ conditional on the stress levels $\mathbf{x}$ and the experimental results (life times) $\mathbf{y}$. c is a normalizing constant. For independent, non–censored observations the likelihood function is

$$L(\mathbf{p}|\mathbf{x},\mathbf{y}) = \left[\prod_{i=1}^{m} f_Y(y_i|x_i,\mathbf{p})\right] \tag{3}$$

where $f_Y()$ denotes the density function of lifetimes Y. The failure probability of the considered structural element is calculated as a total or predictive probability

$$P(\text{failure}|\mathbf{y},\mathbf{x}) = \frac{1}{c}\int_{\Omega_f}\int_{\mathcal{P}} f_Y(y|x_w,\mathbf{p})\, f'(\mathbf{p}) \left[\prod_{i=1}^{m} f_Y(y_i|x_i,\mathbf{p})\right] d\mathbf{p}\, d\mathbf{y} \tag{4}$$

where the parameters $\mathbf{p}$ are defined in the space $\mathcal{P}$, x_w is the working stress and Ω_f is the failure region defined as $\Omega_f = \{Y - y_f < 0\}$. y_f is the required number of load cycles within the lifetime of the structural element. For the planning of additional experiments it is now natural to assume that the cost of each future experiment are proportional to the duration of the test or, if formulation (1) is used, proportional to the expected duration. Furthermore, let the outcomes of the extra experiments be denoted by $\tilde{\mathbf{y}}$ at the selected stress levels $\tilde{\mathbf{x}}$. Then, in abbreviated notation the expected failure probability is

$$E[P_f|\tilde{\mathbf{x}}] = \int_0^{\infty} P(\text{failure}|\mathbf{y},\mathbf{x},\tilde{\mathbf{y}},\tilde{\mathbf{x}})\, f_{\tilde{\mathbf{y}}}(\tilde{\mathbf{y}}|\tilde{\mathbf{x}})\, d\tilde{\mathbf{y}} \tag{5}$$

where $P(\text{failure}|\mathbf{y},\tilde{\mathbf{y}})$ is determined as in eq. (4). This is, in fact, the classical setting of a Bayesian pre–posterior analysis. The new values $\tilde{\mathbf{y}}$ and thus the corresponding density depend on the experimental plan, i.e. on $\tilde{\mathbf{x}}$. From the above it follows that the future outcomes of experiments must obey the distribution derivable from the a priori knowledge. In particular, the parameters are at their expected prior values. Furthermore, since the prior information can be vague a sensitivity analysis must again show whether the experimental plan designed according to eq. (1) is sufficiently robust against variations in the a priori parameters.

In general, the numerical computations implied by eq. (1) can be performed by a FORM/SORM analysis (Guers/Rackwitz [5]) for arbitrary life time models. Other stochastic variables which might affect the reliability of the element can easily be included in the analysis. The integral eq. (5) is best determined by conditional sampling (see Ayub [1] and Karamchandani and Cornell [6]). If the lifetimes are assumed to be lognormally distributed and the relationship between lifetimes and stress levels follows the classical Wöhler–curve all computations are analytic for non–censored experiments. In the following illustrations only this special case is studied.

3. ILLUSTRATIONS

It is assumed that some characteristic of the lifetime (mean, median,...) follows the well known S–N or Wöhler equation

$$N = C\,(\Delta S)^{-k} \tag{6}$$

If a logarithmic transformation of the lifetimes and stress ranges is performed a linear relation exists between the transformed lifetimes and stress ranges. Let $x_i = \ln \Delta S_i$ and $y_i = \ln N_i$. It is assumed that Y follows a normal distribution with mean value

$\mu = \beta_1 + \beta_2 (x - \bar{x})$ and constant standard deviation, σ i.e.

$$y = \beta_1 + \beta_2 (x - \bar{x}) + \sigma I \tag{7}$$

where I is a standard normally distributed variable and

$$\bar{x} = \frac{1}{m} \sum_{i=1}^{m} x_i \tag{8}$$

The likelihood function is

$$L(y \mid x,p) = \sigma^{-m} \exp\left[-\frac{1}{2\sigma^2} \sum_{i=1}^{m} (y_i - \beta_1 - \beta_2 (x - \bar{x}))^2\right] \tag{9}$$

The quantities

$$\hat{\beta}_1 = \frac{1}{m} \sum_{i=1}^{m} y_i , \qquad \hat{\beta}_2 = \frac{\Sigma(x_i - \bar{x})y_i}{\Sigma(x_i - \bar{x})^2} ,$$

$$s^2 = \sum_{i=1}^{m} (y_i - \hat{\beta}_1 - \hat{\beta}_2 (x_i - \bar{x}))^2$$

are sufficient statistics for β_1, β_2 and σ^2, respectively. When a sufficient statistic exists, a conjugate prior exists, too. The parameters β_1 and β_2 given σ^2 are normally distributed with means $\hat{\beta}_1$ and $\hat{\beta}_2$ and covariance matrix

$$\mathrm{Cov}(\beta_1; \beta_2) = \sigma^2 \begin{bmatrix} \frac{1}{m} & 0 \\ 0 & \frac{1}{\Sigma(x_i - \bar{x})^2} \end{bmatrix} \tag{10}$$

Furthermore, $\nu s^2/\sigma^2$ follows a chi–square distribution with $\nu = (m - 2)$ degrees of freedom. The existence of conjugate priors considerably simplifies the problem because the posterior distribution can be determined analytically.

We consider a welded cruciform joint (see Fig. 2) subjected to constant amplitude in plane bending fatigue loads. Due to local stress concentrations or defects, small cracks initiate at the toe of the weld. As mentioned the best estimates of the parameters in a linear model are obtained by performing experiments only at two levels as wide apart as possible. For lifetime experiments the experimental costs decrease with increasing stress level. Therefore, one of the experimental levels is the highest allowable test stress. The lower test level is to be determined by the experimental planning. The optimization variables therefore are:

n_L: Number of experiments at the low stress level.
n_H: Number of experiments at the high stress level.
x_L: The logarithm of the lower stress level.

Eq. (1) can be rewritten as

$$E_Y[C_T] = C_F\, E_Y[P_f] + \sum_{i=1}^{n_L} \left[E_0 + E_Y[N_i]\, C_c \right] + \sum_{i=1}^{n_H} \left[E_0 + E_Y[N_i]\, C_c \right] \tag{11}$$

where E_0 are the experimental cost which are independent of the lifetime of the tested specimen, N_i are the numbers of cycles to failure for the i–th specimen and C_c is the cost per cycle.

At first the sensitivities of experimental plans against changes in the parameters is illustrated. A priori no test results are available, but it is assumed that the logarithm (with base 10) of the lifetimes follow a normal distribution with standard deviation and mean value

$\sigma = 0.13007$ $\qquad\qquad$ $\mu = 16.707 - 5.1262\ x$

Furthermore, the logarithm of the highest allowable stress level is $x_H = 2.5$. We now wish to determine the cost–optimal experimental plan. For the considered tests, rough estimates of the marginal costs are given in Table 1.

Cost of failure, C_F	$10 \cdot 10^6$
Experimental costs, E_0	300
Cost pro cycle	$300 \cdot 10^{-6}$

Table 1 : Marginal costs in ECU.

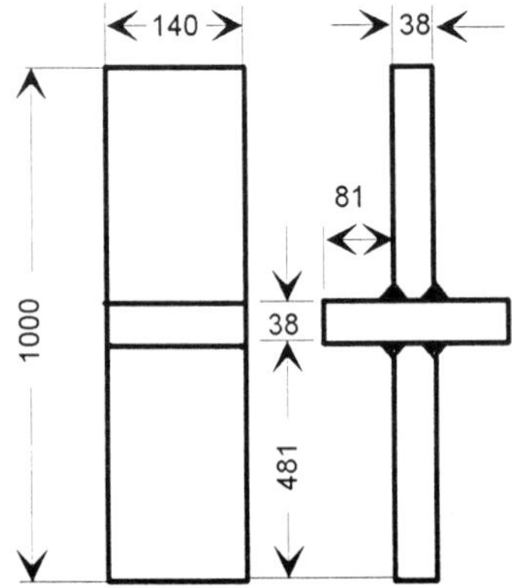

Figure 2: Cruciform joint.

The limit state function is

$$g = N - n_f = C.(\Delta S_w)^{-k} - n_f$$

or equivalently

$$g = \beta_1 + \beta_2 (\log \Delta S_w - \bar{x}) - \log n_f$$

where $C = 10^{(\beta_1 - \beta_2 \bar{x})}$ and $k = -\beta_2$. The working stress amplitude ΔS_w is assumed to be 50 MPa. The number of cycles within the service lifetime is $n_f = 5.10^4$. The following results are obtained:

$n_L = 3$, $n_H = 5$ and $x_L = 2.18$

The expected failure probability is $E[P_f] = 2.58 \cdot 10^{-5}$ and the expected total cost is $E[C_T] = 2980$ ECU.

If σ is assumed to be by 30 % smaller the expected cost of this plan becomes $E[C_T] = 2750$ ECU. The optimal plan, however, now is

$n_L = 3$, $n_H = 4$ and $x_L = 2.20$

The expected cost of this plan is $E[C_T] = 2550$ ECU, which is 7.8 % less than for the originally determined plan. Since a change of σ of 30 % is small relative to the accuracy of the initial "guess" for the parameter it must be concluded that the optimal plan is sensitive towards changes of the parameters.

In many cases experiments are not performed in order to determine the parameters but rather to validate parameter estimates derived from prior information. For example a series of tests have been conducted to assess the fatigue lifetime of the cruciform joint shown in figure 1. The data have been extracted from Lebas and Fauvre [6] (see Tab. 2). The experiments were conducted in air and at 20°C on non protected steel specimens at five different stress ranges, using a frequency of 2 Hz. The following number of load cycles up to detection of a crack with 2 mm depth have been recorded.

Stress range (MPa)	Number of cycles
264	20,800
200	79,300
170	216,000
130	485,000
100	3,670,000

Table 2 : Results from cruciform joint tests

Using the lognormal model the following results are obtained : $\bar{x} = 2.2134$, the mean values of β_1 and β_2 are $\hat{\beta}_1 = 5.3605$ and $\hat{\beta}_2 = -5.1262$ (see Fig. 2). Their estimated variances are $s_{\beta_1}{}^2 = 3.5410\ 10^{-3}$ and $s_{\beta_2}{}^2 = 0.16586$. The estimated variance of the error is $s^2 = 0.0177$.

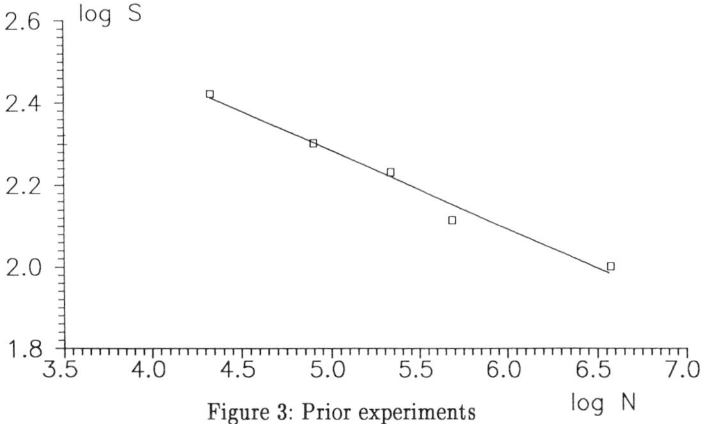

Figure 3: Prior experiments

Under the assumption that the outcomes of the next experiments follow a distribution with the prior parameters the optimal plan is to perform two additional experiments at the highest allowable stress level ($x_H = 2.5$). Due to the large cycle cost any "optimal" low stress level test turns out to be at or very close to the maximum allowable stress level. The sensitivity of the optimal plan towards changes of β_0, β_1 and σ must be investigated. When each of these parameters is changed by its prior standard deviation with the other parameters unchanged the optimal plan remains the same. It can, therefore, be concluded that a robust optimal plan has been found in this case.

It can be shown that plans with less informative prior knowledge (smaller number of prior experiments) become significantly less robust as illustrated by the first example with only vague prior information. This implies that reasonable experiment planning rests to a large extent on the firmness of prior knowledge. If prior knowledge is, in fact, rather vague it may be necessary to design experiments sequentially enabling step by step updating after each experiment.

4. SUMMARY AND CONCLUSION

A Bayesian methodology is suggested which allows the determination of a cost–optimal plan for fatigue tests of structural components. Essential for any planning is some informative prior knowledge about the parameters in the lifetime prediction model. It is suggested to use the expected prior parameters. However, as these are still uncertain, it is necessary to design experimental plans on the basis of sensitivity studies in accordance with classical schemes for the design of fatigue experiments. The method is illustrated at the particular case of crack initiation tests on a cruciform joint from the offshore industry. It is found that without prior experiments the optimal experimental plan depends strongly on initial guesses for the parameters of the distribution of lifetimes. If, on the other hand, informative priors are available and the purpose of the experiments is to validate parameter estimates determined on the basis of prior experiments a robust experimental plan can usually be found.

ACKNOWLEDGEMENTS

This study has in part been supported by Elf Aquitaine which is highly appreciated. We wish to acknowledge useful discussions with H. H. Müller.

REFERENCES

[1] Ayub, B. M., Lai, K.–L., Selective Sampling in Simulation–Based Reliability Assessment, Int. J. Pres. Ves. & Piping, 46, 1991, pp. 229–249.

[2] Chaloner, K., Optimal Bayesian Experimental Design for Linear Models, The Annals of Statistics, 12, 1984, pp. 283–300.

[3] Chernoff, H., Locally Optimal Designs for Estimating Parameters, Ann. Math. Stat., 24, 1953, pp. 586–602.

[4] Escobar, L. A., Meeker, W. Q., Planning Accelerated Life Test with Type II Censored Data, J. Statis. Comput. Simul., 23, 1986, pp. 273–297.

[5] Guers, F., Rackwitz, R., Time–Variant Reliability of Structural Systems Subject to Fatigue, ICASP 5, Vancouver, 1987

[6] Karamchandani, A., Cornell, C. A., Adaptive Hybrid Conditional Expectation Approaches for Reliability Estimation, Structural Safety, 11, 1991, pp. 59–74.

[7] Kielpinski, T. J., Nelson, W., Optimum Censored Accelerated Life Tests for Normal and Lognormal Life Distributions.

[8] Lebas, G., Fauvre, J. C., Collection of data : S–N Tests on welded specimen under constant amplitude loading, BRITE Project 2124, Task 1.2/004, 1989

[9] Mann, N. R., Design of Over–Stress Life–Test Experiments when Failure Times have the Two–Parameter Weibull Distribution, Technometrics, 14, 1972, pp. 437–451.

[10] Meeker, Q. M., Nelson, W., Optimum Accelerated Life–Tests for the Weibull and Extreme Value Distributions, IEEE Transactions on Reliability, R–24, 1975, pp. 321–332.

[11] Nelson, W., Kielpinski, T. J., Theory for Optimum Censored Accelerated Life Tests for Normal and Lognormal Life Distributions.

[12] Nelson, W., Meeker, Q. M., Theory for Optimum Accelerated Censored Life Tests for Weibull and Extreme Value Distributions

[13] Pilz, J., Bayesian Estimation and Experimental Design in Linear Regression Models, Teubner–Text zur Mathematik, Band 55, Leipzig, 1983

[14] Raiffa, H., Schlaifer, R., Applied Statistical Decision Theory, M.I.T. Press, Cambridge, USA, 1961.

[15] Sørensen, J. D., Faber, M. H., Kroon, I. B., Risk Based Optimal Fatigue Testing, Y. K. Lin (Ed.) Probabilistic Mechanics and Structural and Geotechnical Reliability, ASCE, Denver, 1992, pp. 523–526.

Reliability and Optimization of Structural Systems, V (B-12)
P. Thoft-Christensen and H. Ishikawa (Editors)
Elsevier Science Publishers B.V. (North-Holland)
1993 IFIP

Influence of Cumulative Damage on the Seismic Reliability of Multi-Storey Frames

L. Esteva, O. Diaz

Institute of Engineering, National University of Mexico
Apdo Postal 70472, México, D. F. Mexico

1.INTRODUCTION

It is well known that modern methods of earthquake-resistant design lead to structural systems that develop considerable nonlinear behavior when subjected to high-intensity earthquakes occurring in intervals of the order of a few tens of years. Thus, the possibility of significant damage is implicity accepted, provided the probability of collapse is kept sufficiently low. Design intensities and safety factors are determined on the basis of an optimization analysis (either formal or informal) aiming at obtaining a balance between the expected values of repair costs -and other consequences of damage- and the construction cost increments needed to reduce the former.

A large portion of the damage-related costs arise from repair work. On the other hand, in some cases, damage at some locations in a structure is not repaired, either because it is judged to be low or because repairing it implies extremely costly local demolition and reconstruction work. Thus, potential weak links may remain inadverted, thus jeopardizing the future performance of the system. All this leads to accumulation of non-negligible damage associated to the occurrence of sequences of moderate and high-intensity earthquakes. Cumulative damage leads to the degradation of the strength and the stiffness of the structure, and hence to an increase of its proneness to damage and failure during subsequent earthquakes.

The use of energy-dissipating devices provides a promising approach to cope with the foregoing problems. Practically speaking, energy dissipating devices may be dealt with as structural members with special properties, among which are particularly relevant: a) the stability of their hysteretic behavior cycles under the action of a large number of alternating load applications, and b) their possibility of being replaced in case of damage, by means of simple and reliable operations, not requiring the reconstruction of the elements anchoring them to the members of the conventional structure. These advantages are partly overshadowed by the cost increments tied to the fabrication and installation of the mentioned devices; hence the importance of studying the economics of possible applications.

In this paper the process of seismic damage accumulation is formulated as a Markov process, where the states of the system are described by the level and spatial distribution of local damage at the end of each earthquake and by the resulting failure probabilities as functions of earthquake intensities. The transition probabilities between successive states of the system depend on both natural environment (seismicity) and human actions (seismic design decisions; policies and rules for repair of damage and preventive replacement of energy-dissipating devices). The framework for making those decisions and establishing those policies and rules is provided by a formal decision approach based on initial construction costs and expected present values of benefits and damage, the latter including failure consequences, damage repair and device replacement. This framework is developed in the sequel, which also includes the formulation of models to describe the accumulation of damage, the transition probabilities between successive states of a system, and therefore the evolution of system reliability with seismic history. The paper ends with the application of the foregoing concepts to a simple case.

2.GENERAL FORMULATION

Seismic design criteria for conventional structures are expressed by means of a design response spectrum, the ordinates of which are reduced in order to account for nonlinear behavior. When energy-dissipating devices (EDD) are used, these criteria are complemented by a statement concerning the fractions of the lateral force and of the lateral stiffness which are to be taken by each of the systems contributing to the lateral strength: the conventional frame (CF) and the EDD. These fractions may vary along the system's height. For the particular case of a simple (single-bay, single--story) system, the design variables are fully determined through the initial tangent stiffnesses, k_F and k_D, of CF and EDD, respectively, and by the corresponding lateral strengths, R_F and R_D (Fig 1). Typical idealized load-deformation curves for these elements are shown in Figs 1a and b, which distinguish between the deterioration of stiffness and strength typical of reinforced concrete members, and the stability of the hysteretic cycles of the EDD.

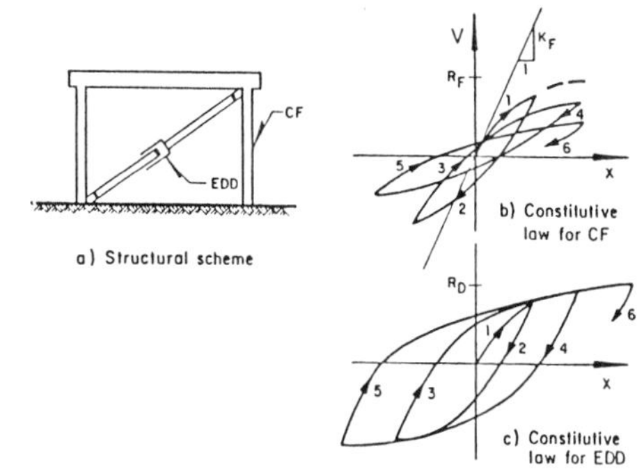

Fig 1 SDOF system with EDD

Every time that a moderate or high intensity earthquake occurs, damage increments take place on both elements. These increments are functions of the amplitudes and numbers of the deformation cycles, and are additive to those accumulated as a consequence of previous events, provided the CF is not repaired and the EDD is not replaced. It is assumed that damage on the CF is visually identified, thus leading to repair actions when it reaches a preestablished level; and that the EDD is replaced when it breaks or, in accordance with a preventive strategy, when it has been subjected to a number of high intensity earthquakes such that the probability of failure in the event of the next potential earthquake is unacceptably high. An optimum strategy for the design, construction and maintenance of such a system implies selecting the structural design parameters, as well as the threshold levels for repair of the CF and the criteria for preventive replacement of the EDD, in such a way as to minimize the sum of initial, maintenance and damage costs, all these updated to a common reference time.

Denoting by T_i, $i = 1,...\infty$ the times of occurrence of earthquakes, by C the initial cost and by L_i, $i = 1,... \infty$ the expenditures associated with failure, damage, repair and maintenance costs immediatly following each earthquake, then the quantity to be minimized is

$$U = C + E\left(\sum_{i=1}^{\infty} L_i\, e^{-\gamma T_i}\right) \tag{1}$$

where E stands for *expected value*, and γ is an adequate discount rate.

The damage accumulated on the CF and the EDD up to and including the j-th event are D_{Fj} and D_{Dj}, respectively. If a repair or replacement operation has taken place after the i-th earthquake, D_{Fi} and D_{Di} are transformed into their updated values, D'_{Fi} and D'_{Di}. The increments corresponding to the (j+1)th earthquake are $\delta_{F(j+1)}$ and $\delta_{D(j+1)}$. In the case of collapse of the structure during the i-th event, followed by immediate reconstruction, L_i equals C + A, where C is the construction cost and A includes all other related costs, including direct and indirect material costs, as well as those arising from human losses, social impact, and so forth. If collapse is not reached, L_i includes the repair costs of structural and infill elements, the replacement of those EDD which have failed or which are estimated to have attained significant fatigue damage, and the losses inflicted by the eventual interruption of the normal operation of the system.

Every time that the damage accumulated on the CF exceeds a preestablished value D_{rF}, the repair must eliminate the accumulated damage; that is, it must restore the initial properties R_F and K_F of the structural frame. Here it is assumed that the damage is visually apparent in the members of a structural frame; but that in the energy dissipators it remains hidden, in the form of partial fatigue, until it reaches so high values that the risk of their failure during the next high-intensity earthquake is excessively high. This justifies the adoption of a preventive policy consisting in replacing the dissipators when they have been subjected to a given number of earthquakes, on the basis of an index of calculated damage, D_{CD}, to be defined in the sequel. Consequently, $D'_{Fi} = D_{Fi}$ if $D_{Fi} < D_{rF}$; $D' = 0$ otherwise; $D'_{Di} = D_{Di}$ if $D_{Di} < 1$ and no preventive replacement takes place; $D'_{Di} = 0$ otherwise.

Because $\delta_{F(j+1)}$ and $\delta_{D(j+1)}$ depend on D_{Fj} and D_{Dj}, the process of damage accumulation is a Markov process, regardless of whether earthquakes occur in accordance with a Poisson, a renewal or a Markov process.

The transition probability matrices between consecutive events are obtained from the probability density functions of $\delta_{F(j+1)}$ and $\delta_{D(j+1)}$ conditional to D_{Fj}

and D_{Dj}. Immediately after the i-th earthquake, the state of the system is expressed in terms of two sets of variables, describing respectively the state of damage on the system and the seismic process. D_{Fi} and D_{Di} belong to the first set. The second is assumed to be described in terms of the vector S_i of seismological variables, if the characteristics of consecutive earthquakes are not independent; otherwise, their probability distribution does not depend on i. In the general case, in order to determine the conditional probability distributions of $D_{F(i+1)}$, $D_{D(i+1)}$ and S_{i+1} given the values corresponding to instant i, it is required, on one hand, to determine the joint p.d.f. of the waiting time and the intensity (T_{i+1}, Y_{i+1}, respectively) of the next earthquake, and on the other, to determine the states of damage D'_{Fi} and D'_{Di} of the system's components, just after the repair of the frame and/or the replacement of some or all the energy dissipators have taken place, if required. The following decision rules are adopted: a) if the damage accumulated in the structural frame exceeds level D_{rF}, the frame is repaired so as to recover the properties it had right after its construction was finished; b) if the energy dissipator fails, or if its calculated damage index exceeds D_{rD}, it is replaced by a new one. The joint p.d.f. of $\delta_{F(i+1)}$ and $\delta_{D(i+1)}$ will be obtained as a function of δ_i, and of D'_{Fi} and D'_{Di}. L_i includes among other costs those associated to the repair and replacement operations mentioned above.

The transition probabilities obtained in accordance with the foregoing paragraphs relate the states of the system at the end of two consecutive earthquakes. These probabilities depend on the seismic activity as well as on the adopted seismic design criteria, and the repair and replacement strategies. Their determination and use in the decision making process are described later.

3.CUMULATIVE DAMAGE AND STRUCTURAL RESPONSE

The following definition of cumulative damage is adopted for both, the structural frame members and the energy dissipating devices.

$$D = k \sum_{j=1}^{n} \xi_j^m \tag{2}$$

where ξ_j is the ratio of x_j, the deformation amplitude of the j-th cycle, to the failure deformation for monotonic load. It is assumed that Eq. 2 applies independently in each loading direction.

D is a measure of the damage determining the degradation of both, stiffness and strength. This degradation is quantitatively expressed by means of Shah and Wang's model (1987), illustrated in Fig 2. In accordance with this figure, the envelope of the cyclic load-deformation curve is bilinear, and the degradation of mechanical properties is represented by means of the evolution of that envelope in each loading direction, as a function of D, and of the peak amplitude reached in any cycle in the direction of interest. The updated envelope is defined by point Q in the figure, having an abscisa equal to x_M, the peak deformation, and an ordinate equal to the product of $(1-\varepsilon)$ by $P_o(x_M)$, where the latter is the load corresponding to x_M in the original envelope and ε is a degradation coefficient related with D as shown in Fig 3, taken from ref 1. Clearly, for reinforced concrete elements the theoretical model predicts excessively large values of ε for D greater than about 0.5. A very good fit to the experimental data of the figure is provided by the equation shown in the figure. For the EDD, Shah and Wang's equation may be used with α = 20, in order to represent the property of keeping a stable hysteretic cycle up to damage levels very close to that leading to failure.

Figures 4 and 5 show means and variation coefficients of damage D_F in terms of intensities, for various values of initial damage states in CF, for a system

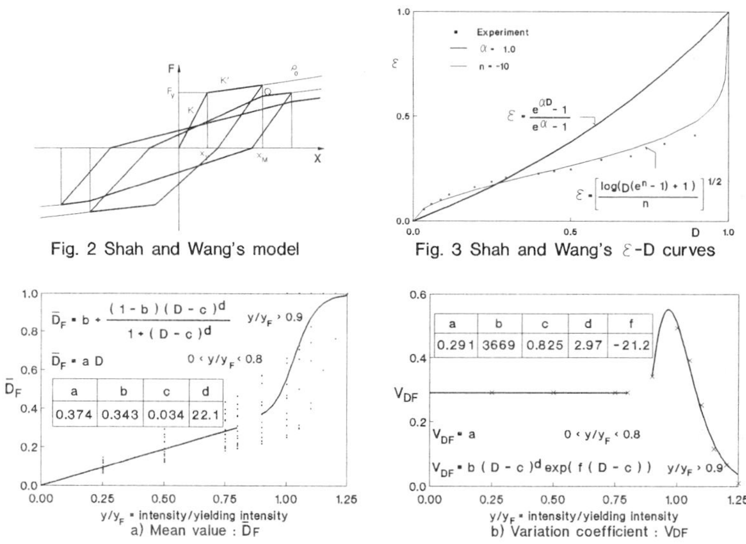

Fig. 2 Shah and Wang's model

Fig. 3 Shah and Wang's ε-D curves

a) Mean value : $\bar{D}_F$

b) Variation coefficient : V_{DF}

Fig. 4 Variation of damage with the normalized intensity y/y_F , for the case of zero initial damage

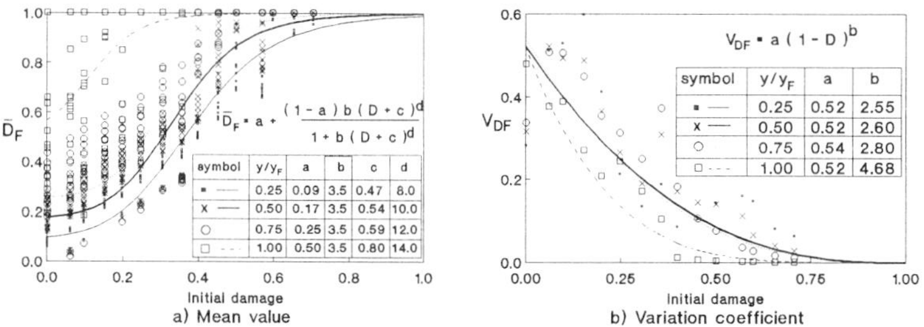

Fig. 5 Final damage in terms of normalized intensity and initial damage

similar to that of Fig 1a, but in this case without EDD. The excitation was a collection of recorded and simulated accelerograms typical of the soft soil area in Mexico City, normalized at different intensitites. The responses were calculated by means of nonlinear step-by-step studies. The intensities are expressed in terms of the maximum ordinate of the linear pseudo acceleration spectrum for damping equal to 0.05 of critical.

4.TRANSITION PROBABILITIES BETWEEN DAMAGE STATES

Let $f_{FD(i+1)}(u,v|D'_{Fi},D'_{Di})$ designate the joint p.d.f. of $D_{F(i+1)}$ and $D_{D(i+1)}$ given D'_{Fi} and D'_{Di}. In order to account for the uncertainty associated to Y_{i+1}, the intensity of the (i+1)th earthquake, $F_{FD(i+1)}$ is calculated as follows:

$$f_{FD(i+1)}(u,\ v|D'_{Fi},\ D'_{Di}) = \int F_{FD(i+1)}(u,v|D'_{Fi},\ D'_{Di},\ y)F_{Y(i+1)}(y)dy \tag{3}$$

The marginal probability density functions of D_{Fi} and D_{Di} are obtained recursively from the initial condition of the system. If the actions of preventive replacement of EDD are neglected, after some algebraic transformations the following recursive relation is obtained.

$$F_{FD(i+1)}(u,v) = \int_o^1 \int_o^{D_{rF}} f_{FD(i+1)}(u,v|u',v')f_{FDi}(u',v')du'dv'$$

$$+ \int_o^1 f_{FD(i+1)}(u,v|0,\ v')f_{Di}(v')P(D_{Fi}>D_{rF}|v')dv'$$

$$+ \int_o^{D_{rF}} f_{FD(i+1)}(u,v|u',\ 0)f_{Fi}(u')P(D_{Di}>1|u')du'$$

$$+ f_{FD(i+1)}(u,v|0,0)P(D_F>D_{rF},\ D_{Di}>1) \tag{4}$$

The recursive computation of F_{FDi} can be done by numerical integration in Eq 4 or by means of Monte Carlo simulation. One drawback of Eq 4 is that it does not account for the possibility of preventive replacement of the EDD's. Incorporating it into the second approach would imply handling a third variable in the mentioned equations and working with the corresponding joint p.d.f. These difficulties are circunvented through the use of the Monte Carlo approach.

5.PREVENTIVE REPLACEMENT OF EDD'S

For s.d.o.f. systems, D_{cD} (calculated damage index) after a given earthquake can be estimated on the basis of two sources of information: the seismic history and the relation between intensity and damage increment. The former is expressed in terms of intensities and times of occurrence of earthquakes, while the latter is contained in Figs 4 and 5. Given this information, D_{cD} equals, by definition, the sum of the expected values of the damage increments δ_{Di} during the earthquakes reported. For the purpose of this analysis, the expected influence of the repair actions on the CF may be accounted for or neglected, depending on the curves used from Figs 4 and 5. The determination of D_{cD} for complex systems is left for future publications.

6.SEISMICITY MODELS

Monte Carlo simulation of values of T_i to be used to obtain the expected value appearing in eq 1 is straightforward when the occurrence of earthquakes is represented by a renewal process, including the Poisson process as a particular case. Accounting for the influence of seismic history on the probabilistic description of future activity can be handled by means of Markov models. In the general case, the joint distribution of the waiting time to the next (large) earthquake and to its intensity is a function of the seismic history (magnitudes, locations and times) covering an interval of the order of one or two centuries, or perhaps smaller. Incorporating this influence on the Monte Carlo simulation of the values of T_i and Y_i offers no problem.

7.MULTISTORY FRAMES

Two parallel studies are underway, trying to understand the process of damage accumulation in multistory frames. One is a comparison of the responses of frames characterized by Takeda's and Shah and Wang's behavior models. The other aims at obtaining incremental damage functions similar to Figs 4 and 5. For this purpose, five three-bay ten-story frames were designed in accordance with Mexico City seismic code of 1987, and subjected to sets of simulated ground motion records with statistical properties similar to those of the SCT (Mexico City) EW accelerogram of 19 September 1985. Mean values of acting loads and mechanical properties of the structural members were derived from the nominal design values, and adopted as deterministic parameters of the systems studied. I to III were conventional frames, having different periods. Systems IV and V were frames provided with EDD's, where the contribution of the latter to the total lateral stiffness was 0.75 and 0.5, respectively, in all stories. D-ε curves defined by the equation shown in Fig 3 were used, with α = 1 for the CF and 20 for the EDD. Design response spectra for linear systems were divided by a factor Q, in order to account for nonlinear behavior. The values of Q adopted for CF's and EDD's ranged from 1.0 to 4.0.

Detailed time histories and maximum values of story deformations are reported in Ref 2. Briefly, ductility demands at stories of frames with Takeda's behavior occasionally reached values as high as 10. Under the assumption of Shah and Wang's behavior, systems designed for low values of Q survived under the action of most of the assumed excitations; the other systems collapsed induced by total loss of strength at one or more members of the CF. For this reason, it was not possible to pass to the next stage in the study, that is, calculating the response of systems having initial damage. The resulting rate of the number of failures to the number of cases studied was considered to be excessively high, and therefore not representative of reality. The authors believe that the very large values of the calculated responses are associated to the inadequacy of the D-ε curves adopted (Fig 3), and to the assumption of instantaneous loss of strength of members of both the CF and the EDD elements when D reaches a value of unity. A new stage in the study is currently in progress, aiming at eliminating the mentioned drawbacks.

8.CONCLUDING REMARKS

The use of energy-dissipating devices is an efficient approach for the reduction of repair costs of structures damaged by earthquakes. Many types of structural members and connections, as well as infill elements, may play the role of those devices, but the success in their use is attached to an approach to seismic design which envisages the reduction of structural (and non-structural) damage and the easiness of its repair as an essential part of the design process, in addition to the determination of the member strengths and stiffnesses required to comply with conventional design rules.

This implies selecting both the location of EDD's in the structural system and the relative values of their strengths, stiffnesses and safety factors with respect to those of the conventional structural members, in such a manner that demands for nonlinear dynamic behavior are concentrated on them. Their fastening to the conventional system must facilitate their replacement in case they are damaged or when a preventive strategy determines it so.

In order to implement the criteria described in the foregoing paragraphs, it will be necessary to count with tools for determining optimum locations and properties of energy dissipators, as well as required stiffness and strengths of the integrated system, and relative contributions of conventional structural members and energy dissipators to those properties. The models proposed in this paper provide a rational framework for reaching the corresponding decisions, as well as for establishing optimum repair/replacement policies. With the aid of those models, simple guidelines and recommendations must be prepared for preliminary design in typical practical cases. Although the results presented here are only preliminary and limited in scope, they show the feasibility of the approach proposed.

REFERENCES

1. Shah, S P, and Wang, M L,"Reinforced concrete hysteresis model based on the damage concept", *Earthquake Engineering and Structural Dynamics, 15* (1987), 993-1003.
2. Rodríguez, J F F, "Respuesta sísmica de edificios con disipadores de energía", Thesis for the degree of Civil Engineer, *School of Engineering, National University of Mexico* (1993).

Reliability and Optimization of Structural Systems, V (B-12)
P. Thoft-Christensen and H. Ishikawa (Editors)
Elsevier Science Publishers B.V. (North-Holland)

Sensitivity Information in Reliability-Based Optimization of Plate Girders for Highway Bridges

D. M. Frangopol, S. Hendawi, M. Tudor

Dept. of Civil, Environmental and Architectural Engineering
University of Colorado at Boulder, Boulder, Colorado 80309-0428, USA

Abstract

Better understanding of design solutions of composite hybrid plate girders for highway bridges is achieved by combining structural reliability analysis and optimization techniques. This paper is based on a reliability-based optimization formulation of composite hybrid plate girders. Nondeterministic cross-sections under random loads are automatically sized for least-weight in full conformance with the American Association of State Highway and Transportation Officials (AASHTO) specification for highway bridges. A design example is presented and sensitivity analyses are carried out.

Keyword Codes: G.1.6; G.3; J.6
Keywords: Optimization; Probability and Statistics; Computer-Aided Engineering

1. INTRODUCTION

Traditionally, structural optimization has been based on deterministic design philosophy. This philosophy represents a narrow view of structural design [1]. In fact, deterministic optimization provides inconsistent reliability levels for different design situations and can lead to optimal designs with less safety and redundancy than normally considered adequate. Therefore, the structural optimization process must be accomplished under realistic conditions. These conditions are provided by the structural reliability theory. The marriage of structural reliability and structural optimization has led to a new design perspective from which reliability-based structural optimization has evolved [1-7].

This paper is based on a reliability-based optimization formulation of composite hybrid plate girders [6,7]. Nondeterministic cross-sections under random loads are automatically sized for least-weight in full conformance with the AASHTO-Standard Specifications for Highway Bridges [8]. A design example is presented and sensitivity analyses are carried out.

2. COMPOSITE HYBRID PLATE GIRDER DESIGN

The design of composite hybrid plate girders (CHPG) is a complex process [9]. The formulation of an automated approach to the deterministic design of CHPG using optimization techniques was presented by Dhillon and Kuo [10]. The investigation of Dhillon and Kuo demonstrated that optimization techniques can be used effectively in an automated approach to design CHPG. In two recent studies, the deterministic formulation of Dhillon and Kuo based on 22 design variables (X_1 to X_{22}) and 13 design parameters (Y_1 to Y_{13}) has been extended to reliability-based structural optimization [6,7].

In this context, eight of the design parameters Y_1 to Y_{13} indicated in [10] are treated as random variables. They represent the yield strength of steel (Y_2,Y_8,Y_9), the dead load (Y_5), the maximum live load shear and moment including impact (Y_6,Y_7), the compressive strength of concrete (Y_{10}), and the superimposed dead load moment (Y_{11}). Using this reliability-based context, the unstiffened and stiffened minimum weight CHPG optimization formulations were given by Hendawi and Frangopol [6,7]. These formulations are in full conformance with the AASHTO-Standard Specifications for Highway Bridges [8]. The constraints on strengths and dimensions considered in these formulations are indicated in [7]. The problem constraints are both reliability-based and deterministic as follows:

$$\beta_i = \bar{g}_i / \sigma_{gi} \geq \beta_i^* \tag{1}$$

$$g_i \leq 0 \tag{2}$$

where i = 1, 2, 3, 4, 6, 7, 11, 12 for the unstiffened case, i = 1, 2, 5, 6, 8, 9, 10, 11, 12 for the stiffened case, β_i = reliability index with respect to limit state i, β_i^* = allowable reliability index, g_i = 0 is the limit-state under consideration i=1,...,12, and $\bar{g}_i$ and σ_{gi} are mean and standard deviation of the performance function g_i. The objective function to be minimized is the weight of the steel plate girder alone, expressed as

$$W_1 = f_1(X_1 \text{ to } X_6, Y_3, Y_{13}) \tag{3}$$

$$W_2 = f_2(X_1 \text{ to } X_6, X_{19} \text{ to } X_{21}, Y_3, Y_{13}) \tag{4}$$

where W_1 and W_2 are the weights of the steel plate girder for the unstiffened and stiffened cases, respectively, and the design variables X_1 to X_6 and X_{19} to X_{21} are specified in [10].

For solving the above reliability-based optimum design problems, the general purpose deterministic optimization program developed by Vanderplaats [11] was linked to the reliability analysis program developed by Lee *et al.* [12].

3. OPTIMUM DESIGN SENSITIVITY

For comparison purposes, the deterministic 80 ft simple span CHPG considered by Dhillon and Kuo [10] is treated herein in a reliability-based context. Results in this section are based on data presented by Hendawi and Frangopol in [6,7]. The random variables are all independent and normal distributed. Furthermore, the allowable reliability indices are all considered to be the same $\beta_1^* = \beta_2^* = \ldots = \beta_6^*$. Sensitivity of reliability optimum solutions with respect to changes in mean values and coefficients of variation of various random variables considered in computations are presented in Tables 1 to 6. The gradient values (i.e., $\partial\beta_i / \partial\bar{Y}$, $\partial\beta_i / \partial V(Y)$), in these tables provide valuable information that can be used to guide trade-off studies in reliability-based optimization of CHPG for highway bridges.

4. CONCLUDING REMARKS

Modern structural reliability-based optimization methodology has reached the stage where it is possible to design structural systems based on reliability constraints with respect to all limit states required by code specifications. Sensitivity analysis plays a very important role in reliability-based analysis, design, and inspection of structural systems. As the structural reliability-based optimization technology moves from the research to practical applications, more emphasis should be placed on (a) including both system reliability and consequences of failure in the optimization process, (b) providing sensitivity information, and (c) moving upward in the design hierarchy towards global optimization.

ACKNOWLEDGEMENTS

This work was supported in part by the United States National Science Foundation under Grant MSM-9013017 and by the Jordan University of Science and Technology for graduate study of the second author. The third author contributed to the results presented in this paper during the preparation of her final report for the graduate course on structural reliability. The senior author thanks Professor A. H-S. Ang for helpful comments and discussions during the 5th IFIP WG 7.5 Conference.

REFERENCES

1. D.M. Frangopol and F. Moses, *Chapter* in Advances in Design Optimization (ed. H. Adeli), Chapman and Hall, (1993) in print.
2. D.M. Frangopol, ASCE J. Struct. Eng., 111-11 (1985) 2288.
3. I. Enevoldsen, Ph.D. Thesis, University of Aalborg, Denmark, (1991).
4. P. Thoft-Christensen, Lecture Notes in Engineering (eds. A. Der Kiureghian and P. Thoft-Christensen), Springer-Verlag, 61 (1991) 387.
5. D.M. Frangopol, *Chapter* in Probabilistic Structural Mechanics Handbook (ed. C. Sundararajan), Van Nostrand Reinhold, (1993), in print.

6. S. Hendawi and D.M. Frangopol, Struct. Safety, (1993) submitted.
7. S. Hendawi and D.M. Frangopol, Proc. ICOSSAR '93, Innsbruck, Austria (1993), in print.
8. AASHTO-Standard Specifications for Highway Bridges, 13th Edition, 1983.
9. C.G. Salmon and J.E. Johnson, Steel Structures, Third Edition, Harper and Row, 1990.
10. B.S. Dhillon and C-H. Kuo, ASCE J. Struct. Eng., 117-7 (1991) 2088.
11. G.N. Vanderplaats, ADS - A Fortran Program for Automated Design Synthesis: Version 1.10, Eng. Design Opt., Santa Barbara, California, 1986.
12. Y-H. Lee, S. Hendawi and D.M. Frangopol, RELTRAN - A Structural Reliability Analysis Program: Version 2.0, Report No. CU/SR-93/6, Univ. of Colorado, Boulder, 1993.

Table 1
Sensitivity of limit states reliabilities to change in mean or coefficient of variation of random variables Y_2 and Y_9

	Reliability index	Allowable reliability index, β_i^*		
	β_i	2.0	3.5	5.0
Mean, $\bar{Y}_2 = \bar{Y}_9$	β_1	0.000	0.000	0.000
		(0.000)	(0.000)	(0.000)
	β_2	0.206	0.170	0.054
		(0.206)	(0.170)	(0.054)
	β_3	0.096	0.089	0.024
	β_4	0.000	0.000	0.000
	β_5	(0.167)	(0.155)	(0.054)
	β_6	0.000	0.000	0.000
		(0.000)	(0.000)	(0.000)
Coefficient of Variation, $V(Y_2) = V(Y_9)$	β_1	0.000	0.000	0.000
		(0.000)	(0.000)	(0.000)
	β_2	-18.468	-33.122	-79.413
		(-18.468)	(-33.122)	(-79.413)
	β_3	-57.775	-61.260	-90.928
	β_4	0.000	0.000	0.000
	β_5	(-11.029)	(-23.453)	(-77.652)
	β_6	0.000	0.000	0.000
		(0.000)	(0.000)	(0.000)

Note: Gradient values for the stiffened case are indicated in parenthesis.

Table 2
Sensitivity of limit states reliabilities to change in mean or coefficient of variation of random variable Y_5

	Reliability index	Allowable reliability index, β_i^*		
	β_i	2.0	3.5	5.0
Mean, $\bar{Y}_5$	β_1	-2.760	-2.201	-0.473
		(-2.822)	(-2.304)	(-0.448)
	β_2	-7.474	-5.867	-1.301
		(-7.613)	(-6.027)	(-1.349)
	β_3	-1.088	-1.011	-0.287
	β_4	-3.135	-3.155	-3.289
	β_5	(-2.094)	(-1.802)	(-0.623)
	β_6	0.000	0.000	0.000
		(0.000)	(0.000)	(0.000)
Coefficient of Variation, $V(Y_5)$	β_1	-0.449	-0.496	-0.052
		(-0.469)	(-0.554)	(-0.047)
	β_2	-3.211	-3.414	-0.394
		(-3.328)	(-3.598)	(-0.422)
	β_3	-0.227	-0.205	-0.022
	β_4	-0.578	-1.103	-3.894
	β_5	(-0.259)	(-0.334)	(-0.093)
	β_6	0.000	0.000	0.000
		(0.000)	(0.000)	(0.000)

Note: Gradient values for the stiffened case are indicated in parenthesis.

Table 3
Sensitivity of limit states reliabilities to change in mean or coefficient of variation of random variables Y_6 and Y_7

	Reliability index	Allowable reliability index, β_i^*		
	β_i	2.0	3.5	5.0
	β_1	-0.003	-0.002	0.000
		(-0.003)	(-0.003)	(0.000)
	β_2	0.000	0.000	0.000
Mean ,		(0.000)	(0.000)	(0.000)
$\bar{Y}_6 = \bar{Y}_7$	β_3	-0.002	-0.002	0.000
	β_4	-0.006	-0.007	-0.016
	β_5	(-0.004)	(-0.003)	(-0.001)
	β_6	-0.001	-0.001	0.000
		(-0.001)	(-0.001)	(0.000)
	β_1	-2.858	-3.471	-0.369
Coefficient of		(-2.832)	(-3.369)	(-0.368)
Variation,	β_2	-0.125	-0.264	-0.290
$V(Y_6) = V(Y_7)$		(-0.106)	(-0.218)	(0.000)
	β_3	-3.629	-3.264	-0.357
	β_4	-9.235	-16.165	-62.160
	β_5	(-4.127)	(-5.320)	(-0.001)
	β_6	-1.498	-1.271	-0.174
		(-1.120)	(-0.849)	(0.000)

Note: Gradient values for the stiffened case are indicated in parenthesis.

Table 4
Sensitivity of limit states reliabilities to change in mean or coefficient of variation of random variable Y_8

	Reliability index	Allowable reliability index, β_i^*		
	β_i	2.0	3.5	5.0
	β_1	0.059	0.051	0.011
		(0.059)	(0.051)	(0.011)
	β_2	0.000	0.000	0.000
Mean,		(0.000)	(0.000)	(0.000)
$\bar{Y}_8$	β_3	0.000	0.000	0.000
	β_4	0.000	0.000	0.000
	β_5	(0.000)	(0.000)	(0.000)
	β_6	0.000	0.000	0.000
		(0.000)	(0.000)	(0.000)
	β_1	-12.446	-24.879	-71.942
Coefficient of		(-12.482)	(-25.065)	(-72.327)
Variation,	β_2	0.000	0.000	0.000
$V(Y_8)$		(0.000)	(0.000)	(0.000)
	β_3	0.000	0.000	0.000
	β_4	0.000	0.000	0.000
	β_5	(0.000)	(0.000)	(0.000)
	β_6	0.000	0.000	0.000
		(0.000)	(0.000)	(0.000)

Note: Gradient values for the stiffened case are indicated in parenthesis.

Table 5
Sensitivity of limit states reliabilities to change in mean or coefficient of variation of random variable Y_{10}

	Reliability index	Allowable reliability index, β_i^*		
	β_i	2.0	3.5	5.0
Mean, $\overline{Y}_{10}$	β_1	0.000	0.000	0.000
		(0.000)	(0.000)	(0.000)
	β_2	0.000	0.000	0.000
		(0.000)	(0.000)	(0.000)
	β_3	0.000	0.000	0.000
	β_4	0.000	0.000	0.000
	β_5	(0.000)	(0.000)	(0.000)
	β_6	0.463	0.412	0.122
		(0.377)	(0.314)	(0.122)
Coefficient of Variation, $V(Y_{10})$	β_1	0.000	0.000	0.000
		(0.000)	(0.000)	(0.000)
	β_2	0.000	0.000	0.000
		(0.000)	(0.000)	(0.000)
	β_3	0.000	0.000	0.000
	β_4	0.000	0.000	0.000
	β_5	(0.000)	(0.000)	(0.000)
	β_6	-19.290	-20.705	-28.118
		(-21.631)	(-23.304)	(-28.114)

Note: Gradient values for the stiffened case are indicated in parenthesis.

Table 6
Sensitivity of limit states reliabilities to change in mean or coefficient of variation of random variable Y_{11}

	Reliability index	Allowable reliability index, β_i^*		
	β_i	2.0	3.5	5.0
Mean, $\overline{Y}_{11}$	β_1	-0.002	-0.002	0.000
		(-0.002)	(-0.002)	(0.000)
	β_2	0.000	-0.001	0.000
		(0.000)	(-0.000)	(0.000)
	β_3	-0.001	-0.001	0.000
	β_4	-0.004	-0.004	-0.004
	β_5	(-0.003)	(-0.002)	(-0.001)
	β_6	-0.001	-0.001	0.000
		(-0.001)	(-0.001)	(0.000)
Coefficient of Variation, $V(Y_{11})$	β_1	-0.018	-0.022	-0.002
		(-0.018)	(-0.021)	(-0.002)
	β_2	-0.001	-0.002	-0.002
		(-0.001)	(-0.002)	(-0.002)
	β_3	-0.020	-0.018	-0.002
	β_4	-0.051	-0.090	-0.345
	β_5	(-0.022)	(-0.030)	(-0.008)
	β_6	-0.010	-0.008	-0.001
		(-0.007)	(-0.005)	(-0.001)

Note: Gradient values for the stiffened case are indicated in parenthesis.

Reliability and Optimization of Structural Systems, V (B-12)
P. Thoft-Christensen and H. Ishikawa (Editors)
Elsevier Science Publishers B.V. (North-Holland)
1993 IFIP

Variance Reduction Technique for Systems of Linear Failure Functions

G. Fu

Structures Research, New York State Department of Transportatation
Albany, NY 12232, USA

ABSTRACT

This paper proposes an importance sampling method for system reliability problems with linearized surfaces of failure mode. It has the advantage of locating all samples in the failure region. Its variance of estimator is estimated by an analytically derived upper bound, which depends on the ratio of the original distribution to the sampling distribution at the design points. Application examples are included for illustration. They show that variance reduction by this method does not depend on the system failure probability to be estimated.

Keyword Codes: F.2.0; G.3; I.6.1
Keywords: Analysis of Algorithms and Problem Complexity, General; Probability and Statistics; Simulation Theory

1. INTRODUCTION

A structural system reliability problem can be formulated as a multi-dimensional integration for system failure probability P_f:

$$P_f = \int_{\mathbf{x}} G(\mathbf{x})f(\mathbf{x})\,dx \tag{1}$$

where $f(\mathbf{x})$ is the distribution function of correlated normal variables $\mathbf{x}$ with mean $\underline{\mathbf{X}}$ and variance matrix $\mathbf{C}$, without loss of generality (since random variables can be generally transferred to normal variables [Hohenbichler and Rackwitz 1981]), and $G(\mathbf{x})$ is an indicator for the structure's limit state:

$$G(\mathbf{x}) = \begin{cases} 0 & \text{if } \mathbf{x} \text{ is in safe region} \\ 1 & \text{if } \mathbf{x} \text{ is in failure region} \end{cases} \tag{2}$$

A series system can be modeled by $G(\mathbf{x})=\text{Min}\{G_j(\mathbf{x})\}$, $j=1,2,\ldots,m$, where

$$g_j(\mathbf{x}) = \begin{cases} 0 & \text{if } z_j(\mathbf{x})>0 \ (\mathbf{x} \text{ is in safe region}) \\ 1 & \text{if } z_j(\mathbf{x})\le 0 \ (\mathbf{x} \text{ is in failure region}) \end{cases} \qquad j=1,2,\ldots,m \tag{3}$$

where $z_j(\mathbf{x})$ is the jth modal limit function. A parallel system can be viewed as a special ca with m=1. The integration in Eq.(1) is often a difficult, if not impossible, task in structu system reliability analysis due to its multiple dimensions and irregular boundaries defined by t indicator function $G(\mathbf{x})$. Thus alternative methods, such as simulation techniques, are sought carry out this integration. The importance sampling method has the potential to reduce t variance of estimator. This method uses a new sampling distribution $p(\mathbf{x})$ for Eq.(1)

$$P_f = \int_{\mathbf{x}} [G(\mathbf{x})f(\mathbf{x})/p(\mathbf{x})]\, p(\mathbf{x})d\mathbf{x} \qquad ($$

For a brief summary of recent advances in this direction, the reader is referred to [e.g., Fu a Moses 1988].

2. PROPOSED IMPORTANCE SAMPLING METHOD

It is assumed that $z_j(\mathbf{x})$ are, or can be approximated by, linear functions:

$$g_j(\mathbf{x}) = \begin{cases} 0 & \text{if } \mathbf{a}_j\bullet(\mathbf{x}-\mathbf{x}_j^*)>0 \\ 1 & \text{if } \mathbf{a}_j\bullet(\mathbf{x}-\mathbf{x}_j^*)\leq 0 \end{cases} \qquad (5$$

where $\mathbf{x}_j^*$ is the design point [Ang and Tang 1984] for mode j, and

$$\mathbf{a}_j = \partial z_j(\mathbf{x})/\partial \mathbf{x}\,\big|\,\mathbf{x}=\mathbf{x}_j^* \qquad (5$$

These problems are referred to as first order problems here, and they are commonly practical interest. A sampling distribution below is suggested for these problems

$$p(\mathbf{x}) = \sum_j w_j\, p_j(\mathbf{x}) \qquad (6$$

where $p_j(\mathbf{x})$ are zero if $\mathbf{a}_j\bullet(\mathbf{x}-\mathbf{x}_j^*)>0$ (in safe region) and twice the normal distribution functi with mean $\mathbf{x}_j^*$ and variance matrix $\mathbf{C}$ if $\mathbf{a}_j\bullet(\mathbf{x}-\mathbf{x}_j^*)\leq 0$ (in failure region), and w_j are weig coefficients defined by

$$p(\mathbf{x}_j^*)/p(\mathbf{x}_1^*)=f(\mathbf{x}_j^*)/f(\mathbf{x}_1^*)\ (j=2,3,\ldots,m) \qquad \text{and} \qquad w_1+w_2+\ldots+w_m=1 \qquad (6$$

This sampling distribution is 1) proportional to the optimal distribution with zero variance [e. Fu and Moses 1988] at the design points, and 2) able to locate all samples in the failure regio

The mean and variance of estimator by the proposed method are

$$E[P_f']=P_f \qquad \text{and} \qquad Var[P_f']=(\int_{\mathbf{x}}[G^2(\mathbf{x})f(\mathbf{x})/p(\mathbf{x})]f(\mathbf{x})d\mathbf{x}-P_f^2)/N \qquad ($$

where N is the number of samples used in the simulation. Although the variance is as diffic as P_f to find as shown, it is bounded as follows:

$$Var[P_f']\leq Max(Var[P_f'])=(Max(f(\mathbf{x})/p(\mathbf{x}))-P_f^2)/N \qquad ($$

here Max(f(**x**)/p(**x**)) occurs at the design points. The analysis of associated simulation error n be found in [Melchers 1984].

VARIANCE REDUCTION

A variance change factor (VCF) is used here to investigate variance reduction of the oposed method. It is defined as the variance ratio of the present method to the conventional onte Carlo simulation method (i.e. using f(**x**) as the sampling distribution). Its upper bound n be derived based on Eq.(8):

$$ax(VCF)=(Max(f(\mathbf{x})/p(\mathbf{x}))-P_f)/(1-P_f) \quad (9)$$

low VCF is desired to reduce the variance. It is found [Fu 1988] that this factor relates ceptable error ϵ, associated confidence level k, and sample size N for these two methods in nple manners:

$$/\epsilon_{MC}=VCF^{1/2}; \quad k_{IS}/k_{MC}=VCF^{-1/2}; \quad N_{IS}/N_{MC}=VCF \quad (10)$$

here subscripts IS and MC denote the importance sampling and conventional Monte Carlo ethods respectively. Similarly, upper or lower bounds of these ratios are given by

$$/\epsilon_{MC}\leq Max(VCF)^{1/2}; \quad k_{IS}/k_{MC}\geq Max(VCF)^{-1/2}; \quad N_{IS}/N_{MC}\leq Max(VCF) \quad (11)$$

nce these relations are analytically derived, they provide a reliable estimation for efficiency d accuracy.

APPLICATION EXAMPLES

1 Example 1

Consider a series system problem with $z_1(\mathbf{x})=x_1-x_2$ and $z_2(\mathbf{x})=61-1.44x_1-x_2$ (m=2), and dependent variables $X_1=N(25,2.5)$ and $X_2=N(10,3.0)$. The modal failure probabilities and eights are found, respectively: $P_{f,1}=.00006126$, $P_{f,2}=.0006850$, $w_1=.0953$, and $w_2=.9047$, ing $\mathbf{x}_1^{*t}=(18.85,18.85)$ and $\mathbf{x}_2^{*t}=(31.15,16.15)$. The exact P_f is computed by $P_f=P_{f,1}+P_{f,2}$ - $_1$*$P_{f,2}=.0007462$, since the two modes are independent of each other.

The simulation is performed using N=4,000, and a confidence level of 95% (k=1.96) is ed to obtain Max(ϵ)=5.7%, with Max(COV)=2.9%. P_f' is obtained to be 0.0007522 with error of 0.8% and COV=1.9%. In comparison with the conventional Monte Carlo method, e sample size of the present method is at least reduced by a factor of Max(VCF)=.002536, thout sacrifice in accuracy (ϵ) or confidence (k). In other words, almost 400 times more mples (about 1.7 million) would have to be used if the conventional Monte Carlo method were nployed.

2 Example 2

This is a problem with combination of series and parallel systems. It is treated as a series of parallel systems as defined earlier. The structural system consists of a deterministic load and two parallel bars made of brittle materials. This results in nondifferentiable failure surfac [Moses 1982] representing two symmetrical sequences of component failures that lead to syste failure. They are equally important and thus equally weighted: $w_1=w_2=0.5$. Axial strengt of bars R_1 and R_2 are assumed normally distributed with a correlation coefficient ρ and equ coefficients of variation of 20%. Their mean values are set equal and given by a safety fact (SF): $\underline{R}_1=\underline{R}_2=0.5*S*SF$. For SF=1.7, 2.2, and 2.7, the results by the present method usi N=4,000 samples are plotted in Fig.1. The exact P_f used for comparison is obtained using integration table [National Bureau of Standards 1959]. A good agreement between the tw methods is observed in Fig.1a. Error ϵ' of these cases is shown in Fig.1b, most being with the range of 2.5% with a maximum of 5.4%. Max(ϵ) shown here is based on a confidence lev of 95% (k=1.96). Fig.1c shows COV that are all within 3.7%. Note that they are n significantly dependent on SF (or equivalently P_f) as the conventional Monte Carlo meth would be, with the same sample size for all cases. This indicates that the standard deviation the estimates is reduced almost proportionally to P_f. Max(VCF) is shown in Fig.1d, whic demonstrates again its decrease with P_f.

4.3 Example 3

This example shows an application to a nonlinear problem with linearized failu surfaces. The original failure function is $z(\mathbf{x})=6+x_1-0.622x_2^2=0$ with $\mathbf{X}^t=(X_1,X_2)$ a independent $X_1=X_2=N(0,1)$. Linearized failure functions are formed according to Eq.5a, usi $\mathbf{x}_1^{*t}=(-0.804,2.890)$ and $\mathbf{x}_2^{*t}=(-0.804,-2.890)$ and $\mathbf{a}_1^t=(1,-3.596)$ and $\mathbf{a}_2^t=(1,3.596)$. T symmetric failure modes give $w_1=w_2=0.5$. Exact $P_f=0.002816$ is computed by a on dimensional integration using the Gamma distribution [Schuëller and Stix 1987].

Listed below are simulation results using N=4,000: Max(VCF)=.008316, $P_f'=.00276$ $\epsilon'=2.0\%$ vs. Max(ϵ)=5.3%, and COV=1.7% vs. Max(COV)=2.7%. Apparently th linearization represents a good approximation. P_f' by the present method based on t linearization is reasonably close to the exact one P_f. The estimated COV and error ϵ' a bounded by the analytically derived upper bounds, respectively. Maximum error Max(ϵ) based on a confidence level of 95% (k=1.96).

5. CONCLUSIONS

A general importance sampling method is introduced. Its sampling distribution proportional to the ideally optimal one at the maximum likelihood points and able to locate samples in the failure domain. This method can be employed in integrals for extremely sm probability when assessing system reliability, expected damage, etc. Upper bounds for varianc error, and COV of estimator are derived analytically, which have been used to evaluate t present method and can be used in general applications to determine required sample size giv an acceptable maximum error associated with a confidence level. Example applications of t suggested method show the assured improvement on efficiency and accuracy for quite gene cases, compared to the conventional Monte Carlo method. These include problems of bo series and parallel systems. They also show that the required sample size is not much depende on the failure probability P_f to be estimated.

. ACKNOWLEDGMENTS

Dr.Jianguo Tang and Mr.Everett M.Dillon with the New York State Department of ransportation assisted in preparing part of the examples. The Federal Highway Administration f the United States provided partial support to the study.

EFERENCES

. Ang,A.H-S. and Tang,W.H. Probability Concepts in Engineering Planning and Design, ol.II-Decision, Risk, and Reliability, John Wiley Sons, 1984

. Fu,G. and Moses,F. "Importance Sampling in Structural System Reliability" Prob. Methods Civil Engineering, (Ed.)by P.D.Spanos, 5th ASCE Specialty Conference, Blacksburg, VA, lay 1988, p.340

. Fu,G. "Error Analysis for Importance Sampling Method" New Directions in Structural ystem Reliability, (Ed.) by D.M.Frangopol, Workshop on Research Needs for Applications f System Reliability Concepts and Techniques in Structural Analysis, Design and ptimization, Boulder, CO, Sept. 1988, p.132

. Hohenbichler,M. and Rackwitz,R. "Non-Normal Dependent Vectors in Structural Safety" SCE J.Eng.Mech., Vol.107,EM6, Dec.1981, p.1227

. Melchers,R.E. "Efficient Monte-Carlo Probability Integration", Civil Engineering Research eport No.7/1984, Monash university, 1984

. National Bureau of Standards, Tables of the Bivariate Normal Dis. Function and Related unctions, Applied Math Series 50, June 1959

. Schuëller,G.I. and Stix,R. "A Critical Appraisal of Methods to Determine Failure robabilities", Structural Safety, Vol.4, 1987, pp.293-309

Fig. 1: Results for Example 2

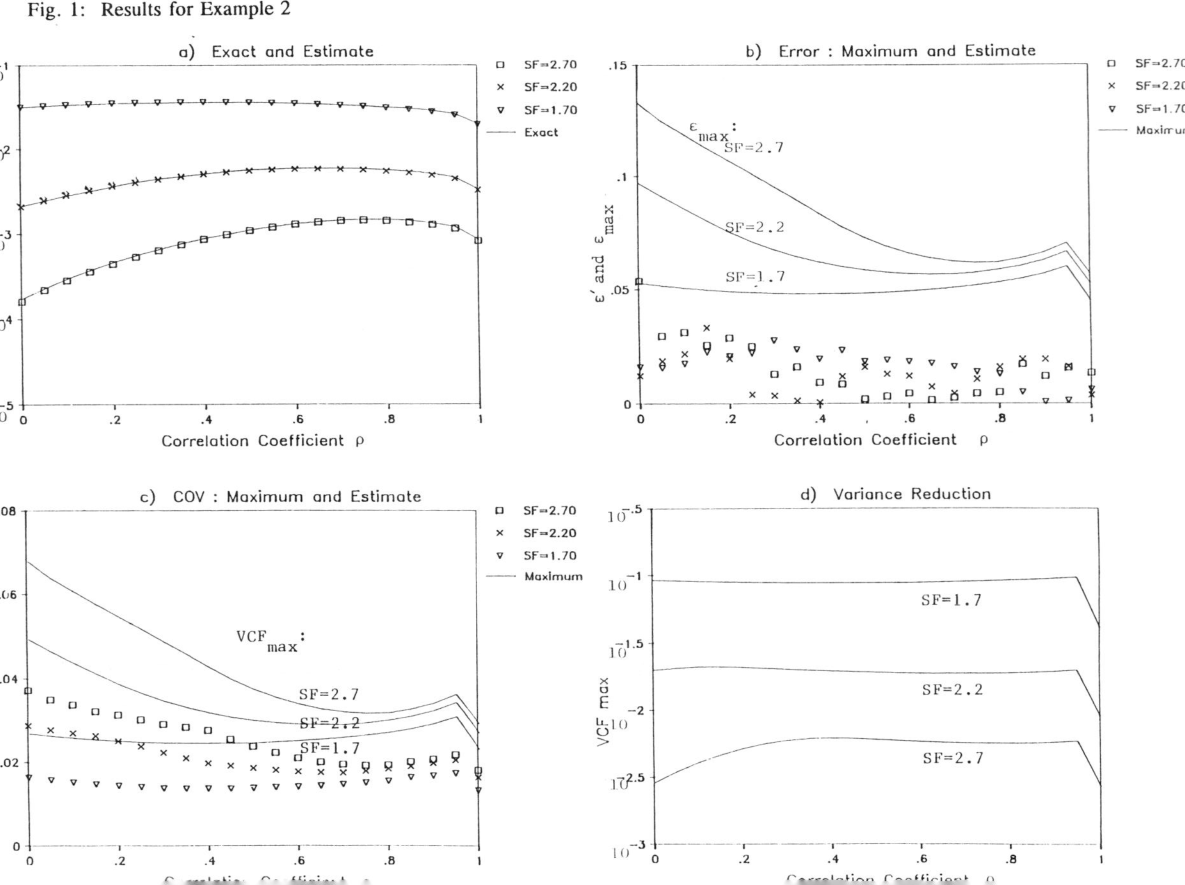

Reliability and Optimization of Structural Systems, V (B-12)
P. Thoft-Christensen and H. Ishikawa (Editors)
Elsevier Science Publishers B.V. (North-Holland)

Knowledge Refinement for Damage Assessment Expert System Based on Neural Network

H. Furuta*, H. Ohtani**, N. Shiraishi*

* Department of Civil Engineering, Kyoto University, Japan

** Kinki Nippon Railway Co., Osaka, Japan

ABSTRACT

For developing a practical expert system, it is a crucial problem to acquire the sufficient amount and good quality of expertise. An attempt is made here to apply the technique of neural network for the knowledge refinement of three types of production rules; ordinal production rule, production rule with certainty factor, and fuzzy production rule.

Keyword Codes: I.1.2.0, I.1.2.1, I.1.2.6
Keywords: Artificial Intelligence, General; Applications and Expert Systems; Learning

1. INTRODUCTION

Recently, many expert systems have been developed in various fields. The authors have been building an expert system for damage assessment of reinforced concrete bridge decks[1-3]. The current version of the system has such a remarkable feature that it includes a fuzzy-set manipulation system to treat fuzzy sets in the process of data handling, rule representation, and inference procedure[3]. However, in order to make the expert system more practical, it is a crucial problem to acquire the sufficient amount and quality of expertise without difficulty.

In this paper, an attempt is made to apply the technique of neural network[4] for the knowledge refinement of production rules. Three kinds of production rules that are ordinal production rules[1], production rules with certainty factors[2], and fuzzy production rules[3] can be modeled as three types of neural networks, respectively. An application of damage assessment of reinforced concrete bridge deck is presented to demonstrate the applicability of the neural network models proposed here.

2. REPRESENTATION OF PRODUCTION RULES BY NEURAL NETWORK MODELS

So far, many expert systems have been developed, most of which are so called production systems that use production rules for knowledge representation. In general, a production rule is expressed as

If X is A then Y is B. (1)

where A and B are the attributes of antecedent and consequent parts, respectively. Evidently, both the antecedent and consequent parts can be extended into multi-attribute cases. Denoting that (Xi Ai) is that Xi is Ai and (Yi Bi) is that Yi is Bi, a general production rule can be written as

If ((X1 A1),(X2 A2), ... ,(Xn An))
then ((Y1B1),(Y2 B2), ... ,(Ym Bm)) (2)

where n and m are the numbers of attributes. ((X1 A1),(X2 A2), ... ,(Xn An)) or (Y1 B1),(Y2 B2), ... ,(Ym Bm)) means that (X1 A1) and (X2 A2) and .... and (Xn An)) or ((Y1 B1) and (Y2 B2) and .... and (Ym Bm), in which "and" indicates "logical and", i.e., intersection.

In order to express Eq. 1, a neural network model is considered, which consists of n input layers , m output layers, and appropriate numbers of intermediate layers. As an example, consider the following production rule.

If ((X1 A1),(X2 A2),(X3 A3)) then ((Y1 B1),(Y2 B2)) (3)

In the neural model, Eq. 3 is expressed in such a way that the input is ((X1 A1),(X2 A2),(X3 A3),(X4 A4),(X5 A5), ... ,(Xn An)) = (1,1,1,0,0, ... ,0) and the output is ((Y1 B1),(Y2 B2),(Y3 B3),(Y4 B4), ... ,(Ym Bm)) = (1,1,0,0, ... ,0). It is possible to refine the knowledge expressed in the form of production rule by learning the past data regarding many pairs of ((X1 A1),(X2 A2), ... ,(Xn An)) and ((Y1 B1),(Y2 B2), (Ym Bm)). If the following data is obtained,

((X1A1),(X2 A2),(X3 A3),(X4 A4), ... ,(Xn An) = (1,0,1,0, ... ,0)
((Y1 B1),(Y2 B2),(Y3 B3),(Y4 B4), ... ,(Ym Bm) = (0.98,0,0,0, ... ,0) (4)

a production rule is derived:

If ((X1 A1),(X3 A3)) then (Y1 B1) (5)

Next, another neural network model is constructed for production rules with certainty factors. Similar to Eq. 2, a production rule with certainty factors is given as

If (p1/(X1 A1),p2/(X2 A2), ... ,pn/(Xn An))
then (q1/(Y1 B1),q2/(Y2 B2), ... ,qm/(Ym Bm)) (6)

where pi and qj are the certainty factors of the i-th factor of the antecedent and the j-th factor of the consequent. To deal with the certainty factors, a neural model shown in Figure 1 is considered. For input and output units, a pair of nodes are prepared, in which one is for the value of the factor and the other is for its certainty factor. Using this neural model, it is possible to refine the production rules with certainty factors.

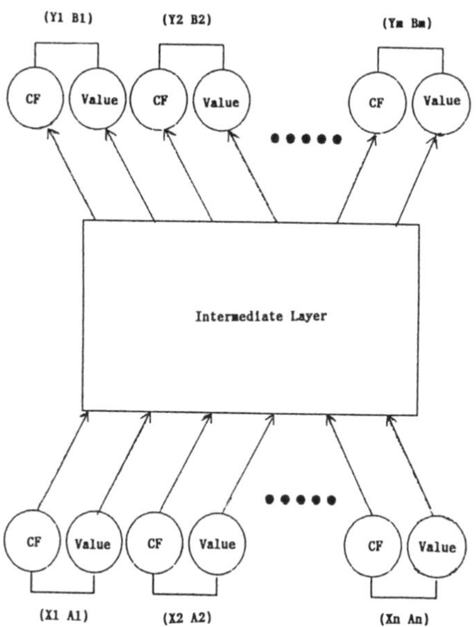

Figure 1 Neural Network Model for Production Rule with Certainly Factors

In usual, a fuzzy production rule[5] can be written as

If(@p1/(X1 @A1),@p2/(X2 @A2), ... ,@pn/(Xn @An))
then (@q1/(Y1 @B1), ... ,@qm/(Xm @Bm)) (7)

where @pi and @qj denote the fuzzy truth values, (Xi @Ai) means that a factor Xi has the value of @Ai in the antecedent, and (Yj @Bj) means that a factor Yj in the consequent has the value of @Bj. Symbols with @ denote fuzzy quantities so that @Ai and @Bj are fuzzy predicates. This fuzzy rule can be expressed by a neural network model shown in Figure 2. For each pair of (@pi, @Ai) or (@qj, @Bj), two units are used, one of which consists of six nodes.

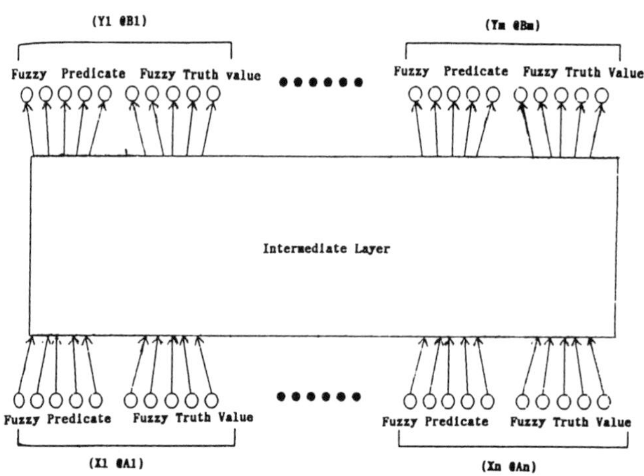

Figure 2 Neural Network Model for Fuzzy Production Rule

3. OUTLINE OF FUZZY EXPERT SYSTEM FOR DAMAGE ASSESSMENT OF RC DECK

In the current version of the damage assessment expert system developed by the authors[3], the past records and inspection results are used as the input data. When the inspection results regarding cracks are firstly input into the system, the matching processes for rules concerning their damage cause, damage degree and damage propagation are executed to provide a solution for the remaining life. At first, based on the inspection results, the damage is classified into cracks, damage of pavement, damage of reinforcing steel, damage of concrete, and structural damage. Followingly, using the design and environmental conditions as well as the inspection data, possible damage causes are estimated. In general, multiple damage causes are estimated, to which "damage mode" is taken into consideration. The damage mode means a group of several damages resulting from the same cause. Identifying a damage mode, damage degree is evaluated for each kind of damage. Based on their evaluation, a damage degree to the damage mode is obtained. Similar to this process, the damage propagation speed is assessed by considering the damage causes estimated. Finally, the remaining life is estimated using the construction year and the results obtained above.

4. AUTOMATIC TUNING OF FUZZY PRODUCTION RULES FOR DAMAGE ASSESSMENT USING NEURAL NETWORK

In order to collect the information to tune several fuzzy production rules which are useful in the damage assessment of RC bridge deck, a survey was conducted, during which several sets of questionnaires were distributed among three experienced engineers of bridge maintenance. Table 1 presents one of the questionnaires. For six cases, the engineers gave truth values to the fact that the extreme wheel loading is one of the damage causes. The first case means that the structural type (i.e., plate girder bridge), the design specification employed, the crack configuration perpendicular to bridge axis, and the short distance between locations of wheel loading and locations of crack occurrence, are closely related to the conclusion that the major damage cause is the extreme wheel load.

Table 1 Questionnaire for Damage Cause (1)

	1	2	3	4	5	6
Structural Type : plate girder	X	X				
Structural Type : arch					X	X
Structural Type : box girder			X			
Design Specification : before 1932 version	X	X	X	X	X	
Crack Configuration : width direction	X	X				
Crack Configuration : bridge axis direction					X	X
Crack Configuration : two directions			X			
Crack Occurrence Time : very recent					X	X
Crack Occurrence Time : recent			X			
Short Distance between Crack Location and Wheel Loading		X	X	X	X	X
Cross Beams Exists			X			
Bridge Deck is Two-directional Reinforcing Slab					X	

For the bridge type, binary expression of "Yes" and "No" can be used. However, because the eighth through tenth factors include such fuzzy sets as "near" and "recent", some appropriate membership functions should be prepared in advance. Here, some membership functions are prescribed for crack occurrence time, distance between the locations of wheel loading and crack occurrence, crack density, thickness of concrete slab, roughness of pavement and so on. Table 2 shows the answers given by the three engineers. For instance, engineer A gave such answers that the truth values for case 1 through case 6 are true, true, very true, very true, true, and more or less true, respectively. On the other hand, engineers B and C gave different answers from that of engineer A.

Table 2 Result of Questionnaire

	Engineer A	Engineer B	Engineer C
Case 1	true	slightly true	-
Case 2	true	true	-
Case 3	very true	hardly true	more or less true
Case 4	very true	hardly true	-
Case 5	true	slightly true	-
Case 6	more or less true	-	-

Here, the symbol - means that the engineer gave no answer. Using the results of the questionnaires, some rules for selecting several damage causes among representative damage causes presented in Table 3 can be derived through the learning process of neural network.

Table 3 Representative Damage Causes

Damage Cause (1)	Extreme Wheel Load
Damage Cause (1)	Impact Effect
Damage Cause (3)	Inadequency of Girder Arrangement
Damage Cause (4)	Short of Deck Depth
Damage Cause (5)	Lack of Main Steel Bar
Damage Cause (6)	Lack of Distribution Bar
Damage Cause (7)	Inadequency of Distributed Cross Beams
Damage Cause (8)	Additional Moment due to Differential Settlement

The learning procedure for the estimation of damage causes is as follows:

1) In order to deal with each factor used in estimating the damage cause, two nodes are given to a crisp quantity with a crisp truth value, and six nodes to a fuzzy quantity and another six nodes to a fuzzy truth value in a neural network model. Through analyzing the questionnaire results, we can construct several sets of data for estimating the damage cause.
2) The output of the neural network model is a set of possible damage causes.
3) Consider a four-layer neural network model, where the intermediate layers consist of 20 nodes, respectively.
4) Using this neural network model, the learning is performed.

For instance, consider the sample data presented in Table 4. In Table 4, the evaluation is done as that 1 denotes "Yes" and 0 denotes "No" for binary variables, the fuzzy evaluation of "Near" is defined by {1/4(m),0.5/6(m)} and that of "fairly true" is defined by {0.5/0.2,1/0.4,0.5/0.6}. Using these data as the input data for the neural computation, it is possible to obtain the learning results. Figure 3 shows that damage cause (1) is the most possible because it has the largest truth value. Applying the same procedure to other input data, it enables to obtain more fuzzy production rules useful in the damage assessment of RC bridge deck.

Table 4 Input for Neural Computation

Factor	Value	Truth V.
Structural Type	1.0	1.0
Design Spec.	1.0	1.0
Crack Config.	1.0	1.0
Occurrence Time of Crack	Very near	Fairly True
Difference between Wheel Load and Crack	Near	True
Crack Density	Medium	True
2-Dimensional Slab	1.0	1.0
Depth of Slab	Large	Slightly True

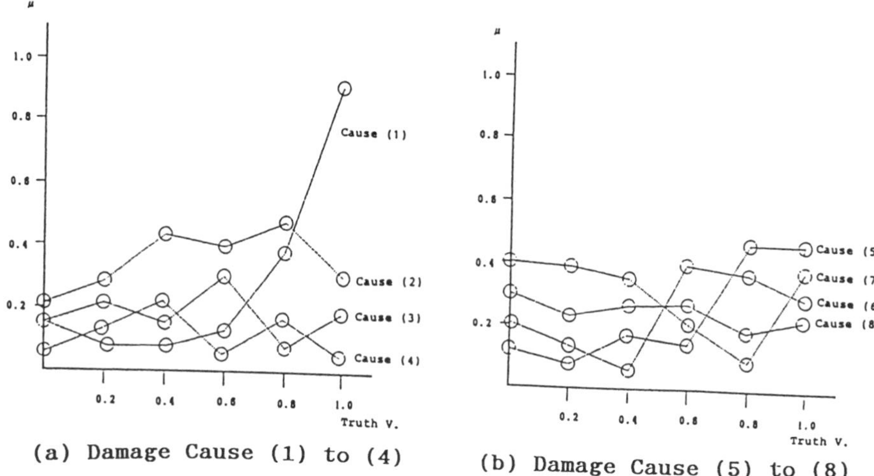

(a) Damage Cause (1) to (4)

(b) Damage Cause (5) to (8)

Figure 3 Output Result For Damage Cause (1)

5. CONCLUSIONS

In order to make expert systems to be actually useful, some improvement is desirable on such issues as the knowledge acquisition and refinement, knowledge representation, treatment of ambiguity or uncertainty, and man-machine interface. By introducing the fuzzy production systems into the inference process, it is possible to solve the above problems to some extent. Namely, it is possible to utilize the knowledge and rules which are expressed in terms of natural language. This enables us to acquire the expertise with ease. Furthermore, the technique of neural network is useful in tuning fuzzy production rules automatically. The three neural network models developed here are applicable to such three production rules as ordinal production rules, production rules with certainty factors, and fuzzy production rules. Through a numerical experiment, it is concluded that the neural network technique is available to generate and refine the production rules for the damage assessment of RC bridge deck. It is, however, desirable to collect the sufficient amount and good quality of data from questionnaires and other sources.

REFERENCES

1. H. Furuta, N. Shiraishi and J.T.P. Yao, Proc. of 5th International Offshore Mechanics and Arctic Engineering Symposium, Vol. 1, pp.11-15, 1986.

2. N. Shiraishi, H. Furuta, M. Umano and K. Kawakami, Fuzzy Sets and Systems, Vol. 44, pp.449-457, 1991.
3. H. Furuta, N. Shiraishi, M. Umano and K. Kawakami, Computers and Structures, Vol. 40, No. 1, pp.137-142, 1991.
4. D. E. Rumelhart et al., Parallel Distributed Processing, MIT Press, 1986.
5. L. A. Zadeh, Information Sciences, Vol. 8, pp.43-80, 1975.

Reliability and Optimization of Structural Systems, V (B-12)
P. Thoft-Christensen and H. Ishikawa (Editors)
Elsevier Science Publishers B.V. (North-Holland)
1993 IFIP

Reliability Analysis of Geometrically Nonlinear Structure by Rigid-Plastic Model

J. M. Johannesen, O. Ditlevsen

Department of Structural Engineering, Technical University of Denmark
Building 118, DK-2800 Lyngby, Denmark

Abstract The self-contradiction in the title of this paper is only apparent because it concerns the application of a particular type of response surface method where the form of the response surface is generated by use of the simple rigid-plastic mechanical theory. The basis is the previously published "model correction factor method". Herein the method is demonstrated to be applicable for efficient and fast reliability analysis of some strongly geometrically nonlinearly behaving frame structures with elastic-plastic constitutive relations.

1. Introduction

Herein the model correction factor method is applied to elastic-plastic frame structures of such slenderness and with such load configurations that geometrically nonlinear effects become important.

The model correction factor method is introduced in Arnbjerg-Nielsen (1991) and is further developed in Ditlevsen and Arnbjerg-Nielsen (1993). It can be classified as a special type of response surface method to be applied to simplify the time consuming reliability analysis when an elaborate mechanical model describes the structural failure behavior. The mathematical class of response surfaces is defined not by some more or less arbitrary class of functions but by a mechanically interpretable model of suitable simplicity. By engineering insight into the actual problem, the simple mechanical model is formulated such that it reflects the most important issues of the elaborate model. However, emphasis is not put on the sophisticated details such as higher order effects that may be built into the elaborate model.

There are several advantages by this type of analysis. One advantage is that the simple model subject to first order reliability analysis directly points at the important domain in the formulation space. Another advantage is that the elaborate fitting procedure in the ordinary response surface method is considerably simplified due to the mechanical information already put into the simple model by its formulation. Moreover the simple model is easily seen through from an engineering point of view making the acceptance of the output from the elaborate model less prone to radical errors or misinterpretations. The method has been succesfully applied in lower bound plastic reliability analysis of damaged concrete decks, Karlsson et al. (1993). In order that this paper can be self-contained the model correction factor method will be summarized in the next section without the argumentations for the validity of the method.

2. Outline of the Model Correction Factor Method

Let (**x**,**y**,**z**) be the total vector of basic variables (input variables) that are contained in the elaborate model. The subvectors **x** and **y** are the vectors of load variables and strength variables respectively. These variables are with sufficient generality defined such that they all have physical units that are proportional to the unit of force. The subvector **z** is the vector of all the remaining basic variables (of type as geometrical and dimensionless basic variables). Two limit state equations $g_r(\mathbf{x},\mathbf{y},\mathbf{z})=0$ and $g_i(\mathbf{x},\mathbf{y},\mathbf{z})=0$ are given representing the elaborate (*r* for "*realistic*") and the simple (*i* for *"idealized"*) model, respectively. It is assumed that for each fixed $(\mathbf{x},\mathbf{y},\mathbf{z})$ the equations

$$g_r(\kappa_r\mathbf{x},\mathbf{y},\mathbf{z})=0 \quad , \quad g_i(\kappa_i\mathbf{x},\mathbf{y},\mathbf{z})=0 \tag{1}$$

can be solved uniquely with respect to κ_r and κ_i, respectively. The solutions are $\kappa_r(\mathbf{x},\mathbf{y},\mathbf{z})$ and $\kappa_i(\mathbf{x},\mathbf{y},\mathbf{z})$. By using the physical property of dimension homogeneity it is shown in Ditlevsen and Arnbjerg-Nielsen (1993) that the two equations

$$g_r(\mathbf{x},\mathbf{y},\mathbf{z})=0 \quad , \quad g_i(\mathbf{x},\frac{\kappa_r(\mathbf{x},\mathbf{y},\mathbf{z})}{\kappa_i(\mathbf{x},\mathbf{y},\mathbf{z})}\mathbf{y},\mathbf{z})=0 \tag{2}$$

are equivalent in the sense that the two set of points they define are identical. The idea of the model correction factor method is to use a suitably simple approximation to the last equation in (2) in the reliability analysis in place of the first equation in (2). The point is to approximate the function $\nu(\mathbf{x},\mathbf{y},\mathbf{z})=\kappa_r(\mathbf{x},\mathbf{y},\mathbf{z})/\kappa_i(\mathbf{x},\mathbf{y},\mathbf{z})$ by a constant or at most by an inhomogeneous linear function of $(\mathbf{x},\mathbf{y},\mathbf{z})$. The approximation is made such that it is particularly good within the region of the space that contributes the most to the failure probability. Let $(\mathbf{x}^*,\mathbf{y}^*,\mathbf{z}^*)$ be a point of this region and let $\nu^*=\nu(\mathbf{x}^*,\mathbf{y}^*,\mathbf{z}^*)$. The equation

$$g_i(\mathbf{x},\nu^*\mathbf{y},\mathbf{z})=0 \tag{3}$$

then defines an approximating limit state in the important region. The problem is now reduced to the problem of how to choose the point of approximation $(\mathbf{x}^*,\mathbf{y}^*,\mathbf{z}^*)$. The answer to this problem is given in the reliability theory. With a judgementally chosen value ν_0 of ν^* a first or second order reliability analysis (FORM or SORM, see eg. Madsen, Krenk and Lind (1986) or Ditlevsen and Madsen (1991)) is made with (3) as limit state. This analysis determines the most central point (the design point) $(\mathbf{x}_1,\mathbf{y}_1,\mathbf{z}_1)$ and an approximate failure probability p_1. Using that $\kappa_i(\mathbf{x}_1,\mathbf{y}_1,\mathbf{z}_1)=1/\nu_0$ an improved value $\nu_1=\nu_0\kappa_{r1}$ of ν^* is calculated where $\kappa_{r1}=\kappa(\mathbf{x}_1,\mathbf{y}_1,\mathbf{z}_1)$. Then a new FORM or SORM analysis is made with (3) as limit state. This gives the most central point $(\mathbf{x}_2,\mathbf{y}_2,\mathbf{z}_2)$ and an approximate failure probability p_2. Proceeding iteratively in this way we get a sequence $(\kappa_{r1},p_1),(\kappa_{r2},p_2),\ldots$ that may or may not be convergent. If the sequence is convergent in the first component it is also convergent in the second component and we have $\kappa_{r1},\kappa_{r2},\ldots\rightarrow 1$, $p_1,p_2,\ldots\rightarrow p$, where p will be denoted as the *zero order approximation* to the probability of the failure event of the elaborate model.

If the sequence is not convergent we still can define the zero order approximation by simple interpolation to the value $\kappa_r=1$ among points (κ_r,β) ($\beta=-\Phi^{-1}(p)$, Φ = standardized normal distribution function) corresponding to the sequence or simply obtained for a series of different values of ν_0.

A check of the goodness of the zero order approximation is made by replacing the function $\nu(\mathbf{x},\mathbf{y},\mathbf{z})$ by its first order Taylor expansion

$$\nu(\mathbf{x},\mathbf{y},\mathbf{z}) \simeq \tilde{\nu}(\mathbf{x},\mathbf{y},\mathbf{z}) = \nu^* + \mathbf{a}^T(\mathbf{x}-\mathbf{x}^*) + \mathbf{b}^T(\mathbf{y}-\mathbf{y}^*) + \mathbf{c}^T(\mathbf{z}-\mathbf{z}^*) \tag{4}$$

at the most central point $(\mathbf{x}^*,\mathbf{y}^*,\mathbf{z}^*)$ corresponding to the limit state (3) with ν^* being the value corresponding to $\kappa_r=1$. The numerical determination of the coefficients **a**, **b** and **c** requires that the values of $\nu(\mathbf{x},\mathbf{y},\mathbf{z})$ are known at least at as many points in the vicinity of $(\mathbf{x}^*,\mathbf{y}^*,\mathbf{z}^*)$ as the number of variables in $(\mathbf{x},\mathbf{y},\mathbf{z})$. These values of ν are obtained by solving the equations (1) with respect to κ_r and κ_i respectively at each chosen point $(\mathbf{x},\mathbf{y},\mathbf{z})$.

With (4) substituted for κ_r/κ_i into the last equation in (2) we get a limit state for which both the probability of failure and the value of κ_r in general will be different from p and the value $\kappa_r=1$ as obtained by the zero order approximation. However, by a unique scaling factor k_r on the load vector $\mathbf{x}$ we can achieve that the limit state $g_i(k_r\mathbf{x},\nu(k_r\mathbf{x},\mathbf{y},\mathbf{z})\mathbf{y},\mathbf{z})=0$ corresponds to the failure probability p. With the Taylor expansion (4) substituted into this equation we get the limit state equation $g_i(k_r\mathbf{x},\tilde{\nu}(k_r\mathbf{x},\mathbf{y},\mathbf{z})\mathbf{y},\mathbf{z})=0$ for which we can determine k_r by iterative application of FORM or SORM analysis such that the corresponding failure probability becomes p. The size of the deviation of k_r from 1 can then be used to judge the accuracy of the zero order approximation. Also the change of the most central point contributes to this judgement.

In case k_r deviates too much from 1 an iterative procedure can be applied just as for the zero order approximation to obtain $k_r=1$ in the limit and a corresponding improved value of p. This value will be called the *first order approximation.*

Remark. In the particular case where the joint distribution of the random load vector **X** is Gaussian it is possible to give a simple interpretation of the deviation of k_r from 1. With **U** being a standard Gaussian vector we have $\mathbf{X}=\mathbf{M}[(\mathbf{M}^{-1}\mathrm{Cov}[\mathbf{X},\mathbf{X}^T],\mathbf{M}^{-1})^{1/2}\mathbf{U}+\mathbf{e}]$ where **M** is the diagonal matrix with the elements of $\boldsymbol{\mu}=E[\mathbf{X}]$ as diagonal elements, $\mathbf{e}^T=[1...1]$, and the symbol $\mathbf{A}^{1/2}$ stands for any matrix that satisfies $\mathbf{A}^{1/2}(\mathbf{A}^{1/2})^T=\mathbf{A}$. If $\mathbf{M}^{-1}\mathrm{Cov}[\mathbf{X},\mathbf{X}^T]\mathbf{M}^{-1}=\{\rho[X_i,X_j]V_{Xi}V_{Xj}\}$ is independent of **M** (in one dimension this simplifies to a fixed coefficient of variation) it follows from the above representation of **X** that the reliability of the limit state $g_r(k_r\mathbf{x},\mathbf{y},\mathbf{z})=0$ is the same as the reliability of the limit state $g_r(\mathbf{x},\mathbf{y},\mathbf{z})=0$ given that the mean of **X** is simply changed from $\boldsymbol{\mu}$ or $k_r\boldsymbol{\mu}$. In this consideration it is assumed that the random vectors **Y** and **Z** are not dependent on $\boldsymbol{\mu}$. This interpretation is used herein for the graphical presentation of the first order approximation results (Figure 3).

3. Carrying Capacity Calculation

The elaborate model is a geometrically nonlinear elasto-plastic FE-model (appendix A). The material is assumed to be linear elastic with strain hardening after the onset of yielding (appendix B). The carrying capacity for the frame structure in the FE-model is defined to be the first load parameter maximum on the load-displacement curve. A load controlled incrementation is capable of tracing the load-displacement curve to a point close to the first load maximum. This point is arbitrarily determined as the point at which the iteration convergence fails. This arbitrariness of the point causes the corresponding numerical derivatives needed for calculating the Taylor expansion of the correction factor $\nu(\mathbf{x},\mathbf{y},\mathbf{z})$ in the reliability analysis calculations to be too unstable to be useful. The problem is eliminated using displacement controlled incrementation, appendix C, allowing load maxima to be calculated more accurately giving sufficiently stable numerical derivatives.

The simple model is a rigid-plastic yield hinge model in which the yield criterion includes moments only. Potential yield hinges are placed at supports, in frame corners and at points with concentrated applied loads. The orthogonal corner frame example considered in the following shows that it can be necessary to have yield hinges at some few other points of the frame. The cross sectional ideal plastic moment capacity is given as 1/4 WD^2 where W and D are the width and depth, respectively, of the assumed massive rectangular beam cross sections.

4. Reliability Analysis

The reliability analyses are carried out on a HP 9000/730 using the general purpose probabilistic analysis program package Proban version 3.0 (Det Norske Veritas Research). In the model correction factor based analysis the reliability calculations are related to the collapse modes of the simple model. Herein only the design point for the collapse mode with lowest reliability as determined by FORM is considered. Thus the obtained reliability estimated in the analysis is the reliability corresponding to the probabilistically most critical collapse mode. Often this represents a good approximation to the system reliability. If the system reliability were to be considered directly it would be necessary to assign a correction factor to the design point for each collapse mode in the simple model or at least to some of the most important of these.

In the examples herein the limit state equations for the elaborate and the simple model get the appearance $-\kappa_r+\lambda_r(\mathbf{x},\mathbf{y},\mathbf{z})=0$ and $-\kappa_i+\lambda_i(\mathbf{x},\mathbf{y},\mathbf{z})=0$ respectively. The functions $\lambda_r(\cdot)$ and $\lambda_i(\cdot)$ are the carrying capacity functions in each of the two models. This particular form of the limit state equations gives the simplification relative to the general problem that solving with respect to κ_r and κ_i only requires one calculation of $\lambda_r(\mathbf{x},\mathbf{y},\mathbf{z})$ and $\lambda_i(\mathbf{x},\mathbf{y},\mathbf{z})$.

5. Examples: Portal Frame and Orthogonal Corner Frame

Reliability analyses are made for a portal frame, Figure 1 (left), L=10.0 m and H=7.0 m. In order to exaggerate geometrically nonlinear effects the frame is made relatively slender. The dimensions of all cross sections (appendix B) are fixed to depth D=0.2959 m and to width W=0.02639 m. The Youngs modulus is fixed to $E=2.10\times10^5$ MPa. The hardening parameter n in (B.1) is fixed to 10. In the elaborate model each of the substructures AB, BC, CD, and DE are subdivided into 6, 5, 5, and 6 beam elements respectively (appendix B). The finite element model has not been verified by a model with more elements. All probabilistic data of the problem are listed in Table 1.

Table 1. Data for reliability analysis of the portal frame (C.o.V. = Coefficient of variation)

Name	Distribution	Description	Mean	C.o.V.
F_{HB}	Normal	Applied horizontal force at point B	μ_F	0.2
F_{VB}	Normal	Applied vertical force at point B	$\eta\mu_F$	0.2
F_{VC}	Normal	Applied vertical force at point C	μ_F	0.2
f_{yPQ}	Lognormal	Yield stress in PQ (=AB, BD or DE)	1500 MPa	0.1

f_{yPQ} are equi-correlated with correlation coefficient 0.3. No other correlation.

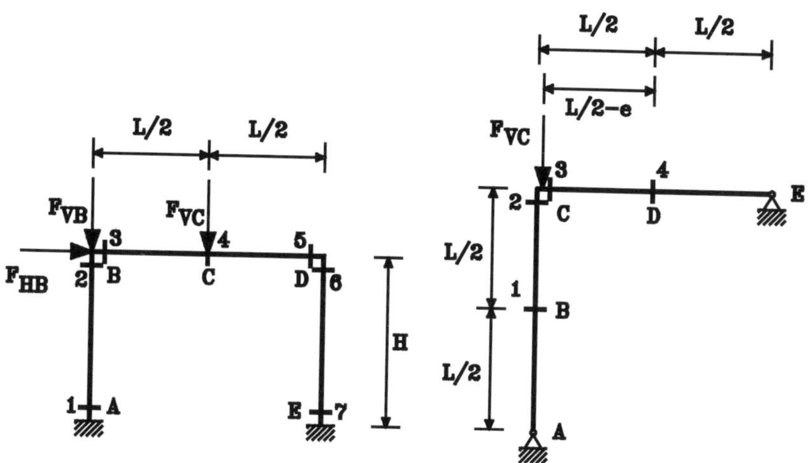

Figure 1. Portal frame (left) and orthogonal corner frame (right). Numbers refer to yield hinge positions

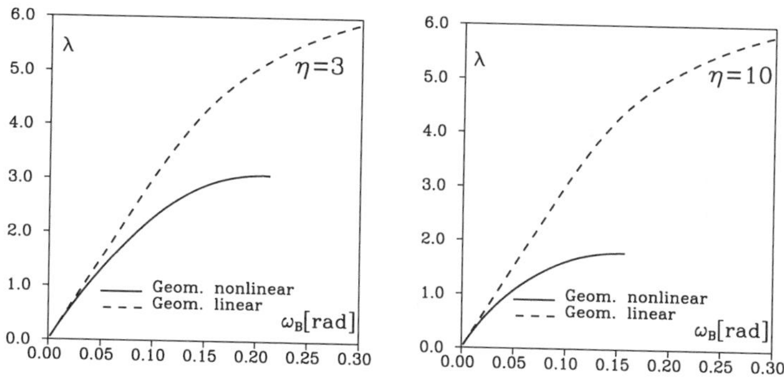

Figure 2. Load parameter λ versus clockwise rotation ω_B of B of the portal frame (controlled displacement, appendix C) for η=3 (left) and η=10 (right)

The reliability analysis is carried out for η=3, Table 1. Different degrees of geometrical nonlinearity can be illustrated by deterministic analyses for different load configuration cases, i.e. different values of η. Load-displacement curves corresponding to geometrically linear and geometrically nonlinear behavior are plotted for η=3 and η=10 in Figure 2. The ordinate λ is the load factor applied to the load configuration such that it balances an imposed rotation ω_B of the frame corner B. For η=0 the effect from the geometrical nonlinearity is very moderate. Otherwise it is seen that the effect from the geometrical nonlinearity is important but that the effect varies relatively slowly with η and thus with F_{VB}. Although F_{VB} has no influence in the simple yield hinge model, the model correction factor method turns out to be able to capture the overall changed mechanical behavior of the structure caused by the rigidity decreasing effect of F_{VB}, Figure 3 (left).

The orthogonal corner frame is shown in Figure 1 (right), L=10.0 m. The cross sectional geometrical properties and the Youngs modulus are as for the portal frame. Ideal plasticity is assumed, i.e. the hardening parameter n in (B.1) is put to infinity (a simplified version of (B.1) is used in this case). In the elaborate model each of the substructures AC and CE are subdivided into 5 beam elements. The finite element model has not been verified by a model with more elements. All probabilistic data of the problem are given in Table 2.

Table 2. Data for reliability analysis of the orthogonal corner frame (C.o.V. = Coefficient of variation)

Name	Distribution	Description	Mean	C.o.V.
F_{VC}	Normal	Applied vertical force at point C	μ_F	0.2
e	Normal	Horiz. eccentricity of F_{VC} (pos. C→D)	0.1 m	0.2
f_{yRS}	Lognormal	Yield stress in RS (=AC or CE)	400 MPa	0.1
f_{yAC} and f_{yCE} are correlated with correlation coefficient 0.3. No other correlation.				

The probabilistically most critical mechanism as determined by FORM (Section 4) is found among 10 and 5 mechanisms for the portal and the orthogonal corner frame respectively. It is noted that the standard formulation of a rigid-plastic yield hinge model for the orthogonal corner frame as considered herein only contains two yield hinges both placed in the horizontal beam, the one at the corner and the other at the load. However, this standard model is unsufficient here because the corresponding beam mechanism is not in any way resembling the elasto-plastic displacement field implied by the elaborate model. In fact one should first establish a reasonable elastic-plastic hinge model that crudely behaves like the elaborate model with respect to geometrical nonlinearity. Thus there obviously should be a reasonable number of separated rotation hinges in the model. Next step is the drastic one to let these hinges be rigid-plastic. The justification of this step relies on the limit state defined by the mechanism equations to be of a reasonable mathematical similarity with the limit state of the elaborate model. At least this is needed in the probabilistically most important region of the formulation space.

6. Results

Results from the correction factor based analyses are compared in Figure 3 to results from direct FORM analyses using the elaborate model. The results from the FORM analyses have been verified by directional simulation (not shown).

It is seen that the first order approximation result based on the model correction factor method gives an excellent approximation to the reliability index. The CPU time used for the FORM analysis was of the order of 1 hour for each point in Figure 3, increasing with increasing reliability index. With a reasonable starting value for the zero order model correction factor, the zero order and the first step of the first order correction factor analysis took only some few minutes in CPU time per point. The iteration to the final first order approximation result was of the order 10 minutes in CPU time.

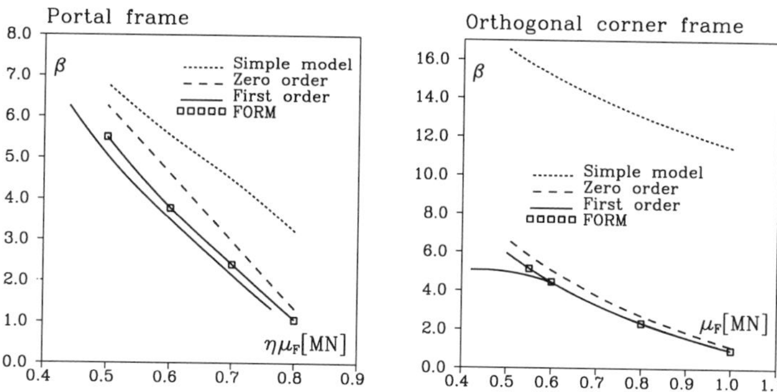

Figure 3. Reliability index $\beta=-\Phi^{-1}(p)$ *versus mean load level. From below the full curves correspond to the first step and the final step of the first order approximation calculation respectively.*

Acknowledgement
This work has been financially supported by the Danish Technical Research Council.

References

Arnbjerg-Nielsen, T (1991): *Rigid-ideal Plastic Model as a Reliability Analysis Tool for Ductile Structures.* Ph.D.-thesis, Department of Structural Engineering, Technical University of Denmark, Series R, No 270.

Byskov, E. (1982-83): *Plastic Symmetry of Roorda's Frame.* Journal of Engineering Mechanics, 10(3), 311-328.

Clarke, M.J. and Hancock (1990), G.J.: *A Study of Incremental-Iterative Strategies for Non-linear Analyses.* International Journal for Numerical Methods in Engineering, Vol 29, 1365-1391.

Ditlevsen, O. and Madsen, H.O. (1990): *Bærende konstruktioners sikkerhed* (in Danish, under translation to English). SBI-rapport 122.

Ditlevsen, O. and Arnbjerg-Nielsen,T. (1993): *Model Correction Factor in Structural Reliability.* Journal of Engineering Mechanics, ASCE.

Forde, B.W.R. and Stiemer, S.F. (1987): *Improved Arc Length Orthogonality Methods for Nonlinear Finite Element Analysis.* Computers and Structures 27(5).

Gierlinsky, J.T. and Graves Smith, T.R. (1985): *A Variable Load Iteration Procedure for Thin-Walled Structures.* Computers and Structures Vol. 21, No. 5, pp. 1085-1094.

Karlsson, M., Johannesen, J.M. and Ditlevsen, O. (1993): *Reliability Analysis of an Existing Bridge.* IABSE Colloquium: Remaining Structural capacity, Copenhagen. IABSE Report Vol. 67, pp. 19-28.

Madsen, H.O., Krenk, S. and Lind, N.C. (1986): *Methods of Structural Safety.* Prentice-Hall.

Powell, G. and Simons, J. (1981): *Improved Iteration Strategy for Nonlinear Structures.* International Journal for Numerical Methods in Engineering, Vol 17, 1455-1467.

Appendix A: Finite Element Formulation

A Lagrangian strain measure is employed. The generalized strain components are the axial strain ε and the curvature κ given by

$$\varepsilon=\frac{du}{dx}+\frac{1}{2}\left(\frac{dw}{dx}\right)^2 \; ; \quad \kappa=-\frac{d^2w}{dx^2} \tag{A.1}$$

where u and w are the axial and transverse displacements, respectively. The fiber strains in a cross section are assumed to vary linearly with depth. The finite element equations are based on an incremental principle of virtual work assuming proportional loading, Byskov (1982-83). The implemented incremental finite element equations (before assembling of elements) are

$$\begin{aligned}&[\int \mathbf{B}^T\mathbf{D}\mathbf{B}dV+\int \mathbf{B}^T\mathbf{d}_1\mathbf{u}_0^T\mathbf{C}dV+\int \mathbf{C}\mathbf{u}_0\mathbf{d}_1^T\mathbf{B}dV+\int d_{11}\mathbf{C}\mathbf{u}_0\mathbf{u}_0^T\mathbf{C}dV+\int \mathbf{C}\sigma^f_0dV\,]\Delta\mathbf{u}\\&=\lambda\int \mathbf{N}^T\mathbf{t}dK-[\int(\mathbf{b}_1+z\mathbf{b}_2)\sigma^f_0dV+\int \mathbf{C}\sigma^f_0dV\mathbf{u}_0]\end{aligned} \tag{A.2}$$

where

0 as subscript refers to the previous state

Δ = operator referring to increment from the previous state to the current state

$\mathbf{D}$ = matrix that defines the tangential constitutive relations between generalized stresses and strains

$\mathbf{d}_1^T$ = first row of $\mathbf{D}$

d_{11} = element in first row and first column of $\mathbf{D}$

σ^f = fiber stress

$\mathbf{N}$ = displacement distribution matrix

$\mathbf{t}$ = load configuration vector

λ = load parameter

$\mathbf{u}$ = displacement vector

$\mathbf{B}$ = linear strain distribution matrix; $\varepsilon_{linear}=\mathbf{u}^T\mathbf{B}\mathbf{u}$; $\varepsilon=\varepsilon_{linear}+\varepsilon_{nonlinear}$; $\varepsilon^T=[\varepsilon\ \kappa]$

$\mathbf{b}_1^T$ = first row of $\mathbf{B}$

$\mathbf{b}_2^T$ = second row of $\mathbf{B}$

$\mathbf{C}$ = nonlinear strain distribution matrix, i.e. $\varepsilon_{nonlinear}=1/2\ \mathbf{u}^T\mathbf{C}\mathbf{u}$

z = coordinate transverse to the element axis

dV = infinitesimal volume element

dK = infinitesimal boundary element

Appendix B: Beam Finite Element and Strain Hardening Material

The finite element is defined as a three-noded element as shown in Figure B.1. In order to get a linear contribution to the axial strain from the axial displacement a node allowing axial displacement is placed at the centre point of the element. The element stiffness matrix and the nodal forces corresponding to given stress distributions are calculated using Gaussian quadrature with integration points along the element axis. Membrane locking phenomena are avoided by only

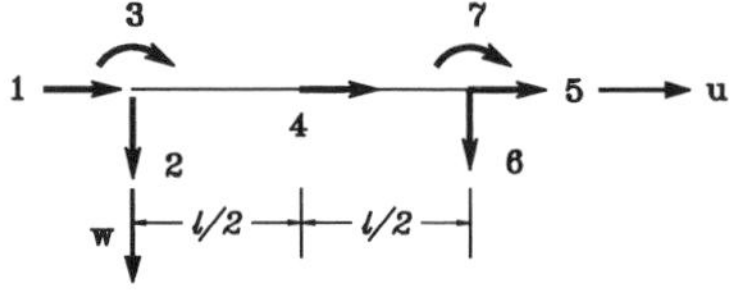

Figure B.1. Beam finite element

using two Gauss points. In the present examples the cross sections are massive rectangular. Integration over the beam cross section is made by Gaussian quadrature. Five integration points along the depth give sufficient accuracy.

The structure is linear elastic up to yielding with Youngs modulus E and initial yielding stress f_{yi}. Strain hardening is assumed according to (B.1). The parameter n=1 corresponds to linear elastic behavior while as n tends to infinity the behavior becomes linear elastic ideal-plastic.

$$\frac{d\varepsilon}{d\sigma} = \frac{1}{E} \qquad \text{for } (|\sigma| \le f_{yi}) \vee [(|\sigma|) \ge f_{yi}) \wedge (\sigma \Delta \sigma \le 0)]$$
$$\frac{d\varepsilon}{d\sigma} = \frac{1}{E}\left(\frac{|\sigma|}{f_{yi}}\right)^{n-1} \qquad \text{for } (|\sigma| > f_{yi}) \wedge (\sigma \Delta \sigma > 0) \tag{B.1}$$

Appendix C: Load Incrementation and Equilibrium Iteration

The loading increment is controlled by a pre-chosen single displacement component in the structure, eg. Powell and Simons (1981). The method requires that the controlling displacement varies monotonously with the loading parameter.

At each loading level the controlling displacement is kept constant while iterations are performed in order to obtain a state of equilibrium. The equilibrium iteration follows a modified Newton-Raphson iteration scheme, except that in order to meet the constant displacement requirement the loading from the unbalanced nodal forces **g** in the system are supplemented with an extra loading proportional to the load configuration **t**. This determines the load parameter increment $\Delta\lambda$ according to the equation

$$\mathbf{h}^{\mathbf{T}}\mathbf{K}^{-1}(\Delta\lambda\mathbf{t}+\mathbf{g})=0 \tag{C.1}$$

The matrix **K** is the tangential system stiffness matrix while the vector **h** relates the controlling displacement $v_{control}=\mathbf{h}^{\mathrm{T}}\mathbf{v}$ to the system displacement vector **v**.

In the present study it is easy to choose displacement components that for relevant realizations of the basic variables fulfill the monotony requirement (eg. rotation of B for the portal frame and rotation of C for the orthogonal corner frame). If such displacement components are difficult to identify there are other possibilities. These are methods like the well known constant arc length method, eg. Forde and Stiemer (1987), or the constant weighted response method, Gierlinsky and Graves Smith (1985), of which the former method can be considered as a special case. For a comprehensive comparative study on a large number of methods reference is made to Clarke and Hancock (1990).

The carrying capacity is defined as the first obtained local maximum of the load parameter during increasing (controlled) displacement from zero. Stop is set after the passage of the maximum as soon as the load parameter becomes below a chosen fraction k ($0 \ll k < 1$) of the obtained maximum value. In order to prevent a break down of the FORM analyses and the simulations using the FE-model directly, a maximum number of displacement steps was set to be used in each carrying capacity calculation. This limitation has caused that the carrying capacity is underestimated in some few cases.

Reliability and Optimization of Structural Systems, V (B-12)
P. Thoft-Christensen and H. Ishikawa (Editors)
Elsevier Science Publishers B.V. (North-Holland)

Extended Fuzzy System Identification Method for Cable Adjustment Work in Daytime

M. Kaneyoshi*, H. Tanaka*, M. Kamei**, H. Furuta***

* Bridge Design Department, Hitachi Zosen Corp., Osaka, Japan

** Public Works Bureau, Osaka City Government, Japan

*** Dept. of Civil Engineering, Kyoto University, Japan

Cable tension adjustment is conventionally carried out in the midnight to eliminate the influence of temperature. Therefore this work becomes very hard. In order to reduce the load, we attempt to develop practical method of estimating the influence of temperature. Fuzzy System Identification method (FSI) developed by the authors is extended to estimate the influence of temperature. First the extended FSI is formulated. Secondly a numerical example of the method is shown using assumptive data. Finally the practical possibility is discussed using the field measured data of Sugahara-Shirokita Bridge. The results confirm that cable tension adjustment in the daytime is theoretically possible by the method.

1. INTRODUCTION

A cable-stayed bridge is a statically high indeterminate structure. Then it is possible to minimize the bending moment in the girder by introducing pre-stresses in the cables to reduce its weight. Even if optimum cable pre-stresses are determined in the design, various kind of errors will be introduced during fabrication or construction [1]. Therefore cable length adjustment, in other words, cable tension adjustment is necessary to alter the stress distribution and the geometrical configuration of the bridge at the construction site. After the measurement of geometrical configuration of girder and tower and of cable tension force, this procedure is executed by adding or removing shim plates at the end of cables in the midnight when temperature of all members of the cable-stayed bridge becomes constant then the influence of temperature is eliminated.

The midnight work of cable length adjustment becomes physically tough especially for the cantilever method erection of the multi-cable type cable-stayed bridge. Moreover if the span of it becomes the longer, the work becomes the harder for its long term construction.

The purpose of this study is to solve this issue by the developing new technology. We attempt to develop practical method of estimating the influence of temperature. Fuzzy System Identification method (FSI) developed by the authors [1], [2] is extended to estimate the influence of temperature.

2. FORMULATION OF EXTENDED FSI

Measured field data $\widetilde{F}$; such as camber values of girders and towers and cable tension are assumed to be fuzzy data. The wave symbol (i.e. ~) denotes fuzzy quantities. Those measured values of standard temperature (i.e. 20° Celsius in Japan) are denoted by $\widetilde{F}_o$ and influence coefficient of camber or cable tension due to unit temperature by K_i. $\widetilde{F}$ is regarded to be dependent on the possibility of coefficients of the model. Thus it is given by the sum of the product of fuzzy coefficient $\widetilde{X}_i$ and K_i as follows.

$$\widetilde{F} = \widetilde{F}_o + \sum_{i=1}^{N} \widetilde{X}_i \cdot K_i \tag{1}$$

Where $\widetilde{X}_i$ is fuzzy variable and is specified in terms of membership function as shown by Fig. 1. The fuzzy variables $\widetilde{X}_i$ and $\widetilde{F}_o$ must be determined by using measured field data. If the fuzzy regression analysis is applied with the introduction of the threshold parameter h ($0 \leqq h < 1$), the fuzzy variables $\widetilde{X}_i$ and $\widetilde{F}_o$ will be given by the following algorithm (i.e. extended FSI):

Find α_i, c_i

$$\text{Maximize} \rightarrow J(c_i) = \sum_{i=1}^{N+M} \sum_{j=1}^{N+M} c_i \cdot | K_{ji} | \tag{2}$$

subject to:

$$F_j + (1-h)\, e_j \geqq (1-h) \sum_{i=1}^{N} c_i \cdot | K_{ji} | + \sum_{i=1}^{N} \alpha_i \cdot K_{ji} + (1-h) \sum_{i=N+1}^{N+M} c_i \cdot | K_{ji} | + \sum_{i=N+1}^{N+M} \alpha_i \cdot K_{ji} \tag{3}$$

$$F_j - (1-h)\, e_j \leqq -(1-h) \sum_{i=1}^{N} c_i \cdot | K_{ji} | + \sum_{i=1}^{N} \alpha_i \cdot K_{ji} - (1-h) \sum_{i=N+1}^{N+M} c_i \cdot | K_{ji} | + \sum_{i=N+1}^{N+M} \alpha_i \cdot K_{ji} \tag{4}$$

$$j = 1, 2, \ldots\ldots, N+M \qquad c_i \geqq 0 \tag{5}$$

Where

- M : the number of field measured items.
- N : the number of zone where the temperature is constant however if there is difference of temperature such as between girder top and girder bottom, the number should be doubled due to the superposition of temperature change and difference.
- F_j : j-component of measured field data
- K_{ji} : influence coefficient of measured field data (j) due to unit temperature change of the member zone (i)
- c_i, α_i : parameters of the membership function $\widetilde{X}_i$
- e_j : measurement error (i.e. fuzzy output)
- h : threshold (fitness) parameter in dealing with fuzzy data ($0 \leqq h < 1$)

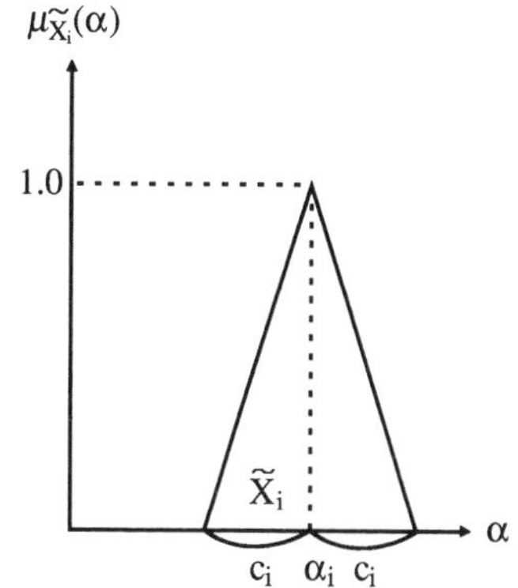

Figure 1 Membership Function of $\widetilde{X}_i$

The third and fourth terms in the right-hand sides of Eqs. (3) and (4) express $\widetilde{F}_o$ in Eq. (1) and influence coefficient K_{ji} is 1 only if i = n+j else it is zero.

If estimated values of $\widetilde{F}_o$ (i.e. measured field data at standard temperature) derived by using above mentioned method are the same as getting from the test performing at midnight, the application of cable tension adjustment method proposed by the authors will become possible and the tough work at midnight could be eliminated.

3. NUMERICAL EXAMPLES

3.1. Model and Numerical Example

A numerical example will be demonstrated by the model of the cable-stayed bridge shown in Fig. 2. The general dimensions as well as nodal points are shown in Fig. 2 and member numbers in Fig. 3. Section properties and dead loads are respectively listed in Tables 1 and 2. Three cases (i.e. Case A ~ C) are assumed according to the different situations of temperature. Assumed temperature (cf. ♦ denotes temperature measuring point) is shown in Table 3. Table 4 shows assumptive measured field data of girder, tower and cable tension. Cambers and member force of tower and girder are given by the extended FSI using the data of case A & B or case A & C as shown in Table 5. Measurement errors of cable tension and camber are assumed to be 1 tf and 1 cm respectively then e_j is 1.0 in the calculation. The fitness parameter h is selected to be 0.5 to use half of fuzzy data. The greatest discrepancy between estimated cable forces in Table 4 and assumed ones in Table 5 is about 12 tf. However the discrepancy becomes only about 1 tf between the estimated cable forces by the extended FSI and assumed ones in Table 5. The girder cambers and horizontal tower displacements are also identical between calculated and assumed values. The results of the estimation by the method are practically satisfactory.

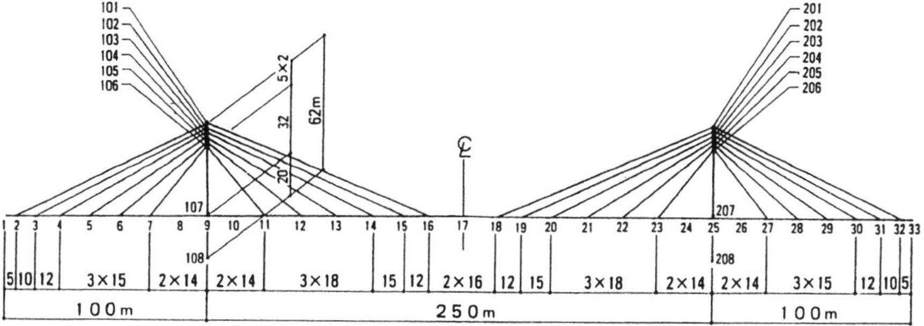

Figure 2 General Dimensions and Nodal Numbers

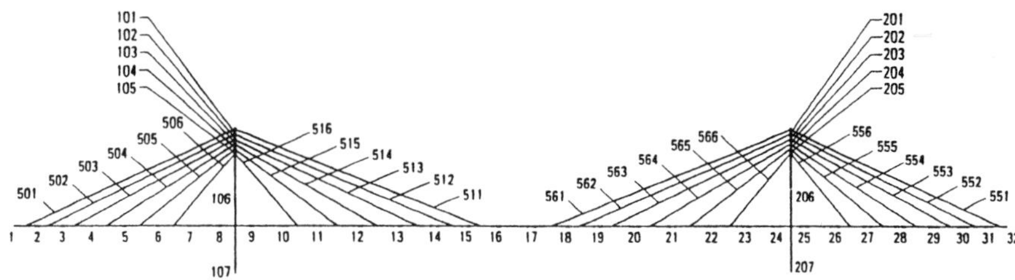

Figure 3 Member Numbers

Table 1
Design Dimensions

	Cross Sectional Area A(m^2)	Moment of Inertia I(m^4)	Young's Modulus E(tf/m^2)
Girder	0.3 ~ 0.4	0.25 ~ 0.35	2.1 x 10^7
Tower	0.3	0.25	2.1 x 10^7
Cable	0.0055 ~ 0.012	————	2.1 x 10^7

Table 2
Dead Load

	W[tf/m]
Girder Pavement, etc.	7.0
Tower	5.0
Cable	0.05 ~ 0.10

Table 3
Temperature Distribution

		Temperature situations	Case A		Case B		Case C	
		Member	Girder top	Girder bottom	Girder top	Girder bottom	Girder top	Girder bottom
Temperature (℃)	Girder	◆ 1,2	30.0	28.0	50.0	55.0	40.0	45.0
		3,4	32.0	30.0	48.0	53.0	40.0	45.0
		◆ 5,6	34.0	32.0	46.0	51.0	40.0	45.0
		7,8	36.0	34.0	44.0	49.0	40.0	45.0
		9,10	38.0	36.0	42.0	47.0	40.0	45.0
		◆ 11,12	40.0	38.0	40.0	45.0	40.0	45.0
		13,14	42.0	40.0	38.0	43.0	40.0	45.0
		◆ 15,16	44.0	42.0	36.0	41.0	40.0	45.0
	Tower	◆ 101 ~ 107	Side Span	Center Span	Side Span	Center Span	Side Span	Center Span
			42.0	40.0	45.0	50.0	40.0	40.0
	Cable	◆ 501,511	44.0		50.0		40.0	
		502,512	42.0		48.0		40.0	
		◆ 503,513	40.0		46.0		40.0	
		504,514	38.0		44.0		40.0	
		◆ 505,515	36.0		42.0		40.0	
		506,516	34.0		40.0		40.0	

Note) Temperature distribution is assumed to be symmetric at the center span.
♦ Selected member for temperature measurement

Table 4
Cable Tension & Deformation

Members \ Temperature situations			Case A	Case B	Case C
Cable tension (tf)	Side span	501	569.13	560.86	569.42
		503	214.95	215.84	215.55
		506	203.04	206.09	199.92
	Center span	511	557.55	558.97	565.27
		513	209.85	209.49	208.33
		516	200.55	200.39	194.43
Girder camber (m)	Side span	Pt.2	0.0036	– 0.0003	– 0.0011
		Pt.6	0.0198	0.0101	0.0030
	Center span	Pt.12	– 0.0120	0.0054	0.0082
		Pt.17	– 0.0348	0.0007	0.0274
Horizontal displacement of tower (m)		Pt.101	– 0.0189	– 0.0393	– 0.0407

Table 5
Comparison of Estimation

Members \ Temperature situations			Case 1	Case 2	Precise estimation at 20℃ (Assumed values)
			Estimation by Case A & B	Estimation by Case A & C	
Cable tension (tf)	Side span	501	574.919	572.941	573.000
		503	212.869	212.703	214.000
		506	192.573	193.349	193.000
	Center span	511	562.303	561.793	562.000
		513	207.649	207.663	209.000
		516	193.705	194.149	193.000
Girder camber (m)	Side span	Pt.2	0.0011	0.0011	0.000
		Pt.6	– 0.0001	0.0004	0.000
	Center span	Pt.12	– 0.0020	– 0.0017	0.000
		Pt.17	– 0.0009	0.0046	0.000
Horizontal displacement of tower (m)		Pt.101	– 0.0011	– 0.0024	0.000

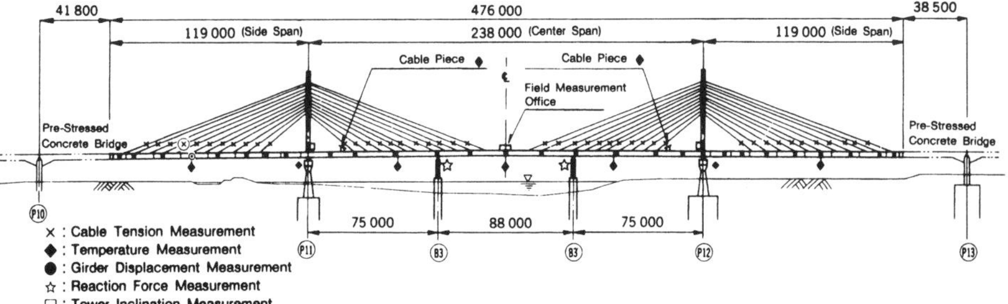

Figure 4 General View of the Sugahara-Shirokita Bridge in Erection Stage

3.2. Calculation with Measured Data

The method is verified using the measured field data of Sugahara-Shirokita Bridge at the time when the cable No. 8 was installed in August 1988 [3]. The general view of the bridge and measurement locations are shown in Fig. 4. Time history of member temperature, girder deformation and cable tension are shown in Fig. 5 and 6, respectively. The hottest season's data (Table 6) are used as inputs. In the analysis the model is divided into several zones where temperature is constant as shown in Fig. 7. Field measurement errors are assumed to be 3.0 tf, 1.0 cm and 3.0°C for cable tension, camber and temperature respectively. The threshold parameter h is fixed to be 0.5 as neutral. The case 1 and 2 in Table 7 are the estimations calculated by the field measured data of August 12 and 13, 12 and 15 in 1988, respectively. These results are compared with the field measured data at 8 p.m. on August 11 in 1988 (Table 7). The temperature was about 29°C at that time and all members of the bridge were almost same temperature. Good agreements are shown in cable tensions and girder cambers except for the cable tension of C4 and C41 and for girder camber point 6. These discrepancies may be due to the lack of cable tension measurement and of temperature measurement. However these results confirm that cable tension adjustment in the daytime is theoretically possible by the extended FSI.

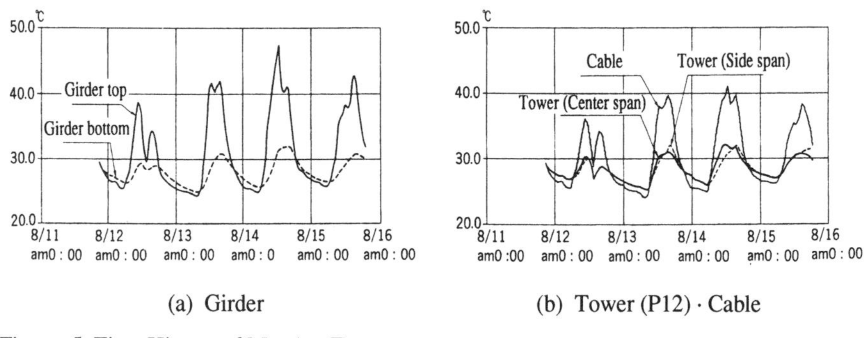

(a) Girder (b) Tower (P12) · Cable

Figure 5 Time History of Member Temperature

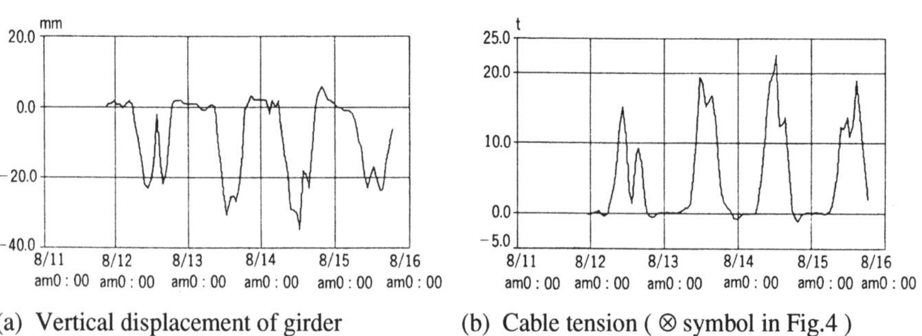

(a) Vertical displacement of girder (⊙ symbol in Fig.4) (b) Cable tension (⊗ symbol in Fig.4)

Figure 6 Time History of Girder Deformation and Cable Tension

Table 6
Example of Field Measured Data at Sugahara-Shirokita Bridge

Field measured data			Measurement date & time		
			August 12 10 a.m.	August 13 15 p.m.	August 15 14 p.m.
Temperature (℃)	Girder point A	Girder top	38.7	44.9	44.0
		Girder bottom	28.4	29.8	30.3
	Girder point D	Girder top	38.9	42.0	43.0
		Girder bottom	28.7	30.7	30.7
	Girder point G	Girder top	38.7	44.9	44.0
		Girder bottom	28.3	29.7	29.9
	Tower of P11	Side span	29.9	30.8	30.8
		Center span	29.1	30.6	30.7
	Tower of P12	Side span	30.3	30.9	30.8
		Center span	29.6	31.3	31.1
	Cable of P11		36.2	40.0	38.2
	Cable of P12		36.1	39.7	38.5
Cable tension (tf)	C4		202.1	203.9	205.9
	C11		189.8	191.1	189.6
	C34		187.0	185.6	183.8
	C41		229.7	231.2	232.9
Girder camber (m)	Side span of P11 side	Pt.6 Pt.8 Pt.12	19.682 19.726 19.814	19.670 19.722 19.807	19.674 19.725 19.810
	Center span	Pt.26 Pt.34 Pt.45	20.008 19.781 19.450	19.999 19.770 19.451	19.998 19.777 19.449
	Side span of P12 side	Pt.56 Pt.60 Pt.62	18.922 18.590 18.401	18.921 18.589 18.395	18.922 18.593 18.398

Table 7
Estimations at 20°C and Field Measured Data at Sugahara-Shirokita Bridge

Field measured data			Case 1	Case 2	Field measured data at 29 ℃ on August 11
			Estimation by Aug. 12 & 13	Estimation by Aug. 12 & 15	
Cable tension (tf)	C4		183.2	182.9	187.0
	C11		202.5	200.6	199.9
	C34		195.6	194.2	198.9
	C41		209.1	213.2	214.1
Girder camber (m)	Side span of P11 side	Pt.6 Pt.8 Pt.12	19.736 19.755 19.819	19.720 19.744 19.813	19.719 19.749 19.815
	Center span	Pt.26 Pt.34 Pt.45	20.007 19.770 19.452	20.007 19.772 19.452	20.001 19.772 19.447
	Side span of P12 side	Pt.56 Pt.60 Pt.62	18.925 18.612 18.445	18.927 18.617 18.448	18.922 18.612 18.444

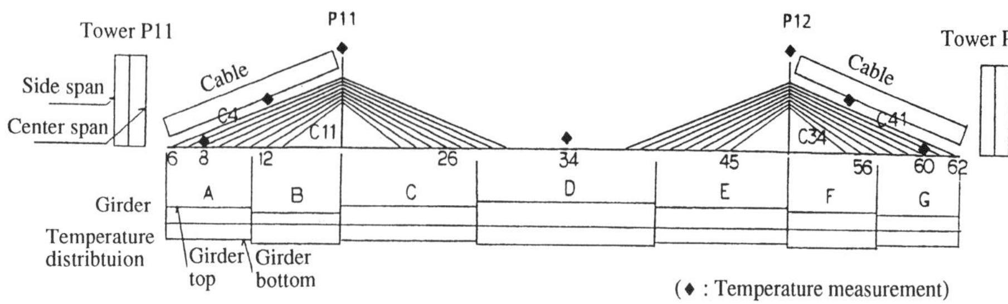

Fig.7 Measurement Points and Temperature Distribution

4. CONCLUSIONS

Through several numerical examples, the following conclusions are derived:

1) The number of measurement place of temperature should be increased and accurate model of the bridge is necessary to obtain accurate estimations.
2) The changes of cable tension and cambers of the girders and towers are large in daytime, therefore automatic measurement system is necessary which can measure aforesaid measurement items in a second.
3) Some temperature measurements of Sugahara-Sirokita Bridge might be interfered by the shadows of the erection machineries. Therefore careful attention should be paid for the selection of field measurement places.
4) The resent estimations are obtained by the only two cases of the field measured data. If many cases are available as the data, the accuracy of the calculation will be improved. The more accurate result will be given if many data under the various temperature are available.

REFERENCES

1. Kaneyoshi, M., Tanaka, H., Kamei, M. and Furuta, H. , Optimum Cable Tension Adjustment Using Fuzzy Regression Analysis, Reliability and Optimization of Structural Systems '90, Proceedings of the 3rd IFIP WG 7.5 Conference Berkely, California, USA, March 26-28, 1990.
2. Kaneyoshi, M., Furuta, H., Kamei, M. and Tanaka, H. , Optimum Cable Tension Adjustment Method by Fuzzy Regression Analysis, Proc. of the Korea-Japan Joint Seminar on Structural Optimization, Seoul, Korea, PP.283-292, May, 1992.
3. Kaneyoshi, M., Tanaka, H., Kamei, M. and Furuta, H. , New System Identification Technique Using Fuzzy Regression Analysis, Proc. 1st Int. Symp. on Uncertainty Modeling and Analysis, Maryland, Dec., 1990.

Reliability and Optimization of Structural Systems, V (B-12)
P. Thoft-Christensen and H. Ishikawa (Editors)
Elsevier Science Publishers B.V. (North-Holland)

Structural Optimization by Neuro-Optimizer

M. Kishi, T. Kodera, Y. Iwao, R. Hosoda

Department of Marine System Engineering, College of Engineering
University of Osaka Prefecture, Sakai, Osaka 593, Japan

1. INTRODUCTION

In design problems of engineering systems some design variables take discrete values on account of the industrial standard and of the design specification. Hence, such design optimization is considered as a combinatorial/discrete optimization. The Branch-and-Bound method is a traditional technique to solve discrete optimization problems; however, without heuristics the algorithm involves time and space complexities for large-scale problems.

In the middle 1980's Hopfield [1] showed that some combinatorial optimization problems can be programmed and solved on artificial neural networks minimizing the quadratic energy function. The basic concept of the Hopfield neural network is a combination of the input-output neuron model and the steepest descent method.

Based on the concept of the Hopfield neural network we have already proposed an optimization method named NEURO-OPTIMIZER [2] which is expected to be able to attain good solutions for general nonlinear discrete optimization problems. The NEURO-OPTIMIZER needs some algorithm to escape from local minima of the energy function. The simulated annealing method [3] is introduced there.

The discrete variable is represented numerically by neurons in the NEURO-OPTIMIZER. The neuron state takes binary values of one or zero. In this paper the number representation of the discrete variable are investigated by a stability analysis, and a mapping technique is proposed as a redundant representation for irregularly discrete variables. Numerical example for a structural system optimization is provided to illustrate the applicability of the NEURO-OPTIMIZER with the mapping technique.

2. NEURO-OPTIMIZER

2.1. Neuron Model

The neural network consists of mutually interconnected neurons. In 1943, McCulloch and Pitts [4] proposed neuronic equations represented by threshold functions as a dynamic model of neural networks. After that various neuron and neural network models have been proposed and studied. The Hopfield neural network [1] minimizing its energy is applicable to quadratic discrete optimization problems. The basic concept of the Hopfield model is a combination of the input-output function and the steepest descent method for the energy function.

In the NEURO-OPTIMIZER based on the concept of the Hopfield model, the system behaviour is formulated with state equations as follows:

$$V_i(t) = \Phi[U_i(t)] \tag{1}$$

$$dU_i(t)/dt = -\partial E/\partial V_i \tag{2}$$

where $V_i(t)$ is the state/output of neuron i at the instant t, U_i is the input from other neurons, and E is the energy function defined for the neural network. Input-output function Φ is a monotonic increasing function which takes values between zero and one, for example, in the following form:

$$\Phi(U) = 1/(1+\exp(-U)) \ . \tag{3}$$

Energy function E must be differentiable with respect to the neuron state V_i. Then the neurons change their state decreasing the energy E, i.e. $dE/dt \leq 0$.

Because of the sigmoid function Φ, the neuron state V_i does not vary proportionally with the input U_i. The state V_i is a macro-variable which varies on a long time scale, and the input U_i is a micro-variable which varies on a short time scale. In general, a self-organizing system is composed of both macro-variables and micro-variables.

2.2. Discrete Optimization

Consider a discrete optimization problem such as:

$$\textit{Find} \quad X$$

$$\textit{Such that} \quad f(X) \rightarrow \textit{minimize} \tag{4}$$

$$\textit{Subject to} \quad g(X) \leq 0 \tag{5}$$

$$h(X) = 0 \tag{6}$$

$$X \in \chi \tag{7}$$

where X is the discrete variable, f is the objective function, g and h are the constraint functions, and solution space χ denotes the finite set of all possible solutions.

The Lagrangean function L is defined as:

$$L(X,\lambda,\mu) = f(X) + \lambda g(X) + \mu h(X) \ . \tag{8}$$

$\lambda(\geq 0)$ and μ are the Lagrange multipliers. The optimal solution X^* can be obtained minimizing L with respect to X $(\in\chi)$. The Lagrange multipliers λ and μ are improved in the following way [5]:

$$\Delta\lambda \propto g(X) \tag{9}$$

$$\Delta\mu \propto h(X) \tag{10}$$

where Δ denotes the increment. Negative λ is treated as zero.

Suppose that the discrete variable $X(\in\chi)$ can be represented by neurons state $\boldsymbol{V}$, and let the energy function E of the neural network be the Lagrangean function L. Then the neurons change their state decreasing the Lagrangean. This means that the above mentioned discrete optimization problem can be solved on the neural network. The equality constraint $V^T(\mathbf{1}-V)=0$, where **1** denotes the vector whose elements are unity, is added for the binarization of V so that each neuron i has a state: $V_i=0$ or $V_i=1$.

2.3. Fluctuating Neuron Model

The discrete optimization problem involves local optima. The energy function of the neural network has a very large number of local minima because of the binarization constraint on the neuron state. Some fluctuating neural network models have been proposed in order to escape from local minima.

One approach to fluctuating neural networks is to use a probabilistic state transition mechanism. The Boltzmann machine [3] is a probabilistic neural network model, where the input-output function Φ gives the probability of the neuron state V_i being one, i.e.:

$$\Phi(U) = 1/(1+\exp(-U/T)) \tag{11}$$

$$\Phi(U_i) = Prob[V_i=1] \ . \tag{12}$$

The parameter T is the so-called temperature which is controlled in an annealing schedule.

The Langevin equation model [2] is another probabilistic neural network model, in which the random noise is added to the input U_i of the form:

$$dU_i/dt = -\partial E/\partial V_i + u_i \ . \tag{13}$$

The noise u_i is the zero-mean normally distributed random variable, and the variance σ_{ui} is given by

$$\sigma_{u_i}^{\ 2} = \alpha T^2 \tag{14}$$

where α is a positive constant, and T is the temperature.

The other approach to fluctuating neural networks is to use a chaotic dynamics. State equations of neural networks have an affinity for the chaos. That approach is attractive; however, at present much investigation should be done in order to control chaotic system parameters.

In the following, we introduce the Langevin equation model because of its simplicity. The temperature is slowly lowered in the annealing schedule, for example:

$$T(q) = T_0/\ln(1+q) \tag{15}$$

where $T(q)$ denotes the temperature during the q-th stage, and T_0 is some positive constant.

3. NUMBER REPRESENTATION OF DISCRETE VARIABLE

3.1. Number Representations

In the NEURO-OPTIMIZER each discrete variable is represented by neurons which take binary values of one or zero. There are various ways of representing discrete variables numerically as follows:

(A) Distributed representation - For the variable X which takes regularly discrete values, e.g. $d, d+c, \ldots, d+(n-1)c$, the distributed representation is introduced of the form:

$$X = d + c\sum_i V_i \tag{16}$$

where the sum is over n-1 values, and each neuron is independent.

(B) Local representation - For the variable X which takes irregularly discrete values, e.g. $d_1, \ldots, d_n$ ($d_i < d_{i+1}$), the local representation is introduced of the form:

$$X = \sum_i d_i V_i \qquad \textit{subject to} \quad \sum_i V_i = 1 \tag{17}$$

where the sum is over n values, and each neuron is exclusive.

(C) Intermediate representation - For the irregularly discrete variable the intermediate representation is also introduced of the form:

$$X = d_1 + \sum_i (d_{i+1} - d_i) V_i \qquad \textit{subject to} \; V_i \geq V_{i+1} \quad (i = 1, \ldots, n-1) \tag{18}$$

where the sum is over n-1 values, and each neuron is dependent.

3.2. Stability Analysis

From Eqs.(1) and (2) we have

$$dV_i / dt = -V_i(1 - V_i)(\partial L / \partial V_i) \triangleq p_i(V) \,. \tag{19}$$

Let $\underline{V}$ be a neurons state vector whose elements take binary values of one or zero. By expanding Eq.(19) in a Taylor series and retaining only the linear terms we obtain

$$\begin{aligned} dV_i / dt &\approx p_i(\underline{V}) + (\partial p_i / \partial V |_{\underline{V}})^{\mathrm{T}} \Delta V = (2\underline{V}_i - 1)(\partial L / \partial V_i |_{\underline{V}}) \Delta V_i \\ &\triangleq q_i \Delta V_i \,. \end{aligned} \tag{20}$$

Thus, it is natural to arrive at the following conclusions for the neural network system subjected very small fluctuations: i) If $q_i<0$ for any i, the system is stable. ii) If $q_i>0$ for some i, the system is unstable.

Simple examples of the stability analysis are now given. Consider an optimization problem such as:

Find X

Such that $f(X) \rightarrow$ ***minimize*** (21)

Subject to $X \in \chi = \{ r_1, \ldots, r_k^*, \ldots, r_n \}$ (22)

where X is a scalar variable, and f is a convex function whose minimal point is $X=r_k^*$. Let $r_1, \ldots, r_n$ be an ordered series with equal difference c. Thus, the distributed representation can be used here. The Lagrangean function is defined as:

$$L(V, \mu_1) = f(r_1 + c\sum_{i=1}^{n-1} V_i) + \mu_1 \sum_{i=1}^{n-1} h_i(V_i) \,. \tag{23}$$

The constraint functions $h_i(V_i)=V_i(1-V_i)$ (i=1,...,n-1) are added for the binarization of neuron state V_i, and μ_1 is the Lagrange multiplier. We want to let the optimal solution $X=r_k^*$ be stable and the other binary solutions be unstable. As a result of the stability analysis, we have the following feasible bounds on the Lagrange multiplier μ_1:

$$0 < \mu_1 < \min_{X \in \chi,\, X \neq r_k^*} |cf'(X)| \tag{24}$$

where f' indicates the derivative of f.

On the other hand the local representation is used for the discrete variable X. Then the Lagrangean function is defined as:

$$L(V,\mu_1,\mu_2) = f(\sum_{i=1}^{n} r_i V_i) + \mu_1 \sum_{i=1}^{n} h_i(V_i) + \mu_2 (\sum_{i=1}^{n} V_i - 1)^2 \tag{25}$$

where μ_2 is the Lagrange multiplier for the constraint function in Eq.(17). As a result of the stability analysis, we found that it is impossible to let only the optimal solution be stable.

3.3. Mapping Technique

Independency and redundancy are the remarkable features of the neural information processing. The industrial standard usually involves irregularly discrete values. Hence, the distributed number representation is desirable even for irregularly discrete design variables in the NEURO-OPTIMIZER. Here we propose a mapping technique in order to transform the sequence of irregularly discrete values to a sequence of regularly discrete values.

Suppose that the variable X takes irregularly discrete values, $d_1,...,d_n$. We introduce a mapping F by which integer i-1 (i=1,...,n) has a one-to-one correspondence to discrete value d_i (see Figure 1). Hence, the optimization problem with respect to X is transformed to the optimization problem with respect to the integer variable Y by the mapping F: $Y \rightarrow X$. The spline function or the Lagrange interpolation can be used in order to represent F. Then, the integer variable Y is represented by Eq.(16) with binary neurons V, and the NEURO-OPTIMIZER can solve the optimization problem.

The mapping technique is applicable when the discrete variable takes linear ordered values. Conversely, for the discrete variable which takes unordered values, the technique is not proper.

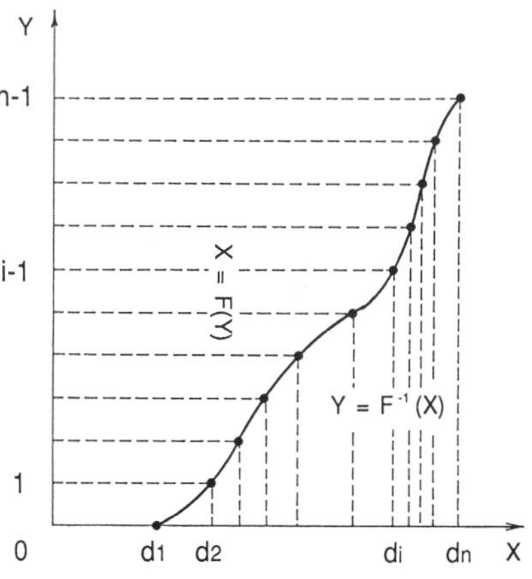

Figure 1. Mapping technique

4. APPLICATION TO STRUCTURAL OPTIMIZATION

Numerical example for a structural optimization problem is presented to demonstrate the applicability of the NEURO-OPTIMIZER. Considered here is an optimal design problem of a truss structure whose members are uniform and homogeneous. The configuration of the structure, the materials to be used, and the loading conditions are assumed to be given. The design problem is to determine the optimal cross-sectional size of the members minimizing the structural weight/mass under deterministic failure criteria. The size of each member is singled out of the candidates.

The design problem is formulated as follows:

$$\textit{Find} \quad X$$

$$\textit{Such that} \quad W(X) \rightarrow \textit{minimize} \tag{26}$$

$$\textit{Subject to} \quad g_y(X) \leq 0 \tag{27}$$

$$g_b(X) \leq 0 \tag{28}$$

$$X \in \chi \tag{29}$$

where X is the design variable vector, W is the structural mass, g_y and g_b are respectively the yielding and buckling constraints [6] of each member, and χ is the set of candidates for X. Let a_k be the cross-sectional area of the k-th member and i_k be the geometrical moment of inertia, then the combination $X_k=(a_k, i_k)^T$ is taken as the design variable.

The optimal design problem is solved for the truss structure as shown in Figure 2 [7]. The number of design variables is eight, considering a symmetric structure, and each design variable has ten candidates (see Table 1). The mapping technique is used to represent the discrete design variables. The state transition of the neural network is simulated by using the Euler method. Neuron state transitions occur synchronously. The temperature and the Lagrange multipliers are renewed at every iteration. The objective function in Eq.(26) is squared in the Lagrangean function, and further, the objective function values and the constraint functions values are scaled to keep the input increment $dU_i/dt\Delta t$ at appropriate levels $(=O(10^{-1}))$. The numerical data are given as $\alpha=1$ in Eq.(14), $T_0=20.0$ in Eq.(15), and the simulation time step $\Delta t=0.05$. Every Lagrange multiplier takes zero-initial-value. The initial condition of the neurons is generated randomly. The material data and the loading conditions are given following the reference [7].

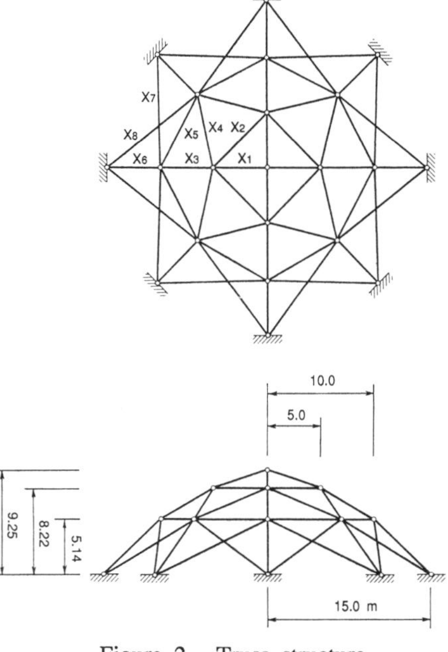

Figure 2. Truss structure

For the optimal design problem fifty simulations are carried out. Figure 3 illustrates the histogram of the solutions by the NEURO-OPTIMIZER. The optimal solution [7], which is shown in Table 2, is found with 48% probability of success. The average number of state transitions of the neural network until convergence is 124.7. On the other hand two thousand feasible solutions are found by the Monte Carlo method. Figure 4 is the histogram of the feasible solutions. Although the NEURO-OPTIMIZER does not guarantee to attain the optimal solution, the quality of the obtained solutions is satisfactory.

Table 1 Member candidates

Type	Area (cm^2)	Moment of Inertia (cm^4)
(1)	1.0	1.0
(2)	5.0	25.0
(3)	13.3	176.9
(4)	14.4	207.4
(5)	17.2	295.8
(6)	24.3	590.5
(7)	27.7	767.3
(8)	29.3	858.5
(9)	34.0	1156.0
(10)	44.0	1936.0

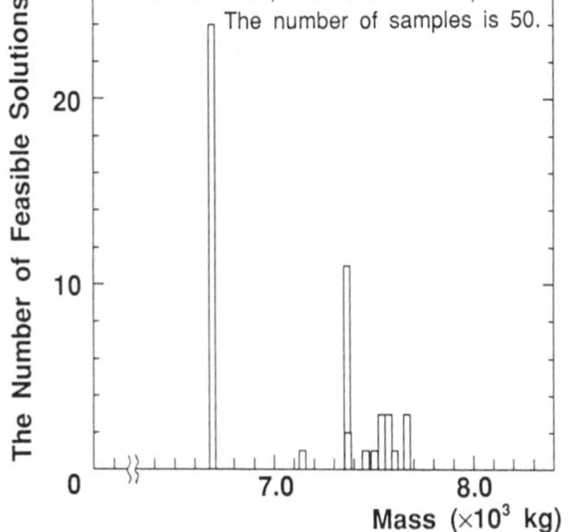

Figure 3. Distribution of solutions by NEURO-OPTIMIZER

The number of samples is 2,000.
The Number of Feasible Solutions
10^2
0
SOLUTIONS BY NEURO-OPTIMIZER
10^4
Mass (kg)

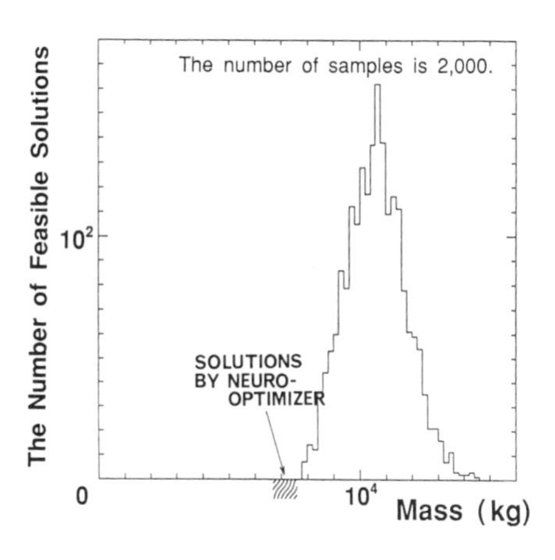

Figure 4. Distribution of feasible solutions

Table 2 Optimal solution

Design Variable	X_1	X_2	X_3	X_4	X_5	X_6	X_7	X_8
Type	(10)	(4)	(6)	(8)	(2)	(7)	(5)	(3)

5. CONCLUSIONS

This paper is concerned with the discrete optimization method, NEURO-OPTIMIZER. The results are summarized as follows:

(1) Combining the input-output neuron model and the steepest descent method, the concept of the NEURO-OPTIMIZER is proposed for general discrete optimization problems.

(2) The number representation of the discrete design variable by neurons is investigated through a stability analysis, and the mapping technique is proposed as a distributed representation.

(3) Numerical examples are provided to verify the applicability of the proposed method. The NEURO-OPTIMIZER with the mapping technique is applied to a discrete structural optimization problem, and the optimal solution is found with a high probability of success.

This paper is silent on scaling mechanisms of the Lagrangean function; however, the scaling is related to the following two quantities: i) the quality of the solution obtained by the NEURO-OPTIMIZER, and ii) the computational time until convergence. Actually the scaling is problem-dependent, the general mechanism should be proposed.

A part of this work is financially supported by a Grant-in-Aid for the Scientific Research, the Ministry of Education, Science and Culture of Japan. All the computations were processed by using the NEC PC-9800 system.

REFERENCES

1. Hopfield, J.J., Electronic Network for Collective Decision Based on Large Number of Connections between Signals, United States Patent, No.4,660,166(1987).
2. Kishi, M., Suzuki, T., and Hosoda, R., Structural Design by Neuro-Optimizer, in Practical Design of Ships and Mobile Units, eds. J.B. Caldwell and G. Ward, Vol.2, Elsevier Applied Science(1992)940.
3. Aarts, E., and Korst, J., Simulated Annealing and Boltzmann Machines: A Stochastic Approach to Combinatorial Optimization and Neural Computing, John Wiley & Sons, 1989.
4. McCulloch, W.S., and Pitts, W., A Logical Calculus of Ideas Immanent in Nervous Activity, Bull. Math. Biophys., 5(1943)115.
5. Platt, J.C., and Barr, A.H., Constrained Differential Optimization, in Neural Information Processing Systems, ed. D.Z. Anderson, American Inst. of Physics(1988)612.
6. Pedersen, P., On the Optimal Layout of Multi-Purpose Trusses, Comput. Struct., Vol.2, (1972)695.
7. Pedersen, P., Optimal Joint Positions for Space Trusses, J. Struct. Div., ASCE, Vol.99, ST10(1973)2459.

Reliability and Optimization of Structural Systems, V (B-12)
P. Thoft-Christensen and H. Ishikawa (Editors)
Elsevier Science Publishers B.V. (North-Holland)

Reliability of Dynamically Excited Linear Structures with Random Properties

H. U. Köylüoğlu*, S. R. K. Nielsen**
* Dept. of Civil Engineering and Operations Research,
** Dept. of Building Technology and Structural Engineering,
University of Aalborg, DK-9000 Aalborg, Denmark

The present paper deals with the reliability of dynamically excited linear structures with random material properties. A stochastic finite element formulation for the dynamic analysis of two dimensional frames is derived utilizing deterministic shape functions in the derivation of element mass and stiffness matrices. Hence, the weighted integral method has been applied. This treatment allows the random stiffness fields to be represented by a few random variables, named weighted integrals, defined as the integrals of the stiffness random field multiplied by deterministic functions. Uncoupling of the random equations of motion in linear structural dynamics is achieved by a normal mode expansion in random eigen values and random eigen vectors. Uncoupled equation for each mode, with modal damping ratios assigned as random variables at this point, are solved under the corresponding modal loading. Probability of failure of a system is considered to be defined in terms of first passage of assigned bounds of a set of response processes which may be selected as characteristic nodal point displacements or internal stresses and section forces. Reliability of the multi-degrees of freedom system is then approximated from a truncated modal expansion of these response processes. In the corresponding modal subspace, the limit state functions from all considered response processes map on hypersurfaces. This treatment leads to extreme decreases in system reliability calculations for multi-degree of freedom structural systems with many failure elements. Moreover, for one, two or three modes, outcrossing rates of the corresponding modal coordinate vectors can be calculated easily. Initially, the probability of failure on condition of a certain realization of the structure is calculated based on independent outcrossings of the modal coordinate vectors assuming Gaussian excitation. Then, the unconditional probability of failure is calculated using simulation techniques.

Keywords codes: J.2; G.3; I.6.3
Keywords: Phisical sciences and engineering; Probability and statistics; Simulation

1. INTRODUCTION

Stochastic structural models, in which system uncertainties in materail properties and/or geometry due to physical imperfections, model inaccuracies and system complexities are modeled using random variables or random processes which may be functions of time and/or space, have been considered by several researchers and consequently a new field, " Stochastic Finite Elements " was coined to stochastic mechanics in 1980's. Although there have been papers on Monte Carlo solutions and reliability considerations, most of the studies done in this field have been on the second order moment analysis of stochastic systems under deterministic loading. The developments in stochastic finite elements are reviewed by Vanmarcke et al. [1], Benaroya and Rehak [2], Der Kiureghian et al. [3] and Brenner [4].

The ultimate goal of stochastic response analysis is reliability considerations with limit states ranging from serviceability and applicability requirements to total collapse. The review given by Der Kiureghian et al. [3] summarizes the evolution, the main problems, e.g. discretization of random fields and gradient computations, and the state of the art on reliability of stochastic structural systems. Most of the literature is on static stochastic systems, e.g. Hisada and Nakagiri [5], Liu and Der Kiureghian [6], Deodatis and Shinozuka [7] and there are only a few studies on dynamic stochastic systems. Spencer and Elishakoff [8] considered the reliability problem of a SDOF system, with random stiffness and with deterministic initial values, subject to Gaussian white-noise excitation. Der Kiureghian et al. [3] introduced the conditional derivative method for computing gradients and an algorithm for finding design points in time-variant reliability problem so that FORM and SORM could be applied.

In this paper, reliability of linear structures with random stiffness and random damping properties subject to random dynamic excitation is studied. For this purpose, first a new and practical method for the calculation of system reliability of linear structures subject to random dynamic excitation is introduced.

Next, stochastic structural systems are considered and the above mentioned reliability method is applied to a recently developed stochastic finite element formulation for the dynamic analysis of linear two dimensional elastic frames with random stiffness and random damping properties, Köylüoğlu and Nielsen [9], based on the weighted integral method of Deodatis [10], Deodatis and Shinozuka [7] and Takada [11], in which Galerkin finite elements with deterministic shape functions is applied to stochastic differential equations. It should be noted that in this way of discretization, the results are mesh independent and random fields can be accurately represented by a small number of random variables.

Unconditional outcrossing rate and unconditional reliability of a stochastic system are computed using simulations. The theory is applied to a three storey one bay framed subject to horizontal earthquake excitation modeled as filtered Gaussian white-noise. The bending moments at the end sections of all beam elements and the top storey horizontal displacement form the considered response processes in the reliability problem. A parametric study to find the relationship between unconditional outcrossings rate, unconditional system reliability versus correlation length and coefficient of variation of the element random fields and coefficient of variation of random modal damping ratios is performed.

2. CROSSING RATES AND RELIABILITY MEASURES

For a scalar stationary Gaussian random process $Y(t)$, the mean rate of outcrossings ν of a time-wise constant interval $[-a, b]$ is well known, e.g. Vanmarcke [12].

$$\nu = \frac{1}{2\pi}\frac{\sigma_{\dot{Y}}}{\sigma_Y}\left(\exp\left(-\frac{(b-\mu_Y)^2}{2\sigma_Y^2}\right) + \exp\left(-\frac{(-a-\mu_Y)^2}{2\sigma_Y^2}\right)\right) \tag{1}$$

where μ_Y, σ_Y and $\sigma_{\dot{Y}}$ are self explanatory.

For a stationary vector process $\mathbf{Y}(t)$, the outcrossings rate of a safe domain S bounded by the time-wise constant surface Γ has been derived by Belyaev [13].

$$\nu = \int_\Gamma \left(\int_0^\infty \dot{y}_n f_{\mathbf{Y}\dot{Y}_n}(\mathbf{b}, \dot{y}_n)d\dot{y}_n\right)d\Gamma \tag{2}$$

$$\dot{Y}_n(t) = \mathbf{n}^T(\mathbf{b})\dot{\mathbf{Y}}(t) \tag{3}$$

where the surface element $d\Gamma$ of Γ is specified by the position vector $\mathbf{b}$. The unit vector of the area element in the outward direction is denoted by $\mathbf{n}(\mathbf{b})$ and $\dot{Y}_n$ signifies the velocity of $\dot{\mathbf{Y}}(t)$ relative to the surface in the outward directed normal vector. $f_{\mathbf{Y}\dot{Y}_n}(\mathbf{y}, \dot{y}_n)$ denotes the joint probability density function (jpdf) of $\mathbf{Y}$ and $\dot{Y}_n$. The innermost integral in (2) calculates the outcrossings through the surface element $d\Gamma$ and the outhermost integral sums the local outcrossings from all parts of the boundary.

If $\mathbf{Y}(t)$ is a stationary Gaussian process, then (2) can be written as

$$\nu = \int_{\Gamma} \left(\sigma\varphi(\frac{\mu}{\sigma}) + \mu\Phi(\frac{\mu}{\sigma}) \right) f_{\mathbf{Y}}(\mathbf{b}) d\Gamma \tag{4}$$

where $f_{\mathbf{Y}}(\mathbf{y})$ is the jpdf of $\mathbf{Y}(t)$ and $\varphi(\cdot)$ and $\Phi(\cdot)$ are the pdf and distribution function of a standardized normal variable. $\mu = \mu(\mathbf{b})$ and $\sigma = \sigma(\mathbf{b})$ are the mean value and the standard deviation of $\dot{Y}_n(t)$ on condition of $\mathbf{Y}(t) = \mathbf{b}$. Let $\mathbf{C}_{\mathbf{YY}}$, $\mathbf{C}_{\mathbf{Y}\dot{Y}}$ and $\mathbf{C}_{\dot{\mathbf{Y}}\dot{Y}}$ specify the covariance and cross-covariance matrices of $\mathbf{Y}(t)$ and $\dot{\mathbf{Y}}(t)$. Then, these quantities are

$$\mu(\mathbf{b}) = \mathbf{n}^T(\mathbf{b}) \mathbf{C}^T_{\mathbf{Y}\dot{Y}} \mathbf{C}^{-1}_{\mathbf{Y}Y} (\mathbf{b} - E[\mathbf{Y}]) \tag{5}$$

$$\sigma^2(\mathbf{b}) = \mathbf{n}^T(\mathbf{b})(\mathbf{C}_{\dot{\mathbf{Y}}\dot{Y}} - \mathbf{C}^T_{\mathbf{Y}\dot{Y}} \mathbf{C}^{-1}_{\mathbf{Y}Y} \mathbf{C}_{\mathbf{Y}\dot{Y}}) \mathbf{n}(\mathbf{b}) \tag{6}$$

In most cases, the surface integral (4) can only be evaluated numerically.

Then, assuming independent outcrossings, the reliability $R([0,t])$ that the response will be in the safe domain S in the time interval $[0,t]$ can be approximated as

$$R([0,t]) \simeq \exp(-\nu t) \int_S f_{\mathbf{Y}}(\mathbf{y}) d\mathbf{y} \tag{7}$$

3. TECHNIQUE FOR SYSTEM RELIABILITY

In this section, based on a reduction to the modal space, a new and practical method for the calculation of system reliability of linear structures subject to random excitations is proposed.

Failure criteria in structural mechanics are normally defined in terms of internal forces and displacements (or stresses and strains). Physical bounds for the response quantities are yielding or cracking for the internal forces and serviceability limits for the displacements. Typical safe domains for the ith nodal element internal bending moment value $M_{b,i}(t)$ and for the jth nodal displacement $U_j(t)$ of a structure subject to dynamic excitation have the following form.

$$-M^-_{b,y,i} < M_{b,i}(t) < M^+_{b,y,i} \quad , \quad i = 1, ..., I \tag{8}$$

$$-U^-_j < U_j(t) < U^+_j \quad , \quad j = 1, ..., J \tag{9}$$

where the limit values $M^-_{b,y,i}$, $M^+_{b,y,i}$ and U^-_j, U^+_j are taken as positive quantities. It should be noted that the inherent moments and displacements due to statical loading must be considered in these bounds at the assigning step. I and J denote the number of control points for the displacements and the bending moments respectively. The internal bending moments are assembled in the I-dimensional vector $\mathbf{M}_b(t)$.

For linear dynamic systems with linear viscous damping, the equation of motion reads

$$\mathbf{M}\ddot{\mathbf{U}}(t) + \mathbf{C}\dot{\mathbf{U}}(t) + \mathbf{K}\mathbf{U}(t) = \mathbf{F}(t) \tag{10}$$

where $\mathbf{M}$, $\mathbf{C}$ and $\mathbf{K}$ are the mass, the damping and the stiffness matrices respectively. Equation (10) can be solved using modal decomposition for which circular eigenfrequencies ω_k and eigenvectors Φ_k are obtained from

$$(\mathbf{K} - \omega_k^2 \mathbf{M})\Phi_k = 0 \qquad k = 1, ..., n \tag{11}$$

Then, $\mathbf{U}(t)$ may alternatively be expanded in the base of eigenvectors $\Phi_1, ..., \Phi_n$.

$$\mathbf{U}(t) = \Phi_1 Y_1(t) + \Phi_2 Y_2(t) + ... + \Phi_n Y_n(t) \tag{12}$$

where $Y_k(t)$ denotes the kth modal coordinate.

A similar modal expansion is valid for the element internal bending moments $\mathbf{M}_b(t)$.

$$\mathbf{M}_b(t) = \mathbf{M}_{b,1}Y_1(t) + \mathbf{M}_{b,2}Y_2(t) + \ldots + \mathbf{M}_{b,n}Y_n(t) \tag{13}$$

where $\mathbf{M}_{b,k}$ are the bending moment at the considered sections obtained when the structure is deformed according to the kth eigen vector. Let Φ_k^e be the nodal point displacements of element e in the kth eigenmode in the local coordinates, and $\mathbf{V}_k^e$ be the conjugated modal end section reaction forces. These are related as

$$(\mathbf{k}_e - \omega_k^2 \mathbf{m}_e)\Phi_k^e = \mathbf{V}_k^e \tag{14}$$

where $\mathbf{k}_e$ and $\mathbf{m}_e$ are the stiffness and mass matrices of the element e. The components of $\mathbf{M}_{b,k}$ belonging to the element e are then contained as a subvector of $\mathbf{V}_k^e$.

It is well known that summation in (12) and (13) can be truncated after a few terms since the main contribution comes from the lower modes. This proposes carrying system reliability computations to a modal subspace. By this, not only system reliability computations are decreased but also a good approximation for the system reliability is obtained. In the following, truncation after the first term and the second terms are covered for the reliability of the system.

When the first mode totally governs the motion, (12) and (13) can be truncated after the first modal contribution. In this case, (9) becomes

$$-U_j^- < \Phi_{1,j}Y_1(t) < U_j^+ \quad , \quad j = 1, \ldots, J \tag{15}$$

$\Phi_{1,j}$, the jth component of the first eigenvector, can be positive or negative. Hence

$$\min\left(\frac{-U_j^-}{\Phi_{1,j}}, \frac{U_j^+}{\Phi_{1,j}}\right) < Y_1(t) < \max\left(\frac{U_j^+}{\Phi_{1,j}}, \frac{-U_j^-}{\Phi_{1,j}}\right) \quad , \quad j = 1, \ldots, J \tag{16}$$

Similar inequalities can be derived from the bending stiffness inequalities (8). The smallest right hand side and the largest left hand side among all such inequalities should be selected as the bounds. This leads to

$$a_1 < Y_1(t) < b_1 \tag{17}$$

$$a_1 = \max_{\substack{i=1,\ldots,I \\ j=1,\ldots,J}} \left(\min\left(\frac{-M_{b,y,i}^-}{M_{b,1,i}}, \frac{M_{b,y,i}^+}{M_{b,1,i}}\right), \min\left(\frac{-U_j^-}{\Phi_{1,j}}, \frac{U_j^+}{\Phi_{1,j}}\right)\right) \tag{18}$$

$$b_1 = \min_{\substack{i=1,\ldots,I \\ j=1,\ldots,J}} \left(\max\left(\frac{M_{b,y,i}^+}{M_{b,1,i}}, \frac{-M_{b,y,i}^-}{M_{b,1,i}}\right), \max\left(\frac{U_j^+}{\Phi_{1,j}}, \frac{-U_j^-}{\Phi_{1,j}}\right)\right) \tag{19}$$

where $M_{b,k,i}$ is the ith component of $\mathbf{M}_{b,k}$. The outcrossings rate then follows from (1).

$$\nu = \frac{1}{2\pi}\frac{\sigma_{\dot{Y}_1}}{\sigma_{Y_1}}\left(\exp\left(-\frac{(b_1 - \mu_{Y_1})^2}{2\sigma_{Y_1}^2}\right) + \exp\left(-\frac{(a_1 - \mu_{Y_1})^2}{2\sigma_{Y_1}^2}\right)\right) \tag{20}$$

When (12) is truncated after the second term, (9) becomes

$$-U_j^- < \Phi_{1,j}Y_1(t) + \Phi_{2,j}Y_2(t) < U_j^+ \quad , \quad j = 1, \ldots, J \tag{21}$$

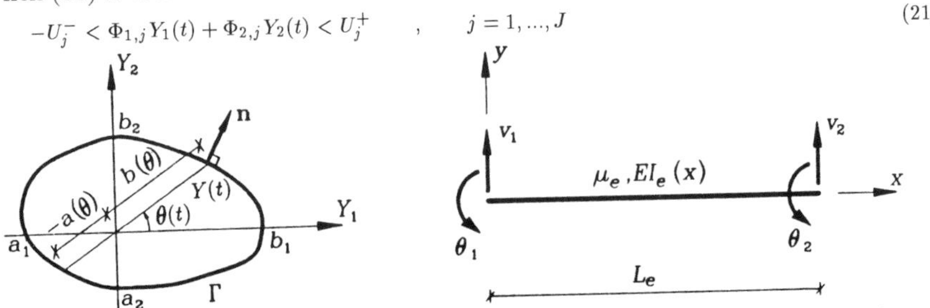

Figure 1. Safe domain in two modal coordinates and mapping from $(Y_1(t), Y_2(t))$ to $(Y(t), \theta(t))$.
Figure 2. Beam element with 4 local degrees of freedom.

Using a mapping from $(Y_1(t), Y_2(t))$ to $(Y(t), \theta(t))$, $\theta(t) \in [-\frac{\pi}{2}, \frac{\pi}{2}]$, see Figure 1, where

$$Y_1(t) = Y(t)\cos(\theta(t)) \tag{22}$$

$$Y_2(t) = Y(t)\sin(\theta(t)) \tag{23}$$

and introducing the quantity $\Phi_j(\theta)$ as

$$\Phi_j(\theta) = \Phi_{1,j}\cos(\theta) + \Phi_{2,j}\sin(\theta) \quad , \quad j = 1, ..., J \tag{24}$$

reduce (21) to

$$\min\left(\frac{-U_j^-}{\Phi_j(\theta)}, \frac{U_j^+}{\Phi_j(\theta)}\right) < Y(t) < \max\left(\frac{U_j^+}{\Phi_j(\theta)}, \frac{-U_j^-}{\Phi_j(\theta)}\right) \quad , \quad j = 1, ..., J \tag{25}$$

Unifying conditions from displacements and the similar conditions from the bending moments provides the following safe domain for $Y(t)$.

$$a(\theta) < Y(t) < b(\theta) \tag{26}$$

$$a(\theta) = \max_{\substack{i=1,...,I \\ j=1,...,J}} \left(\min\left(\frac{-M_{b,y,i}^-}{M_{b,i}(\theta)}, \frac{M_{b,y,i}^+}{M_{b,i}(\theta)}\right), \min\left(\frac{-U_j^-}{\Phi_j(\theta)}, \frac{U_j^+}{\Phi_j(\theta)}\right)\right) \tag{27}$$

$$b(\theta) = \min_{\substack{i=1,...,I \\ j=1,...,J}} \left(\max\left(\frac{M_{b,y,i}^+}{M_{b,i}(\theta)}, \frac{-M_{b,y,i}^-}{M_{b,i}(\theta)}\right), \max\left(\frac{U_j^+}{\Phi_j(\theta)}, \frac{-U_j^-}{\Phi_j(\theta)}\right)\right) \tag{28}$$

where

$$M_{b,i}(\theta) = M_{b,1,i}\cos(\theta) + M_{b,2,i}\sin(\theta) \quad , \quad i = 1, ..., I \tag{29}$$

For $\theta = 0$, (27) and (28) reduce to (18) and (19) for the first mode. At $\theta = \frac{\pi}{2}$, the corresponding limits for the second mode are obtained, i.e. $a_2 = a(\frac{\pi}{2})$, $b_2 = b(\frac{\pi}{2})$.

The limit state surface Γ is constructed numerically from (27) and (28) when θ is varied in the interval $]-\frac{\pi}{2}, \frac{\pi}{2}[$. Then, the outcrossings rate is numerically computed from (4) using the trapezoidal rule so is the reliability function from (7). It should be noted that after a polar coordinate transformation, the integral in (7) can be obtained first analytically in the radius direction for each angle increment, next numerically for all the angles.

4. STOCHASTIC FINITE ELEMENT FORMULATION

In this section, a stochastic finite element formulation for the dynamic analysis of two dimensional frames is derived applying the Bernoulli-Euler theory.

The displacement field in the transverse direction $v(x,t)$ is determined from 4 local degrees of freedom for each element which are selected as the displacements $v_1(t)$, $v_2(t)$ and rotations $\theta_1(t)$, $\theta_2(t)$ of the end sections as shown in Figure 2. It is assumed that the assumptions of Bernoulli-Euler beam theory are valid even when μ, E and I are modeled as random fields. For simplicity mass per unit length μ is considered to be constant and deterministic, whereas the bending stiffness of each element is assumed to be a sum separable homogeneous random field.

$$(EI)_e(x) = \overline{(EI)_e}\,(1 + f_e(x)) \tag{30}$$

where $\overline{(EI)_e}$ is the mean value of the bending stiffness of element e, and $f_e(x)$ is a one-dimensional, zero-mean, homogeneous random field of the eth element. Since, the bending stiffness cannot take on negative values, $f_e(x)$ is assumed to be bounded from below as $f_e(x) > -1$ with probability 1.

Extending the random displacement field as a linear combination of the indicated nodal degrees of freedom and using deterministic cubic interpolation functions, the element stiffness matrix can be evaluated as

$$\mathbf{K}^e = \mathbf{K}_0^e + X_0^e \Delta K_0^e + X_1^e \Delta K_1^e + X_2^e \Delta K_2^e \tag{31}$$

where X_i^e, $i = 0, 1, 2$ are random variables defined as

$$X_i^e = \int_0^{L_e} x^i \; f_e(x) \; dx \qquad i = 0, 1, 2 \tag{32}$$

The expansion in (31) is termed the weighted integral method and the explicit expressions for the coefficient matrices ΔK_0^e, ΔK_1^e and ΔK_2^e are given in Deodatis [10]. $\mathbf{K}_0^e$ is the deterministic part or the expected value and $X_0^e \Delta K_0^e + X_1^e \Delta K_1^e + X_2^e \Delta K_2^e$ is the stochastic part of the element stiffness matrix $\mathbf{K}^e$. The integrals X_i^e defined by (32) are called " weighted integrals ". These are the basic random variables uniquely representing the random field, $f_e(x)$. Joint statististical moments of arbitrary order among these variables are easily derived from (32).

The stochastic global stiffness matrix and the consistent global mass matrix of a stochastic system can be obtained after the usual assembling operation of the finite element method. The differential equations of motion with proportional viscous damping in the global coordinate system are given by (10), where $\mathbf{M}$ is deterministic consistent mass matrix and random matrices $\mathbf{K}$ and $\mathbf{C}$ are assumed to be positive definite with probability 1.

The random matrix $\mathbf{K}$ can be written as

$$\mathbf{K} = \mathbf{K}_0 + \Delta\mathbf{K} = \mathbf{K}_0 + \sum_{p=1}^{3N_e} X_p \Delta\mathbf{K}_p \tag{33}$$

where $\mathbf{K}_0$ is the global deterministic stiffness matrix, X_p are the weighted integrals from all N_e elements and $\Delta\mathbf{K}_p$ are the corresponding coefficient matrices in global coordinates.

The eigenvectors Φ_k now become random. Using modal decomposition, the following stochastic differential equation with random coefficients are obtained for each mode.

$$\ddot{Y}_i(t) + 2\zeta_i \Omega_i \dot{Y}_i(t) + \Omega_i^2 Y_i(t) = F_i(t) \tag{34}$$

where Ω_i is the ith random eigen frequency, $F_i(t) = \Phi_i^T \mathbf{F}(t)$ is the modal loading and ζ_i denotes the ith modal damping ratio. ζ_i are assigned as random variables independent of the weighted integrals. The random system is then defined by a vector $\mathbf{X}$ of dimension m, made up of the $3N_e$ weighted integrals and the modal damping ratios.

Assuming the loading vector $\mathbf{F}(t)$ is proportional in all components, i.e.

$$\mathbf{F}(t) = \mathbf{A} F(t) \tag{35}$$

where $\mathbf{A}$ is a deterministic vector and $F(t)$ is a stationary process with zero mean value function and autospectral density function $S_{FF}(\omega)$, the cross-covariance of the modal coordinates $R_{jk}(\mathbf{x}) = Cov[Y_j(t) Y_k(t) | \mathbf{X} = \mathbf{x}]$, the cross-covariance of the modal velocities $S_{jk}(\mathbf{x}) = Cov[\dot{Y}_j(t) \dot{Y}_k(t) | \mathbf{X} = \mathbf{x}]$ and the cross-covariance of the modal coordinates and velocities $T_{jk}(\mathbf{x}) = Cov[Y_j(t) \dot{Y}_k(t) | \mathbf{X} = \mathbf{x}]$ can be explicitly obtained.

$$R_{jk}(\mathbf{x}) = P_j(\mathbf{x}) P_k(\mathbf{x}) \int_{-\infty}^{\infty} H_j^*(\omega; \mathbf{x}) H_k(\omega; \mathbf{x}) S_{FF}(\omega) d\omega \tag{36}$$

$$S_{jk}(\mathbf{x}) = -P_j(\mathbf{x}) P_k(\mathbf{x}) \int_{-\infty}^{\infty} H_j^*(\omega; \mathbf{x}) H_k(\omega; \mathbf{x}) \; (i\omega)^2 S_{FF}(\omega) d\omega \tag{37}$$

$$T_{jk}(\mathbf{x}) = P_j(\mathbf{x})P_k(\mathbf{x}) \int_{-\infty}^{\infty} H_j^*(\omega;\mathbf{x})H_k(\omega;\mathbf{x})\,(i\omega)S_{FF}(\omega)d\omega \tag{38}$$

where * means complex conjugation and

$$P_j(\mathbf{x}) = \sum_{i=1}^{N} \Phi_{ij}(\mathbf{x})A_i \tag{39}$$

$$H_j(\omega;\mathbf{x}) = \frac{1}{M_j(\Omega_j^2 - \omega^2 + 2\zeta_j\Omega_j\omega i)} \tag{40}$$

$P_j(\mathbf{x})$ and $H_j(\omega;\mathbf{x})$ are respectively the modal participation factor and the modal frequency response function on condition of $\mathbf{X} = \mathbf{x}$. $M_j = \Phi_j^T\mathbf{M}\Phi_j$ is the jth modal mass, which is random, because of the random eigenvector. If $S_{FF}(\omega)$ is rational, (36), (37) and (38) can be obtained analytically using the method of residues. As an example consider the case of Gaussian white noise excitation, $S_{FF}(\omega) = S_0$. Then, (36), (37) and (38) become

$$R_{jk}(\mathbf{x}) = P_j(\mathbf{x})P_k(\mathbf{x}) \frac{4\pi S_0(\zeta_j\Omega_j + \zeta_k\Omega_k)}{M_jM_k[4\Omega_j\Omega_k(\zeta_j\Omega_k + \zeta_k\Omega_j)(\zeta_j\Omega_j + \zeta_k\Omega_k) + (\Omega_j^2 - \Omega_k^2)^2]} \tag{41}$$

$$S_{jk}(\mathbf{x}) = P_j(\mathbf{x})P_k(\mathbf{x}) \frac{4\pi S_0\Omega_j\Omega_k(\zeta_j\Omega_k + \zeta_k\Omega_j)}{M_jM_k[4\Omega_j\Omega_k(\zeta_j\Omega_k + \zeta_k\Omega_j)(\zeta_j\Omega_j + \zeta_k\Omega_k) + (\Omega_j^2 - \Omega_k^2)^2]} \tag{42}$$

$$T_{jk}(\mathbf{x}) = P_j(\mathbf{x})P_k(\mathbf{x}) \frac{2\pi S_0(\Omega_j^2 - \Omega_k^2)}{M_jM_k[4\Omega_j\Omega_k(\zeta_j\Omega_k + \zeta_k\Omega_j)(\zeta_j\Omega_j + \zeta_k\Omega_k) + (\Omega_j^2 - \Omega_k^2)^2]} \tag{43}$$

Then, with the limit states at hand, using the technique developed in the previous section system reliability can be computed for each of the realizations. Finally, unconditional outcrossing rates ν and unconditional reliability of the system $R([0,t])$ must be obtained using the total probability theorem, i.e.

$$R([0,t]) = \int_{R^m} R([0,t]|\mathbf{X} = \mathbf{x})f_{\mathbf{X}}(\mathbf{x})d\mathbf{x} \tag{44}$$

where $R([0,t]|\mathbf{X} = \mathbf{x})$ is the conditional reliability conditioned on the realization $\mathbf{X} = \mathbf{x}$ and $f_{\mathbf{X}}(\mathbf{x})$ is the jpdf of $\mathbf{X}$. In general, the expectation (44) cannot be evaluated analytically. Even numerical evaluation is out of the question if the dimension m of $\mathbf{X}$ is large, e.g. larger than 10. Taylor expansion of $R([0,t]|\mathbf{X})$ from the mean value $E[\mathbf{X}]$ cannot be used, because $R([0,t])$ is a non-analytical function of $\mathbf{X}$, due to the non-analytical boundaries defined by (18), (19) and (27), (28). Moreover, the gradients will be proportional to time t and the Hessian matrix proportional to t^2 and the expansion will blow up with time. Various approximate integration methods such as FORM and SORM based calculations, see e.g. Wen and Chen [14], Madsen and Tvedt [15], Laplace asymptotic methods, see e.g. Bryla et al. [16], cannot be used as they require the gradient and Hessian matrix to be known. The only method left behind is to evaluate (44) by means of Monte-Carlo simulation. Although some kind of variance reducing simulation technique such as importance sampling, see e.g. Melchers [17], or stratified sampling, see e.g. Rubinstein [18], could be addressed for less simulations, ordinary Monte Carlo simulation technique, in which Gaussian realizations of the structure, i.e. random stiffness matrix and uniform realizations of the damping ratio are generated, is applied.

5. NUMERICAL EXAMPLES

Two numerical examples are studied in detail to illustrate the theory. In both examples eigenvectors are normalized to unit mass. The limit-state surface Γ is constructed numerically from (27) and (28) and the numerical integrations of (4) and (7) are performed using a trapezoidal rule with angular increments of $\frac{\pi}{180}$. The first example considers a deterministic system to illustrate the proposed technique to calculate the system reliability in the modal space. The second example covers the stochastic finite element analysis defined herein.

Consider a deterministic 3 DOF system of Figure 3 subject to horizontal mean-zero white noise excitation with intensity S_0. The system parameters and prescribed displacement limit-states of the system are given under the figure. The results of the eigen and the stationary vibration analyses are also noted under the figure.

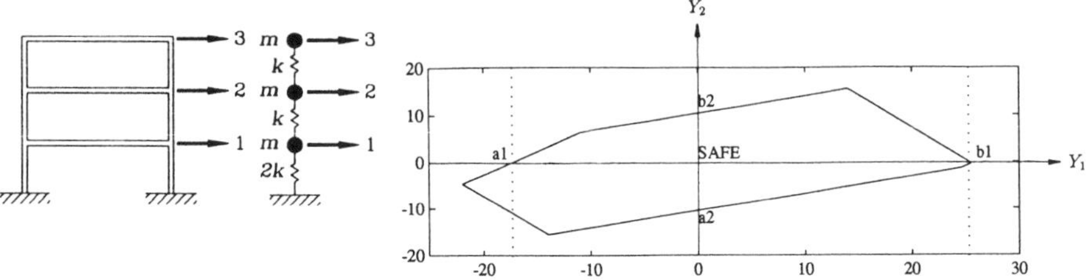

Figure 3. 3 DOF deterministic system. $m = k = 1$, $S_0 = \frac{1}{\pi}$, $\zeta_1 = \zeta_2 = \zeta_3 = 0.03$, $\mathbf{A}^T = [1,1,1]$, $U_1^- = U_1^+ = 6$, $U_2^- = 10$, $U_2^+ = 15$, $U_3^- = U_3^+ = 20$. $\Omega_1^2 = 2-\sqrt{3}$, $\Omega_2^2 = 2$, $\Omega_3^2 = 2+\sqrt{3}$, $\sigma_{Y_1}^2 = 298.969$, $\sigma_{Y_2}^2 = 1.964$, $E[Y_1(t)Y_2(t)] = 0.0701$, $\sigma_{\dot{Y}_1}^2 = 80.108$, $\sigma_{\dot{Y}_2}^2 = 3.928$, $E[\dot{Y}_1(t)\dot{Y}_2(t)] = 0.0513$, $E[Y_1(t)\dot{Y}_2(t)] = -1.049$.

Figure 4. Safe domain S for the 3DOF example.

For one-mode coordinate, the safe domain is found as $a_1 = -17.3205 < Y_1(t) < b_1 = 25.3588$ with an outcrossing rate of $\nu = 0.0780$ and the reliability function of $R([0,t]) = 0.7705\exp(-0.0780t)$. The safe domain has been indicated with dots in Figure 4. For two-modal coordinates case, the safe domain is indicated with the unbroken line in Figure 4. The outcrossing rate ν and the reliability function $R([0,t])$ are evaluated as 0.0603 and $0.7620\exp(-0.0603t)$, respectively. Although the first mode governs the motion, second mode do influence the reliability.

Next, the stochastic 3 storey one bay frame of Figure 5 with 9 DOF which is subject to horizontal mean-zero white noise excitation with intensity S_0 is analyzed using the derived stochastic finite element formulation. System parameters are listed under Figure 5 with some results from the eigen analysis of the frame. The yielding limits for the beams and columns are assigned as $\min_{n=1,\ldots,N}[M_{b,n,s} - 2.5\sigma_{M_{b,n}}]$ and $\max_{n=1,\ldots,N}[M_{b,n,s} + 2.5\sigma_{M_{b,n}}]$ where $M_{b,n,s}$ denotes the static bending moment at the nth failure element and $\sigma_{M_{b,n}}$ denotes the stationary standard deviation of the bending moment due to dynamic loading for the same node. N runs for 6 for the beams and for 12 for the columns in this example. Top storey displacement limit is assigned as $\mp 2\sigma_{U_7}$ where σ_{U_7} is the standard deviation of the top storey displacement. Hence, totally 19 failure elements have been considered in the reliability model.

All bending stiffness fields $(EI)_e(x)$ are assumed to be Gaussian with small coefficient of variation V_{EI}, i.e. $V_{EI} < 0.4$ and the autocovariance function of the zero-mean random field $f^e(x)$ of each element is assumed to be in the following form.

$$R_{ff}(x_1, x_2) = V_{EI}^2 \exp\Big(-(\frac{x_1 - x_2}{bL_e})^2\Big) \tag{45}$$

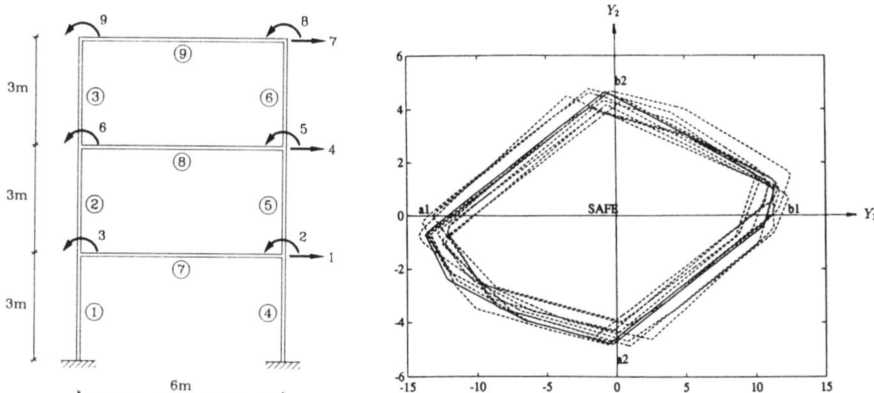

Figure 5. Three storey one bay frame with 9 DOF. All elements $\overline{E} = 3.5 \times 10^{10}\ N/m^2$, columns $\overline{I} = 4.0 \times 10^{-4}\ m^4$, $\mu = 300\ kg/m$, beams $\overline{I} = 3.0 \times 10^{-3}\ m^4$, $\mu = 2800\ kg/m$ (for beams 7 and 8), $\mu = 1900\ kg/m$ (for beam 9), $S_0 = 0.02\ m^2/s^3$, $\mathbf{A}^T = [1,0,0,1,0,0,1,0,0]$, beams $M^-_{b,y} = 450000Nm$, $M^+_{b,y} = 348000Nm$, columns $M^-_{b,y} = 236000Nm$, $M^+_{b,y} = 249000Nm$, $U^-_7 = U^+_7 = 0.0864m$. $\omega_{1,0} = 11.000\ rad/s$, $\omega_{2,0} = 31.789\ rad/s$, $\omega_{3,0} = 47.353\ rad/s$.

Figure 6. The safe domain for the deterministic structure $V_{EI} = V_\zeta = 0$ and for 10 realizations of the random structure with $b = 0.3$ and $V_{EI} = V_\zeta = 0.3$ in the two mode subspace.

where V_{EI} is the standard deviation of the random field, thus the coefficient of variation of the random bending stiffness field. b is a nondimensional correlation parameter such that $b \to \infty$ and $b \to 0$ denote fully correlated and perfectly uncorrelated random fields, respectively. Then, the covariance structure of the weighted integrals which is a monotonously increasing function of b and V_{EI} can be determined easily using (32). For simplicity, the random fields for each element are assumed to be mutually independent and both b and V_{EI} are taken as the same for all elements. All the damping ratios are taken as mutually independent, uniformly distributed as $U[0.02(1-\sqrt{3}V_\zeta), 0.02(1+\sqrt{3}V_\zeta)]$, where the mean value is seen to be 0.02 and the variational coefficient is V_ζ.

The bending moments in the end-sections of all the beam elements and the top-storey horizontal displacement form the considered response processes in the reliability problem. The safe domain for the deterministic structure, i.e. $V_{EI} = V_\zeta = 0$ has been shown with the unbroken line in Figure 6, besides with the safe domain of 10 other realizations of the structure.

System reliability for the deterministic structure with one-modal coordinate is calculated as $R([0,t]) = 0.8854\exp(-1.0071t)$, with two-modal coordinates is found as $R([0,t]) = 0.8805\exp(-0.6751t)$.

In what follows, a parametric study to evaluate the relationships between expected outcrossings rate $E[\nu]$ and $R([0,t])$ versus the non-dimensional correlation length b common to all element members and the variational coefficients V_{EI} and V_ζ is performed and the results with two-modal consideration are plotted in Figures 7-12. All results are obtained from 1000 Monte Carlo simulations in which Gaussian weighted integrals of each element and uniform damping ratios are generated and the reliability problem is solved for each of these realizations.

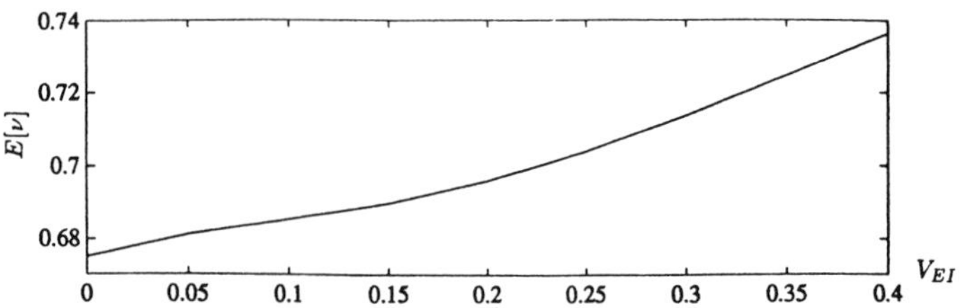

Figure 7. Unconditional outcrossings rate $E[\nu]$ versus the coefficient of variation of the bending stiffness V_{EI} for $V_{EI} \in [0, 0.4]$, $V_\zeta = 0.00$ and $b = 0.15$.

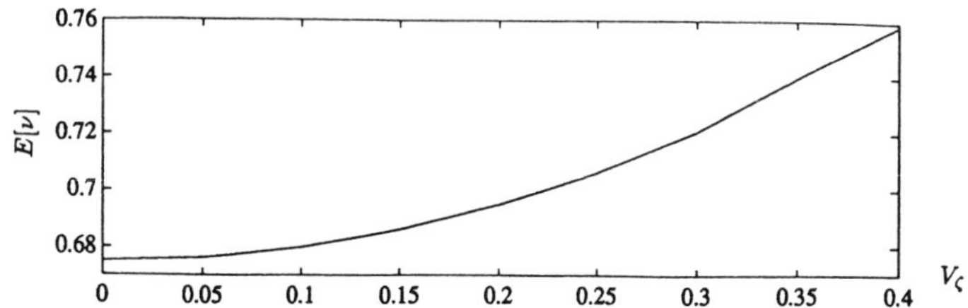

Figure 8. Unconditional outcrossings rate $E[\nu]$ versus the coefficient of variation of the damping ratios V_ζ for $V_\zeta \in [0, 0.4]$ and $V_{EI} = 0.00$.

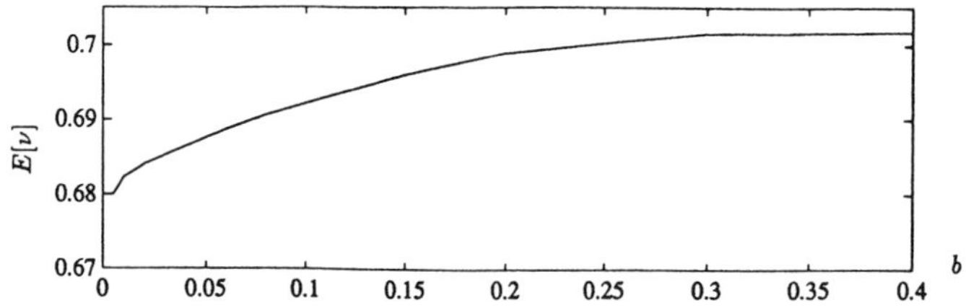

Figure 9. Unconditional outcrossings rate $E[\nu]$ versus the correlation parameter b for $b \in [0.005, 0.4]$, $V_{EI} = 0.20$ and $V_\zeta = 0.00$.

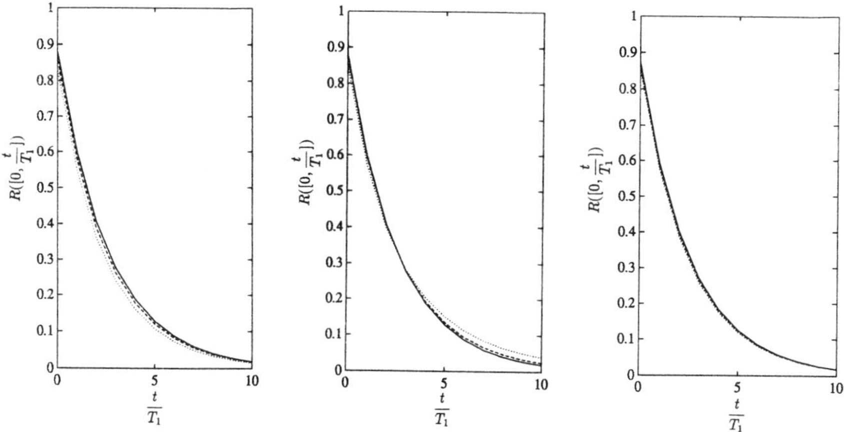

Figure 10. Unconditional system reliability $R([0, \frac{t}{T_1}])$ versus nondimensional time $\frac{t}{T_1}$, where T_1 is the first eigen period of the deterministic system, for V_{EI} =0.00 (—),0.20 (- - -),0.40 (· · ·), $V_\zeta = 0.00$ and $b = 0.15$.

Figure 11. $R([0, \frac{t}{T_1}])$ versus $\frac{t}{T_1}$ for V_ζ =0.00 (—),0.20 (- - -),0.40 (· · ·) and $V_{EI} = 0.00$.

Figure 12. $R([0, \frac{t}{T_1}])$ versus $\frac{t}{T_1}$ for b =0.01 (—),0.10 (- - -),0.40 (· · ·), $V_{EI} = 0.20$ and $V_\zeta = 0.00$.

From Figures 7, 8, 9, it is concluded that the unconditional outcrossings rate increases with an increase in the randomness of the system via an increase in the coefficient of variation V_{EI}, an increase in the correlation length b of the random fields, an increase in the coefficient of variation of the damping ratios V_ζ. Figure 9 shows that only changes when the correlation length is small, affects the results. If b is increased beyond $b \simeq 0.3$, no further increase in $E[\nu]$ is noticed. The coefficient of variation of the damping ratios are found to be the most critical among these three. Because in order to compute outcrossing rate via (4), random damping ratios and random eigenfrequencies are used directly through (36), (37) and (38) and the variability of the random eigenfrequencies is less than the variability of the system which is defined using V_{EI} and b (Köylüoğlu and Nielsen [9]).

On the other hand, Figure 10, 11 and 12 show that the system reliability function of the deterministic system is an upper bound at lower excitation intervals and a lower bound at larger excitation levels. This is clearly seen in Figure 11 and the same is observed for V_{EI} at later times. This result matches with the results of Spencer and Elishakoff [8].

6. CONCLUSION

Based on a reduction to the modal subspace, a new and practical method for the calculation of reliability of structural systems subject to random excitations is developed. This method leads to extreme decreases in the calculation of system realiability for MDOF systems with large number of failure elements. A stochastic finite element formulation is outlined and the proposed theory for system reliability is applied to both deterministic and stochastic structures. A parametric study is performed to find the relationship between the outcrossings rate and system reliability versus the correlation structure and the coefficient of variation of the element random fields and the modal damping ratios.

7. ACKNOWLEDGEMENT

The present research was supported by the Danish Technical Research Council under grant no. 16-5043-1.

8. REFERENCES

1. E. Vanmarcke, M. Shinozuka, S. Nakagiri, G. Schueller and M. Grigoriu, Structural Safety, 3 (1986) 143-166, Amsterdam, The Netherlands.
2. H. Benaroya and M. Rehak, Appl. Mech. Rev., 40 (1988) 201-213.
3. A. Der Kiureghian, C.C. Li and Y. Zhang, Lecture Notes in Engng. IFIP 76, Proc. Fourth IFIG WG 7.5 Conference, Germany, Eds. : R. Rackwitz and P. Thoft-Christensen, Springer-Verlag, 1991.
4. C. Brenner, Internal Working Report No. 35-91, Institute of Engng. Mech., University of Innsbruck, Austria, 1991.
5. T. Hisada and S. Nakagiri, Proc. Third ICOSSAR (1981) 395-408, Elsevier.
6. P.L. Liu and A. Der Kiureghian, J. Engng. Mech., ASCE, 117 (1989) 1806-1825.
7. G. Deodatis and M. Shinozuka J. Engng. Mech., ASCE, 117 (1991) 1865-1877.
8. B.F. Spencer and I. Elishakoff, J. Engng. Mech., ASCE, 114 (1988) 135-149.
9. H.U. Köylüoğlu and S.R.K. Nielsen, Proc. of EURODYN'93, Second European Conf. on Structural Dynamics, June, 1993, Trondheim, Norway.
10. G. Deodatis, J. Engng. Mech., ASCE, 117 (1991) 1851-1864.
11. T. Takada, J. Prob. Engng. Mech., 5 (1990) 146-156.
12. E. Vanmarcke, Random Fields, MIT Press, Cambridge, Massachusetts, USA, 1983.
13. Y.K. Belyaev, Theory of Probability Applications, 13 (1968) 320-324.
14. Y.K. Wen and H.C. Chen, Prob. Engng. Mech., 2 (1987) 156-162.
15. H.O. Madsen and L. Tvedt, Probabilistic Methods in Civil Engng., Edited by Spanos, P.D., 432-435, ASCE, NewYork, USA, 1988.
16. P. Bryla, M.H. Faber and R. Rackwitz, Proc. Tenth Int. Conf. Offshore Mechanics and Arctic Engng., Edited by Soares, C.G. et al., 143-150, ASME, NewYork, USA, 1991.
17. R.E. Melchers, Structural Safety, 6 (1989) 3-10.
18. R.Y. Rubinstein, Simulation and Monte Carlo Method John Wiley & Sons, NewYork, USA, 1981.

Reliability and Optimization of Structural Systems, V (B-12)
P. Thoft-Christensen and H. Ishikawa (Editors)
Elsevier Science Publishers B.V. (North-Holland)

Optimization of RC Structures - Theory and Applications

M. B. Krakovski

Dept. of Naval Architecture & Ocean Engineering
University of Ulsan, Republic of Korea, 680-749

Abstract
A general approach to optimum design of RC structures is presented. The approach is based on the principle of divided parameters. The theoretical concepts of the principle are described. A numerical example is given.

Keyword Codes: G.1.6
Keywords: Optimization

1. INTRODUCTION

The problem of optimum design of RC structures is very complex because of the difficulties to combine in one and the same algorithm rigorous methods of mathematical programming and mostly empirical methods for conventional design of RC structures. The paper presents a generalized approach to RC structures optimum design. The approach is general in the sense that it permits to optimize any RC structure under arbitrary loading taking into account all practical requirements. The approach is called the "principle of divided parameters" [1,2].

2. STATEMENT OF THE PROBLEM

The general problem of RC structures optimum design can be formulated in the following way. It is required to minimize the objective function y of variable parameters $x_1,...,x_l$

$$\min (x_1,...,x_l) \tag{1}$$

subject to constraints

$$f_i (x_1,...,x_l) \leq 0; \; i = 1,...,L. \tag{2}$$

Any criterion of quality (total cost, wholesale price, steel and cement consumption etc.) can be used as an objective function. The variable parameters can include geometrical dimensions of the structure, cross sectional dimensions of its members, properties of materials, types of reinforcement etc. The constraints are the requirements for the structure written in the form of equalities and inequalities.

3. DIVISION OF GENERAL OPTIMIZATION PROBLEM INTO INTERNAL AND EXTERNAL PROBLEMS

All variable parameters of the structure $x_1,...,x_l$ are divided into two groups – the external parameters $x_1,...,x_n$ ($n<l$) and internal parameters $x_{n+1},...,x_l$. The external parameters are usually associated with geometrical dimensions and material properties, whereas the internal parameters are usually associated with reinforcement. Out of the total number L of constraints (2) N constraints containing only the external parameters are selected:

$$f_i\,(x_1,...,x_n) \leq 0;\ i = 1,...,N. \tag{3}$$

These constraints are termed external. The remaining M = L – N constraints contain both the external and internal parameters:

$$f_i\,(x_1,...,x_l) \leq 0;\ i = N+1,...,L. \tag{4}$$

These constraints are termed internal.

The minimum d_{i1} and maximum d_{i2} values for the external variable parameters can be fixed with due regard for current codes as well as for geometrical requirements. Then

$$d_{i1} \leq x_i \leq d_{i2};\ i = 1,...,n. \tag{5}$$

Inequalities (5) provide an example of external constraints (3) and in some cases they are the only external constraints.

The general problem of optimum design (1), (2) is divided into two subproblems – internal and external.

The internal optimization problem is a problem of determining

$$\min_{x_{n+1},...,x_l} y(x_1,...,x_l) \tag{6}$$

having regard to internal constraints (4), the external variable parameters being fixed. Only the internal variable parameters can vary in the course of solving the internal optimization problem.

The external optimization problem is a problem of determining

$$\min_{x_1,...,x_n}\ \min_{x_{n+1},...,x_l} y(x_1,...,x_l) \tag{7}$$

having regard to external constraints (3). Only the external variable parameters can vary in the course of solving the external optimization problem. As a result, such values of the external parameters are obtained, which select the optimum solution out of all solutions to the internal optimization problem. In this case external (3) as well as internal (4) constraints are satisfied, i.e. general problem (1), (2) is solved.

In the process of optimization tests are carried out. A test is defined as a set of the following operations: values of the external parameters are fixed; the external constraints are checked (if they are not satisfied the test is stopped); the internal optimization problem is solved; the value of the objective function is computed.

4. THE PROCESS OF SOLVING THE INTERNAL OPTIMIZATION PROBLEM

The division of parameters into external and internal ones is most commonly governed by available methods of conventional design and optimization of RC structures. These methods provide a basis for solving the internal optimization problem. Consider three typical cases.

1. A method of conventional design is known and minimum cross sectional area of reinforcement is determined directly from the current Code so as to resist all load effects. In this case the following parameters are also determined from the Code: the depth of concrete cover, the number of reinforcing steel bars (longitudinal and transverse), spacings between the bars etc. It is good practice to assume these parameters as internal ones. Thus the internal optimization problem can be solved on the basis of the Design Code.

2. A method of conventional design is known, but minimum cross sectional area of reinforcement cannot be determined directly from the Design Code. There exists only a possibility to check whether internal constraints (4) are satisfied or not. Here the internal parameters are assumed in the same way as in the previous case. The internal optimization problem is solved on the basis of the Design Code in combination with a method of one-dimensional search.

3. A method of optimization with respect to some parameters is known. The rest of the parameters that could be varied in the course of optimization are fixed. The proposed principle makes it possible to assume that all the parameters are variable by dividing them into the external and internal ones. The internal parameters are the parameters of the existing optimization method, whereras the internal parameters are those previously fixed. Thus the optimization method can be used to solve the internal optimization problem. The external variable parameters are fixed in each test in accordance with the procedures developed to solve the external optimization problem.

Several load combinations and the requirements of different limit states are taken care of in the process of handling the internal optimization problem.

5. CASES WITH VIOLATED CONSTRAINTS. THE PROCESS OF SOLVING THE EXTERNAL OPTIMIZATION PROBLEM

Occasionally a test cannot provide values of external variable parameters satisfying internal constraints (4), i.e. the internal optimization problem has no solution. This situation can be taken care of in solving the external optimization problem in one of the following ways:

1. A penalty function is used, i.e. in this test the value of the objective function is increased according to certain rules.
2. The order in accordance with which the external variable parameters are changed in the process of solving the external optimization problem is replaced by another order.

To solve the external optimization problem special algorithms have been developed. They are based on the theory of optimum experimental design and methods of unconstrained optimizations (e.g. the methods of steepest descent, of coordinate descent, of conjugated gradients, of pattern search, of flexible polyhedron search etc.). The algorithms should fix values of external parameters so as to satisfy external constraints (3). It is easy for a feasible domain in the form of hyperparallelepiped (5). In other cases external constraints are simply verified. If they are not satisfied in a test, an algorithm solving the external optimization problem deals with the situation in the same manner as in the absence of a solution to the internal optimization problem.

To apply some of the above algorithms solving the external optimization problem the space of external variables should be normalized, i.e. transformed into a dimensionless form. Different approaches to this transformation have been developed.

The discreteness of the external variable parameters is accounted for in the algorithms solving the external optimization problem: continuous values of the parameters are rounded off to the nearest descrete values. Generally, this procedure is not applicable because a solution can fall outside the feasible domain. But for the principle of divided parameters tests are carried out at the external space points resulting from the rounding off. Therefore it is always known whether the internal constraints are satisfied or not.

Algorithms to solve multiextremum problems and to carry out multiobjective optimization have been developed.

6. AN EXAMPLE OF OPTIMIZATION OF A RC FLEXURAL MEMBER WITH RECTANGULAR CROSS SECTION

Many structures (beams, columns, foundations, roofing shells etc.) have been optimized on the basis of the principle of divided parameters. To illustrate an application of the principle of divided parameters here an example of optimization of RC flexural member witn rectangular cross section under a constant bending moment M = 85 kN.m is considered.

The external variable parameters are: x_1 - width of the section (b); x_2 - its depth (h); x_3 - class of concrete (R_c). The member is reinforced with class A-II steel (according to the Russian Design Code) and this class does not vary. The distance from the upper face to the centroid of the compression reinforcement is constant and equal to 3.5 cm. The internal parameters are: number and diameter of bars defining cross sectional areas of tension A_s and compression A_s' reinforcement; number of reinforcing cages; the effective depth; spacing of the bars.

The internal constraints are the requirements of the Russian Design Codes for the flexural strength of the member. Internal optimization problem solved according to the above Code for any fixed set of external parameters is to determine the minimum values of A_s and A_s' providing flexural strength of the member.

960 tests have been carried out for all possible combimations of the following discrete values of the external parameters:

$x_1(b)$ = 10, 12, 14, 15, 16, 18, 20, 22, 24 cm;

$x_2(h)$ = 25, 30, 35, 40, 45, 50, 55, 60, 65, 70, 75, 80 cm;

$x_3(R_c)$ = 12.5, 15, 20, 25, 30, 35, 40, 45 MPa.

Thus the only external constraints are the inequalities:

10 cm $\le x_1 \le$ 25 cm; 25 cm $\le x_2 \le$ 80 cm; 12.5 MPa $\le x_3 \le$ 45 MPa.

The value of the objective function in each test is the sum of costs of concrete and steel for a member 1m long. The unit price of the steel was assumed to be 1020 roubles per cu.m. The unit prices of concretes of classes B12.5, 15, 20, 25, 30, 35, 40, 45 were 36, 38.5, 40, 41, 42.5, 44.5, 46, 47 roubles per one cu. m respectfully.

The exhaustive search thus made the example serve as a test to verify different algorithms used to solve the external optimization problem.

Below is presented an algorithm based on the theory of optimum experimental design. The external optimization problem is solved in the following order.

A. Set a point of the initial trial $x^{(1,0)}$. Assume k = 1.

B. At the k-th step of the algorithm perform the following operations:

1. Select a linear plan. Perform tests at the point $x^{(k,0)}$ and at the points $x^{(k,j)}$ defined by the linear plan.

2. Find the gradient of the objective function in the space of the external variables and move along the antigradient.

3. Denote the point with the lowest value of the objective function by $x^{(k+1,0)}$. Go to the (k+1)-th step.

C. Stop optimization process when the objective function cannot be reduced any more.

The linear plans are represented as design matrices. Plans of Box and Wilson [3] and of Placket and Burman [4] are used as linear plans. The matrix elements are normalized values of external parameters. A normalized value $\bar{x}_i^{(k,j)}$ of the external parameter $x_i^{(k,j)}$ is given by

$$\bar{x}_i^{(k,j)} = (x_i^{(k,j)} - x_i^{(k,0)})/a_i \tag{8}$$

where a_i is a step between descrete values of x_i.

The process of optimization is presented in Table 1. The following values of the external parameters were taken as the initial trial: $x_1(b)=20$ cm, $x_2(h)=45$ cm and $x_3(R_c)=15$ MPa. According to the procedure a of the first step of the algorithm (k=1) the tests followed a Box-Wilson linear plan, namely fractional factorial 2^{3-1}. The plan is defined by a matrix of normalized values of external parameters (+1 of −1). Each row of the matrix defines a set of the external parameters for a test, each column is related to one of the external parameters. From the tests carried out in accordance with the design matrix the following approximation of the relationship between the objective function and the external parameters can be obtained:

$$y=b_0 + \sum_{i=1}^{n} b_i\bar{x}_i \tag{9}$$

where b_i are coefficients given by

$$b_i = \frac{1}{T} \sum_{u=1}^{T} \bar{x}_{iu} y_u \quad (i = 0,1,...,n). \tag{10}$$

Here T is the number of tests according to the design matrix; x_{iu} is an element of the design matrix; y_u is the value of the objective function in the u-th test; i,u are the numbers of a column and of a row of the design matrix respectively.

Approximation (9) is used in the procedure b which is a movement along the antigradient: the values of the external parameters vary directly as the coefficients b_i, the values are rounded off and the tests are carried out.

For k=1 approximation (9) takes the form:

$$y = 4.52 + 0.355\bar{x}_1 + 0.165\bar{x}_2 + 0.135\bar{x}_3. \tag{11}$$

Moving along the antigradient the lowest value of the objective function (y=3.82 rbls) was obtained for j=8. The values of the external parameters were assumed as the initial trial for the second step, k=2. Here the minimum value of the objective function was y=3.14 rbls for j=9 and it couldnot be reduced for k=3. The exhaustive search has shown that this value is really the global minimum.

Table 1

k	$x^{(k,o)}$	Procedure	j	Normalized parameter values x_1	x_2	x_3	Dimensional parameter values x_1(b)cm	x_2(h)cm	$x_3(R_c)$	y rbls
–	–	–	0	0	0	0	20	45	15	4.57
1	20 45 15	a	1	−1	−1	+1	18	40	20	4.24
			2	+1	−1	−1	22	40	12.5	4.68
			3	−1	+1	−1	18	50	12.5	4.30
			4	+1	+1	+1	22	50	20	5.28
		b	5	−1	0	0	18	45	15	4.48
			6	−2	−1	−1	16	40	12.5	4.26
			7	−2.5	−1	−1	15	40	12.5	3.85
			8	−3	−1	−1	14	40	12.5	3.82
			9	−4	−2	−1	12	35	12.5	3.84
2	14 40 12.5	a	1	−1	−1	0	12	35	12.5	3.84
			2	+1	−1	0	16	35	12.5	4.17
			3	−1	+1	0	12	45	12.5	3.63
			4	+1	+1	0	16	45	12.5	4.03
			5	0	0	+1	14	40	15	3.76
		b	6	−1	0	0	12	40	12.5	3.68
			7	−2	+1	+1	10	45	15	3.37
			8	−2	+2	+2	10	50	20	3.33
			9	−2	+3	+2	10	55	20	3.14
			10	−2	+4	+3	10	60	25	3.34
3	10 55 20	a	1	0	−1	−1	10	50	15	3.39
			2	0	+1	−1	10	60	15	3.19
			3	0	−1	+1	10	50	25	3.38
			4	0	+1	+1	10	60	25	3.34
			5	+1	0	0	12	55	20	3.52
		b	6	0	+1	−1	10	60	15	3.19

7. CONCLUSION

In conclusion it would be well to consider the following advantages of the principle of divided parameters:

1. This principle represents a generalized approach to optimum design of RC structures subject to a variety of external actions with due regard for all design requiremetns. It enables to use any objective function and the latest achievements in conventional structural design.

2. The principle allows to reduce the dimension of the problem replacing one problem of l variables by two simpler problems: the first is of n external variables and the second is of $m=l-n$ internal variables. When solving the

internal optimization problem the majority of constraints should be satisfied first and foremost, whereas an optimization algorithm is relatively simple. On the other hand, in the solution of the external optimization problem the external constraints are weak and the primary consideration is the optimization procedures. Thus the principle of divided parameters separates two difficulties of optimization – the satisfaction of constraints and the search for the optimum solution – relating them to the internal and external optimization problems respectfully.

3. The algorithm to solve the internal optimization problem may comprise methods of conventional design of RC structures as well as methods of their optimization with respect to some parameters. Thus the optimum solution to the internal optimization problem meets all the requirements which were satisfied in the process of conventional design or optimization. In addition, the discreteness of internal variable parameters can be dealt with when necessary.

4. The algorithms to solve the external and internal optimization problems are independent of one another making it possible:

a) to apply one and the same algorithm for the external optimization problem in combination with different algorithms for the internal optimization problem;

b) to use different algorithms for the external olptimization problem in combination with one and the same algorithm for the internal optimization problem. The possibility to use combinations of different algorithms enables the optimum solution to be found for all practical purposes.

5. It is a relatively easy matter to solve the external optimization problem since the search is conducted as a rule along the feasible domain border without moving inside the domain.

6. The principle of divided parameters makes it possible to apply without trouble the most sophisticated methods of conventional design of RC structures and methods of their optimization with respect to some parameters modifying algorithms for the internal and external optimization problem respectfully.

7. The optimization methods based on the principle of divided parameters are universal, simple and rigorous.

The above advantages permit to make wide use of the principle of divided parameters in practical design of RC structures.

REFERENCES

1. M.B.Krakovski, Transactions of NIIZhB Analysis and Design of RC Structures, NIIZhB, Moscow, Stroyizdat Publishers, 1977 (in Russian).
2. M.B.Krakovski, Recommendations on Optimum Design of RC Structures, Moscow, NIIZhB, 1981 (in Russian).
3. G.E.Box, K.B.Wilson, J. of the Royal Statist. Soc., Ser. B, No 1, 13, (1951).
4. R.L.Plackett, J.P.Burman, Biometrica, No 4, 33 (1946).

Reliability and Optimization of Structural Systems, V (B-12)
P. Thoft-Christensen and H. Ishikawa (Editors)
Elsevier Science Publishers B.V. (North-Holland)

Improvement of Russian Codes for RC Structures Design on the Basis of Reliability Theory

M. B. Krakovski

Dept. of Naval Architecture & Ocean Engineering
University of Ulsan, Republic of Korea, 680-749

Abstract
An algorithm and computer program for structural reliability estimation are presented. Monte Carlo approach with subsequent approximation of the simulation results by Pearson's curves is used. The above approach is applied for reliability analysis of RC flexural member with rectangular cross section. Cases with insufficient and excessive reliability are revealed. The results of the analysis have made it possible to formulate the proposals for improvement of Russian Codes for RC structures design with the aim of reliability regulation. The suggested and existing approaches are compared.

Keyword Codes: G.3; J.6
Keywords: Probability and Statistics; Computer-aided Engineering

1. INTRODUCTION

The current Russian Codes for RC structures design are based mostly on empirical equations. These equation were derived with individual reliability considerations for each case. Therefore even the reliability of one and the same type of structure varies with geometrical dimensions, reinforcement ratio, strength of materials etc. The scatter of reliability level for different structures is rather wide, but reliability does not depend on the importance of structures.

The aim of the presernt investigation is to develop an approach to reliability regulation of RC structures designed according to Russian Codes.

2. ALGORITHM AND COMPUTER PROGRAM FOR RELIABILITY ESTIMATION OF RC STRUCTURES

All variable parameters of the structure $X_1,...,X_n$ and of the loads $Q_1,...,Q_s$ are assumed to be random values with known laws of distribution. The deterministic method for the design of the structure is also assumed to be known. Failure of the structure is defined as violation of at least one of the requirements. The task is to find the probability of non-failure of the structure.

The problem is solved in the following order.

1. Using Monte Carlo simulation methods and the known laws of distribution obtain m realization of random values - parameters of the structure and loads - $x_{1i},...,x_{ni}$ and $q_{1i},...q_{si}$ (i=1,...,m).

2. Carry out m deterministic designs of the structure by the selected method. The designs determine m times the values of k output parameters of the structure $y_{1i},...,y_{ki}$ (i=1,..,m), such as stresses, deflections, crack widths, bearing capacity etc.

3. The values of each output parameter thus obtained are assumed to be observed values of the random variable representing this output parameter. Using the values choose an appropriate probability density function out of the family of Pearson's curves for each output parameter $y_1(z),...,y_k(z)$.

4. Find the ultimate possible value of the output parameters $y_1^0,...,y_k^0$.

5. Determine the reliability of the structure for each output parameter:

$$P_i = \int_{y_i^0}^{\infty} y_i(z)dz \tag{1}$$

$$P_i = \int_{-\infty}^{y_i^0} y_i(z)dz \tag{2}$$

Use (1) if the value of y_i^0 limits the value of the i-th output parameter from the bottom and (2) if the limit is from the top.

The algorithm was implemented in the form of a computer program written in Fortran 77. The inverse problem can also be solved: knowing the probability density function $y_i(z)$ such values of y_i^0 can be found that a prescribed level of reliability P_i is obtained.

Accuracy of the program results was proved by test examples [1]. The possibility of using statistics of extreme values in designs was demonstrated [2].

3. ANALYSIS OF RELIABILITY OF RC FLEXURAL MEMBER WITH RECTANGULAR CROSS SECTION

According to the Russian Building Code for design of RC structures the characteristic strength of materials is defined with a reliability of 0.95. It means that 95% of all possible strength measurements are expected to exceed the characteristic strength. To obtain the design strength it is necessary to divide the characteristic strength by the partial coefficient for properties of materials which is greater than or equals unity for strength and serviceability limit states respectively. In this case the reliability P_d of design strength of materials for the strength limit state varies with the coefficient of variation v of the strength of material: as v increases or decreases P_d decreases or increases accordingly. As a result, changes in the values of v cause changes in the reliability of quantities characterizing the bearing capacity of RC structures.

In the reliability analysis of a RC flexural member with rectangular cross section the task was to investigate how the reliabilities of quantities characterizing the bearing capacity of RC structures is affected by the reliability of the concrete and reinforcement strength values specified in the Code.

The analysis was made using the above approach. The prism strength of concrete and the strength of reinforcement were assumed to be random normally distributed variables with the following characteristics: class of concrete B25 (from Design Code SNiP 2.03.01-84), characteristic prism strength R_{pr}^{ch} = 18.5 and design prism strength R_{pr} = 14.5 MPa, class of steel reinforcement A-III, characterisic strength R_s^{ch} = 390 and design strength R_s = 365 MPa.

According to the provisions of the Russian Design Code SNiP 2.03.01-84 mean values of the prism strength of concrete $\bar{R}_c$ and the strength of steel $\bar{R}_s$ as well as their standard deviations σ_c and σ_s are determined by the equations:

$$\bar{R}_c = R_{pr}^{ch} /(1.07(1-2v_c)); \quad \sigma_c = \bar{R}_c v_c \tag{3}$$

$$\bar{R}_s = R_s^{ch} /(1-1.64v_s); \qquad \sigma_s = \bar{R}_s v_s \tag{4}$$

where v_c and v_s are coefficients of variation of prism strength of concrete and strength of steel respectfully.

To determine the reliability of design values for the strength of reinforcement R_s and that of concrete R_c the values

$$t_s = (\bar{R}_s - R_s)/\sigma_s \text{ and } t_c = (\bar{R}_c - R_c)/\sigma_c \tag{5}$$

were calculated.

The reliabilities P_s and P_c corresponding to t_s and t_c were found according to the normal dinstribution law.

In addition to the characteristics of the normal distribution law for the strength of materials, the reinforcement ratio μ = $100A_s/bh_0$ was fixed for calculations (where A_s is cross sectional area of reinforcement, b and h_0 are width and effective depth of the member respectfully). As a result the probability density function in the form of one of the Pearson's curves, type 1 to 7, for bearing capacity of the member characterized by the relative moment $M' = M/bh_0^2$ was obtained. The value of the relative moment M_0' with a definite reliability P_m was taken for further analysis.

The results of the calculations are presented in Fig. 1 where the M′ versus μ curves were plotted for different values of the coefficients of variation of the concrete prism strength v_c, of the steel reinforcement strength v_s and also for different values of P_m. A total of five curves numbered 1,2,...,5 are

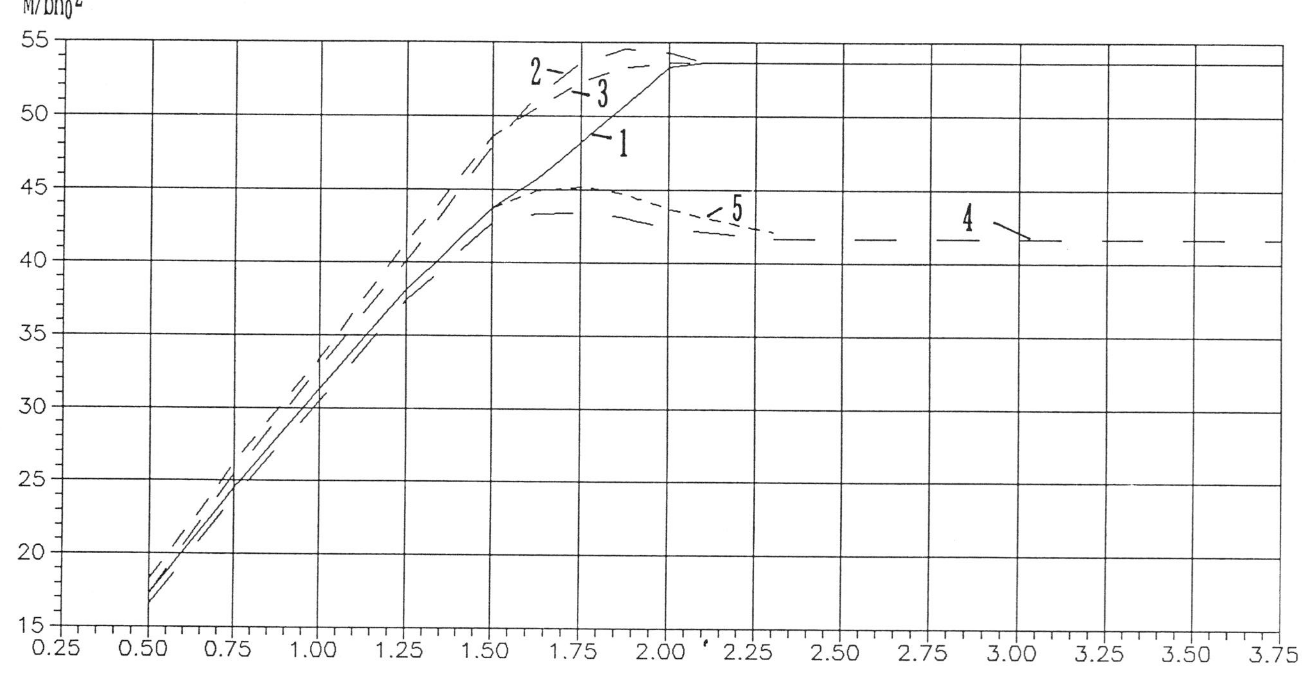

Fig. 1. Investigation of the reliability of a flexural reinforced concrete member with rectangular cross section.

presented in Fig. 1. The moments corresponding to the curves are denoted M_1',M_{20}',...,M_{50}'. The subscript "0" in the designations M_{20}',...,M_{50}' means that these values were obtained with the definite reliability P_m.

The moments M_1' were calculated following the procedure of the Building Code for design values of the strength of materials. Further, curve 1 was used as a reference for comparison with other results.

To plot curve 2 the coefficients of variation v_s and v_c were assumed so as to obtain $t_s = t_c = 3$ and $P_m = P_s = P_c = 0.9986$ from equations (1) for inverse problem, (3),(4),(5). It appears that M_1' and M_{20}' are practically the same within the zone of overreinforcement ($\mu > 2.1$) but for $\mu < 2.1$ the values of M_{20}' are somewhat greater than M_1'.

To plot curve 3 the maximum possible value for the coefficient of variation of concrete prism strength $v_c = 0.2$ and a relatively high coefficient of variation for the strength of steel $v_s = 0.1$ were assumed. In this case $t_c = 2.37$, $t_s = 0.2$, $P_c = 0.9910$ and $P_s = 0.9938$. The reliability P_m of the moments M_{30}' was assumed to be equal to the lowest value of the material reliability values P_c and P_s, i.e. $P_m = 0.9910$. Fig. 1 shows that curves 2 and 3 are close to each other for $\mu < 1.5$ and $\mu > 2.25$. And only in the zone close to overreinforcement ($1.5 < \mu < 2.25$) M_{20}' is higher than M_{30}'.

In curve 4 the reliabilities of the strength of the materials were the same as in curve 3 but a greater reliability $P_m = 0.9986$ was assumed for M_{40}'. Curves 1 and 4 were found to be practically the same in the zone of underreinforcement (for $\mu < 1.5$). But the moment M_{40}' proved to be much lower than M_1' near the zone of overreinforcement and within this zone ($\mu > 1.5$). For instance the difference was 9.4 and 18.6 percent for $\mu = 1.75$ and $\mu = 2$ respectively, while it was about 20% for $\mu > 2.5$.

To plot curve 5 the initial data were the same as for curve 4 but a lower coefficient of variation was assumed for the strength of reinforcement: $v_s = 0.6$. The calculations were made for $1.5 \leq \mu \leq 2.25$. The moments M_{50}' were found to be slightly greater than M_{40}' for the indicated above μ values, i.e. the reduction of the bearing capacity ($M_{20}' < M_1'$) was caused mainly by the high value of the coefficient of variation of the concrete prism strength.

It is also of interest to observe the change in the reliability of M_1' (calculated in accordance with the Design Code) when μ is close to the zone of overreinforcement and within this zone. In this case the reliability of M_1' is the probability that the bearing capacity of the flexural member with random material strength exceeds M_1'. When the uniformity of material strength is high (the coefficients of variation v_c and v_s are low) the reliability of M_1' calculated from the probability density function of M_2' is about 0.999. But when the uniformity is low for both materials or even only for concrete (random variables M_3' and M_4') the reliability of the moment M_1' decreases down to about 0.99. This reliability is too low and steps should be taken to increase this value.

The results have a clear physical meaning. Within the overreinforcement zone the strength of the flexural member is governed only by that of concrete. The ultimate bending moment of the member can be computed from the Design Code (the design prism strength is used in the equations). If the reliability of the design prism strength low then the reliability of the ultimate moment is low as well. The reliability of the bearing capacity may be insufficient in all cases where the strength of a structure is governed mainly by that of one material, and the reliability of the design strength of this materal is low. To avoid these dangerous situations some changes in the concepts of limit state design of RC structures are suggested. These proposals are described below.

4. PROPOSALS TO IMPROVE LIMIT STATE DESIGN OF RC STRUCTURES

As has been noted above the characteristic strength of a material is specified in the Code with a reliability of 0.95 and the design strength is obtained by dividing the characteristic strength by the partial coefficient for material properties γ which is higher than or equals unity for strength and serviceability limit states respectfully. The coefficient γ takes into account various non-statistical factors. In addition it is proposed to establish requirements for the reliability of design strength of materials specifying that it cannot be lower than 0.9986. This reliability is specified in the Russian Building Code SNiP II-21-75 for the design strength of concrete with the mean coefficient of variation $v_c = 0.135$.

Below the reliability of RC structures is estimated considering only the variability of the strength of the materials - concrete and reinforcing steel. The loads are assumed to be equal to their design values. The reliability is defined as the probability that the bearing capacity of a member with random strength characteristics of the materials exceeds the bearing capacity calculated according to the Design Code, when strength of the materials equal their design values. Thus the particular values of the loads are ignored and only the bearing capacity of the member is taken into account.

Since the reliability of the design strength of the materials is at least 0.9986 the reliability of the bearing capacity is also close to 0.9986 when the strength of the member depends mainly on one material and is higher than that when several materials interact in a structure (concrete, tension and compression reinforcement, prestressed and non-prestressed steel etc.). To make all structures equally reliable with a reliability of 0.9986 the material property combination factor can be introduced:

$$k_c = M_1/M_0 > 1 \tag{6}$$

where M_1 and M_0 are the bearing capacity with a reliability of 0.9986 and that determined from the Design Code respectfully.

M_1 is computed on the basis of the above approach and M_0 is found by conventional design methods using the design strength of materials.

For instance, calculations made with the help of the above computer program have shown that for a flexural member with rectangular cross section k_c = 1.05 if

$$0.45 \leq \xi/\xi_R \leq 0.7 \tag{7}$$

and k_c = 1 in all other cases. Here ξ and ξ_R are a relative depth and the maximum relative depth of concrete compressioln zone respectfully. Thus if conditions (7) are satisfied the design bearing capacity of a member obtained from the Design Code can be increased by 5%.

The importance factor k_0 may be also introduced to reduce the design bearing capacity for more important structures. Thus the basic equation for limit state design of RC structures takes the form:

$$\sum_i n_i S_i \leq k_c k_0 m \Phi(\lambda_1 R_1, \lambda_2 R_2, \ldots) \tag{8}$$

where S_i are characteristic loads applied to a structure;
- n_i are load factors;
- R_i are design strength of materials;
- λ_i are service behaviour factors:
- m is an importance factor for the building which the structure in question belongs to;
- Φ is the bearing capacity of the structure.

5. COMPARISON WITH THE EXISTING APPROACH

In [3] another approach to structural reliability regulation is introduced. It is suggested to determine partial coefficients used in structural codes so as to minimize the quantity S given by

$$S = \sum_{i=1}^{m} \omega_i \Delta(P_{fi}(\bar{\gamma}), P_{ft}) \tag{9}$$

subject to the constraint

$$\sum_{i=1}^{m} \omega_i P_{fi}(\bar{\gamma}) = P_{ft} \quad \text{with} \quad \sum_{i=1}^{m} \omega_i = 1.0 \tag{10}$$

and where

$\Delta(P_{fi}(\bar{\gamma}),P_{ft})$ is an agreed function of the quantities $P_{fi}(\bar{\gamma})$ and P_{ft};
$P_{fi}(\bar{\gamma})$ is the failure probability of the i-th structural component designed using the set of partial coefficients $\bar{\gamma}$;
P_{ft} is the target failure probability;
$\omega = (\omega_1,...,\omega_m)$ is a set of weighting factors indicating the relative importance of each of m structural components included in the partial factors evaluation.

Thus, as indicated in [3], the aim of this approach is to minimize the deviations of the probabilities P_{fi} from the target probability of failure P_{ft} whilst maintaining the *average* probability of failure at the target level.

In contrast to this the above approach makes it possible to regulate not only the *average* reliability, but to obtain a prescribed level of reliability for different structures and to change reliability in accordance with the importance of structures.

These two approaches can be also used in combination: first the approach described in [3] is used and partial coefficients are calculated; then the approach described in this paper is applied for more detailed regulation of structural reliability.

6. CONCLUSION

The proposed approach makes it possible to regulate reliability of RC structures by two factors which are used in addition to any existing system of partial coefficients. The material properties combination factor k_c makes all identical structures equally reliable, the importance factor k_0 regulates reliability of RC structures according to their importance. Thus a prescribed level of reliability for any group of structures or even for individual structure can be obtained.

REFERENCES

1. M.B.Krakovski, Stroitelnaya Mekhanika e Raschet Sooruzheniy. No. 2 (1982), in Russian.
2. M.B.Krakovski and L.I.Stolipina, Problems of Optimization and Reliability in Structural Mechanics, Abstracts of the Reports to the National Conference. Moscow, Stroyizdat Publishers, 1983, in Russian.
3. P.Thoft-Christensen and M.J.Baker, Structural Reliability Theory and Its Application, Springer-Verlag, Berlin, Heidelberg, New York, 1982.

Reliability and Optimization of Structural Systems, V (B-12)
P. Thoft-Christensen and H. Ishikawa (Editors)
Elsevier Science Publishers B.V. (North-Holland)

Load Model for Bridge System Reliability

J. A. Laman, A. S. Nowak
Department of Civil and Environmental Engineering
University of Michigan, Ann Arbor, Michigan, 48109, USA

Efficient evaluation of existing bridges requires knowledge of the load and resistance parameters. This paper concerns the development of a data base for load modeling using weigh-in-motion (WIM) technology. Presented are results of steel highway girder bridge testing using WIM and a strain data acquisition system. The ten girder, three simple span, two lane steel girder bridge is instrumented in its 33'- 6" entry span for all data collection. The bridge is located on US23 which is a major route between Detroit, Michigan and Toledo, Ohio with no weigh station on the route. Vehicle weight statistics are collected simultaneously with strain data. Distribution functions are generated for gross vehicle weight (GVW) and measured girder strains allowing an analysis and comparison of the theoretical stress spectra based on WIM data to the actual measured stress spectra experienced by the bridge girders. Girder response to heavy vehicle multiple presence is presented as well as lateral load distribution functions and live load fatigue spectra for the girders. From these statistics conclusions are drawn regarding stress histories of the girders as a function of time and position of the girder in the bridge which is important for the system reliability load model.

Keyword Codes: J.2, J .6

Keywords: Physical Sciences and Engineering. Computer Aided Engineering.

1. INTRODUCTION

Reliability evaluation of bridges as a system is an increasingly important topic in the effort to assess the deteriorating infrastructure. Accurate and inexpensive methods are needed to determine the actual strength of the bridge, its remaining life, and the actual load spectrum. The major factors which have contributed to the present situation are: the age, inadequate maintenance, increasing load spectra and environmental contamination. As a consequence, the deficient bridges are posted, repaired or replaced. This disposition of bridges must consider economical and safety implications. To avoid high costs of unnecessary replacement or repair, the evaluation must accurately determine the present load carrying capacity of the structure and predict loads and any further changes in the applicable time span.

The objective of this study is to develop an approach for evaluation of the existing bridge fatigue load spectra. The truck loads are determined on the basis of weigh-in-motion (WIM) measurements. Fatigue analysis is based on the rainflow algorithm and Miner's rule. Direct measurements of strain histories in bridge girders allows for the development of component specific fatigue load models. As a result, critical girders can be identified.

Field measurements and analysis were performed on the bridge located in Monroe County south of Milan, Michigan. The bridge carries US23 southbound over the Saline River.

2. BRIDGE DESCRIPTION

The bridge is a composite steel, ten girder structure. Exterior girders are W36X150 and the eight interior girders are W27X102. Diaphragms consist of three-eighth inch plate riveted to L3X3X$^3/_8$ at the top and bottom flange forming a "C" shape. Slab thickness is 7" with standard New Jersey type barriers. Measurements were taken on the southbound entrance span which is 33'- 6" in length and is simply supported between the abutment and the first pier. The cross section is shown in Fig. 1. The site has a moderate average daily truck traffic (ADTT) of approximately 2,500 with a relatively high density of heavily loaded trucks.

3. INSTRUMENTATION

Dynamic strains were obtained with a data acquisition and analysis system from the SoMat Corporation. The system consists of a field computer, memory module, communications module, and 4 strain gauge modules. Strains are collected in either attended or unattended mode. A second portable computer communicates with the SoMat system to control the data acquisition mode, calibration, initialization, data display, and data downloading. The system is specifically configured to collect highway bridge dynamic strain histories and perform statistical analysis of the data. This is possible due to the modular component arrangement of the system.

The data acquisition system is powered by a 12 volt battery with sufficient power for approximately three weeks of continuous data collection. Data is sampled simultaneously at a sampling rate of 200 Hz. The SoMat field computer executes the rainflow algorithm as the data is sampled, allowing collection of approximately 4.5 billion cycles per channel. Following collection, data is reviewed and downloaded to the portable computer hard drive for storage, processing, and plotting.

WIM statistics are collected with bridge weigh-in-motion (BWIM) equipment from Bridge Weighing Systems, Inc. The system is designed to collect static truck axle weights, axle spacing, GVW, vehicle speed, lane of travel, and arrival time. The system is constructed with a central processing unit, an analog front end capable of eight channels, and data storage. The system is powered by a 12 volt battery. A portable computer is used for communication with the BWIM system.

4. ANALYTICAL PROCEDURE

To study the effect of multiple presence of vehicles on the bridge as well as the effect of vehicle and bridge dynamics, the truck data captured by WIM was used to analytically determine the statistics of the static strain history. These statistics were then compared to the dynamic strain history collected simultaneously with the WIM data. Differences in the strain histories can generally be attributed to a dynamic component and multiple presence occurrences.

A program was devised to read the WIM data files and effectively "drive" the captured truck files analytically across the bridge. As the truck progresses across the bridge in one foot increments, the program analyzes the bridge using the semi-continuum method by Jaeger and Bakht. The resulting girder moments calculated at the same locations where strain transducers were placed are stored and analyzed using the rainflow algorithm to determine strain hysteresis loops.

The semi-continuum method idealizes the bridge girders as an assembly of parallel beams interconnected by the slab as a transverse semi-continuum. This is a special case of the grillage analogy in which the number of transverse beams approaches infinity. The method considers torsional stiffness in both the transverse and longitudinal directions and applies a harmonic analysis of externally applied loads. For this study five harmonics were calculated at each increment of the truck. The transverse elements are discrete and the longitudinal are continuous, thus semi-continuum. The semi-continuum method has been shown to be very accurate with the added advantage over other methods in that the computing requirements are much lower.

Stress histories which are wide band in nature do not allow for simple cycle counting. The cycles are irregular with variable frequencies and amplitudes. Several cycle counting methods are available for the case of wide band and non-stationary processes, each successful to a degree in predicting the fatigue life of a structure. The rainflow method is preferred and is utilized in this study for both the static and dynamic stress histories due to the identification of strain ranges within the variable amplitude and frequency stress histogram that are associated with closed hysteresis loops. (See Committee, Doweling, Wirsching and Shehata)

5. APPLICATION OF DATA TO RESISTANCE MODELS

Fatigue resistance test data has been shown to fit a log-log linear curve of the form:

$$Log\, N = m\, Log\, S + b \qquad (1)$$

where N = number of cycles to failure; S = stress level; b = intercept at S = 1.0. and m = slope of the S-N curve.

Equation (1) reflects behavior for a constant stress range value. Actual bridge components are subjected to varying amplitude stress cycles. Commonly occurring stress histories in fatigue analysis are often categorized as either narrow band or wide band processes. Narrow band processes are characterized by an approximately constant period. Wide band processes are characterized by higher frequently small excursions superimposed on a lower, variable frequency process. For the case of steel girder highway bridges, where the loading is both random and dynamic in nature, the stress histories are wide band.

Loading, and therefore the response of a highway girder bridge, is not a truly stationary process in which a mean and a standard deviation can be calculated, however the effect of passenger vehicles and other low gross weight vehicles may be filtered to extract the stress cycles which contribute to the fatigue damage. By neglecting the very low amplitude stresses the stress cycles producing fatigue damage will be acquired for the system reliability load model

Wide band random fatigue load effect is generally accounted for by utilizing a cumulative damage formula. The most common and widely accepted of the cumulative damage theories is the Palmgren-Miner's hypothesis, or Miner's Rule. Miner's Rule, as does others, assumes a relationship between variable amplitude loading and results of constant amplitude fatigue testing. The damage of each cycle in the complex stress history adds incrementally to the total fatigue damage based on the assumption that the accumulation of fatigue damage proceeds linearly over time for a stationary process. Miner developed a relationship between the fatigue life N (number of load cycles to failure) and eqivalent stress range, S_r,

$$N = A \,/\, S_r \qquad (2)$$

where A = constant depending on category of detail, and

$$S_r = (\, \Sigma\, S_i^3\, \gamma_i)^{1/3} \qquad (3)$$

$$\gamma_i = n_i \,/\, n \qquad (4)$$

where n_i = applied number of load cycles at stress level S_i; n= total number of cycles. The equivalent stress range was calculated for the girders in the tested bridge and are presented in Fig. 6.

6. MEASUREMENT RESULTS

Statistical data is presented in the form of cumulative distribution functions (CDF). This special scale is used to present and compare the critical upper tails of CDF's. The distributions are plotted on the normal probability paper. Construction and use of the normal probability paper is explained in basic textbooks (e.g. Benjamin and Cornell). Only straight lines on normal probability scale correspond to normal distributions. Therefore, the shape of plotted CDF's exhibited by the measurement data can be used for verification of the distribution type (normal, log normal, and others). When plotting the CDF's, it is convenient to use the inverse normal distribution as the vertical scale.

For comparison, the inverse normal distributions of GVW for October 21 and October 22 truck traffic on southbound US23 are shown in Fig. 2. 797 trucks were observed during 3.6 hours on the first day and 998 trucks were observed during 5.6 hours on the second day. Considering only trucks with GVW greater than 10,000 pounds, 502 trucks were observed on the twenty first, averaging 7.4 axles per truck and 680 on the twenty second, averaging 6.9 axles per truck. It is important to note that the legal limit in Michigan is about 150 kips for 11 axle trucks, which is exceeded by a significant number of vehicles.

Strain test results are presented in Fig 3 through Fig. 5. The strain distributions for girders 3 to 9 are presented in Fig. 3 based on dynamic strain collection for a two week period following the WIM data collection. Girder 5 experiences the largest strains. Girders 3, 4, 6, and 7 experience moderate strains, and girders 8 and 9 are the least loaded girders. Equivalent stress, S_r, is plotted in Fig. 6 for the data collected on the twenty-first and second as well as for the two week period following.

Results of the dynamic strain data collection and comparison with the static strain statistics derived by analysis from the WIM results for October 21 and October 22 are shown in Fig. 4 and Fig. 5 respectively. It can be seen that the distribution of test strains is significantly to the right of the static, analytically derived distribution of strains.

Strains were also measured for a one week duration in the central diaphragms to determine the extent of its participation in live load distribution. The diaphragm strains were measured a distance of one foot from the girder center-line at the bottom flange. CDF's for the diaphragm between girders 5 and 6 are presented in Fig. 7. For comparison, the corresponding CDF's for girder 5 and 6 are also plotted in the same figure.

7. CONCLUSIONS

A procedure is presented for evaluation of the fatigue load spectra. The measurements are carried out on a steel composite girder bridge in Southeast Michigan. The load model is derived on the basis of strain records and the rainflow algorithm.

Fatigue load spectra are determined for individual girders. The resulting stress spectra are considerably different and are dependent on girder position with respect to roadway lanes. These differences verify component-specific live load spectra. This observation is important for the system reliability load model.

Effective stress range values are calculated for girders in the considered bridge. There is a considerable difference depending on girder location. The dynamic and multiple presence of vehicles has a significant effect on the resulting distribution of strains.

Stresses were measured in diaphragms. The observed values are much smaller than girder stresses which indicates that diaphragms participate to only a small extent in the distribution of live load forces.

8. ACKNOWLEDGMENTS

The presented research has been sponsored by the Michigan Department of Transportation (MDOT) and The Great Lakes Center for Truck Transportation Research (GLCTTR) which is gratefully acknowledged. The authors thank Glenn Bukoski and Leo DeFrain of MDOT and Tom Gillespie of GLCTTR for their support. Thanks are also due to Tadeusz Alberski for his generous assistance and suggestions in conducting the field measurements.

REFERENCES

Benjamin, J.R. and Cornell, C.A., *Probability, Statistics, and Decision for Civil Engineers*, McGraw Hill Book Company, New York, 1970.

Committee on Fatigue and Fracture Reliability of the Committee on Structural Safety and Reliability of the Structural Division, (1982), "Fatigue Reliability: Variable Amplitude Loading," *Journal of the Structural Division*, ASCE, Vol. 108, No. ST1, January 1982, pp. 47-69.

Dowelling, N.E., (1972), "Fatigue Failure Predictions for Complicated Stress-Strain Histories," *Journal of Materials*, JMLSA, Vol. 7, No. 1, March 1972, pp. 71-87.

Jaeger, L.G. and Bakht, B., *Bridge Analysis by Microcomputer,* McGraw Hill Book Company, New York, 1989.

Laman, J.A. and Nowak, A.S., (1993), "Live Load Spectra for Bridge Evaluation," Symposium on Practical Solutions for Bridge Strengthening and Rehabilitation, DesMoines Iowa, April 1993.

Wirsching, P.H. and Shehata, A.M., (1977), "Fatigue Under Wide Band Random Stresses using the Rainflow Method," *Journal of Materials and Technology*, Trans of ASME, July 1977, pp. 205-211.

FIGURES

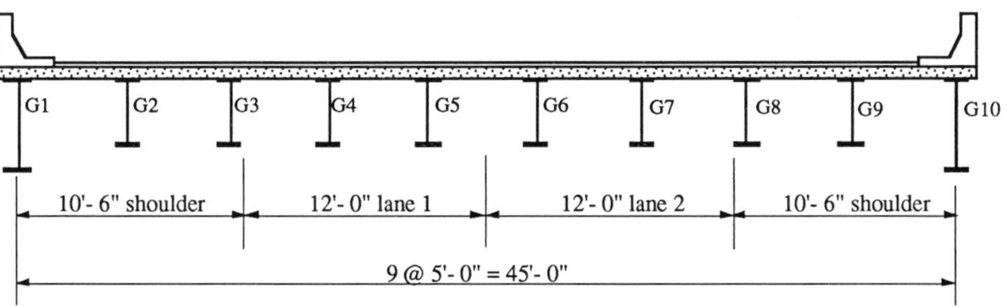

SECTION LOOKING NORTH

Fig. 1, Section of Entrance Span.

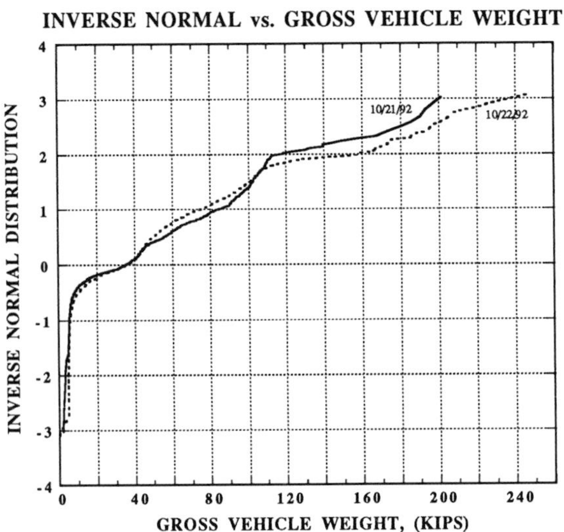

Fig. 2, Inverse Normal Distribution vs. GVW.

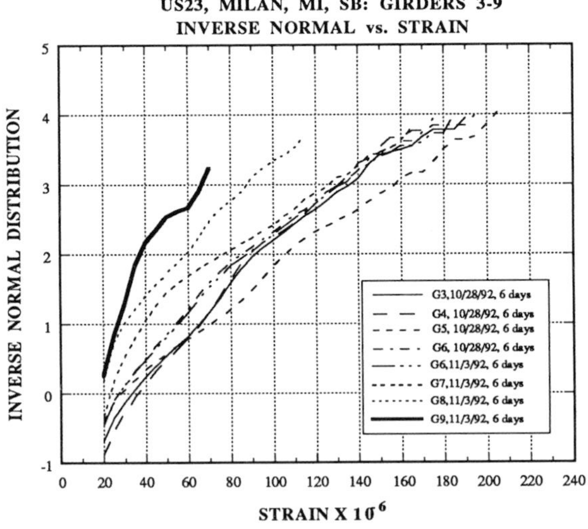

Fig. 3 , Inverse Normal Distribution vs. Strain - Girders 3 - 9.

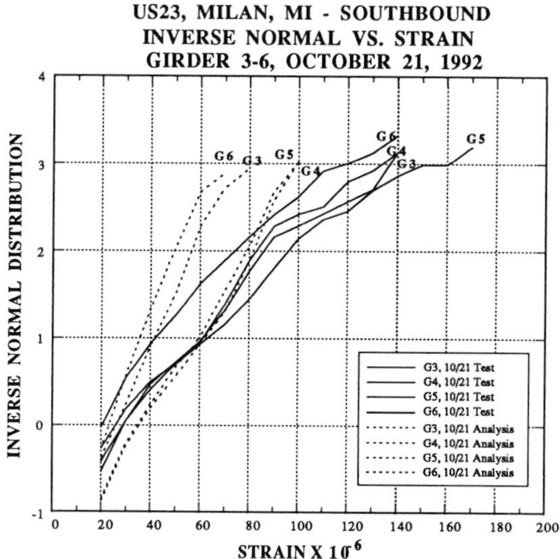

Fig. 4, Inverse Normal Distribution vs. Strain.

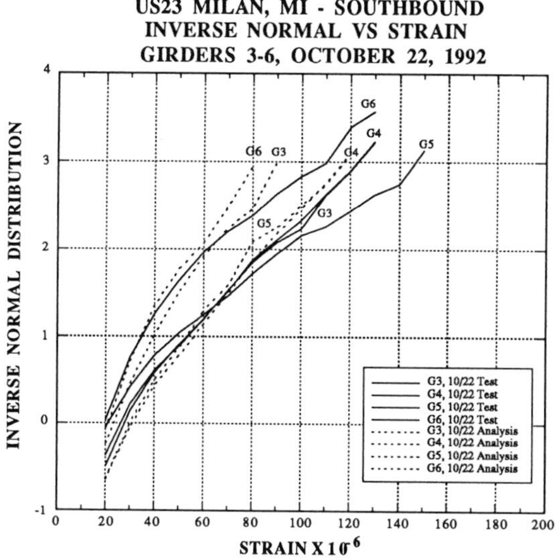

Fig. 5, Inverse Normal Distribution vs. Strain,.

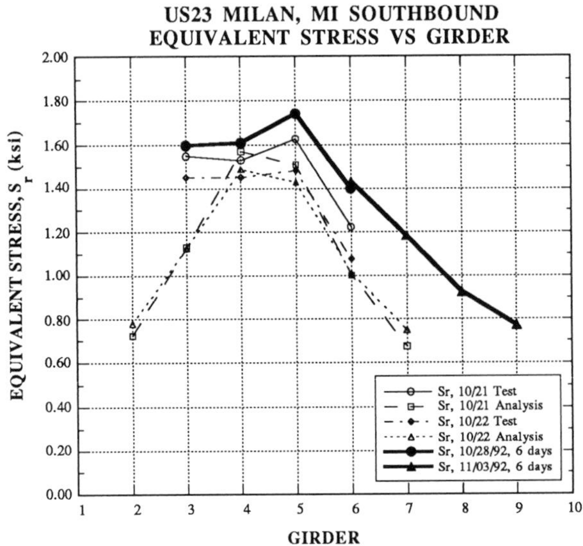

Fig. 6, Equivalent Stress vs. Girder.

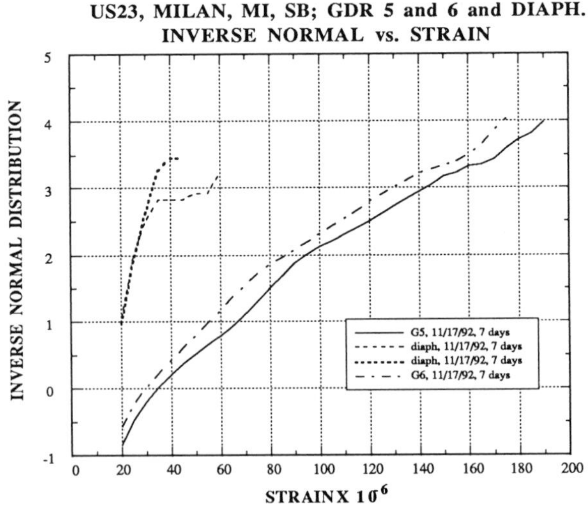

Fig. 7, Inverse Normal Distribution vs. Strain.

Reliability and Optimization of Structural Systems, V (B-12)
P. Thoft-Christensen and H. Ishikawa (Editors)
Elsevier Science Publishers B.V. (North-Holland)

Estimation of the Failure Probability of Building Based on Observed Earthquake Records

M. Matsubara, M. Takeda, Y. Kai

Nuclear Power Division, Shimizu Corporation, Tokyo, 105-07, Japan

Abstract

To estimate the failure probability of a Nuclear Power Plant in Japan, authors have established the seismic Probabilistic Risk Analysis (PRA) procedure. A modification to the method is suggested in this report, and the difference between the results of two kinds of PRA methods is studied. Although the previous method and the modified method show little difference in the evaluation of the annual flexural cracking frequency of the ordinary building, the previous method gives an irregularly larger result than the modified method in the evaluation of the annual flexural cracking frequency of the base-isolated building.

Keyword codes : I.6.1; I.6.3; I.6.4
Keywords : Simulation and modelling, Simulation Theory ; Application ; Model Validation and Analysis

1. Introduction

To estimate the failure probability of a Nuclear Power Plant in Japan, authors have established the seismic Probabilistic Risk Analysis (PRA) procedure [1],[2],[3]. In our recent study, it is suggested that some variations in the seismic hazard analysis and the fragility analysis may correlate in the previous PRA procedure. This may cause improper failure probability as a result. A modified PRA method is suggested in this study, and the difference between the results of two kinds of PRA methods is compared. The variation β consists of the variations caused by uncertainty (βu) and by the randomness (βr). The βr is the focus of this study.

2. Modification of the PRA method

In the previous method, the variation of the attenuation relationship was obtained by the variation of the maximum acceleration (βa_{max}). The variation of the spectral was obtained by the variation of normalized response spectrum, which is the response spectrum normalized by its maximum acceleration ($\beta S_A/a_{max}$).

From our recent study [4], the statistical characteristics of input motion and that of soil property are highly correlated. However the variation of input motion is only considered here and that of soil property is omitted. The sum of the variation of spectral shape and that of the maximum acceleration is larger than the variation of response spectra by the dynamic analysis using the input motion sample directly. Therefore the variation of the spectral shape and the variation of the maximum acceleration must be considered simultaneously in the fragility analysis. For this purpose, the variation of the attenuation relationship is divided into two kinds of variation: the variation of seismic energy dissipation along the way from the seismic center to the site and the variation of response of the ground near the site. According to ref. [5], the variation of the peak acceleration is greater than that of the square root of the total energy of the earthquake records ($\sqrt{E_T}$). E_T is obtained as below.

$$E_T = \int_0^T (a^2_{NS}(t) + a^2_{EW}(t) + a^2_{UD}(t))dt \quad (1)$$

where $a_{NS}(t)$:acceleration of NS component
$a_{EW}(t)$:acceleration of EW component
$a_{UD}(t)$:acceleration of UD component

This difference is assumed to mean the variation of response of the ground. The variation of the $\sqrt{E_T}$ (β_{ET}) is assumed to stand for the variation of the seismic energy dissipation. The variation of the response of the ground (β_R) can be obtained as the difference of the variation of response spectrum (β_{SA}) and that of the seismic energy dissipation (β_{ET}).

$$\beta_R = \sqrt{\beta_{SA}^2 - \beta_{ET}^2} \quad (2)$$

3. Statistic analysis of ground motion records

3.1 Outline of the observation system of the ground motion

Two test buildings were constructed side by side on a relatively hard loam layer with gravel [6]. These buildings are actual 3-story reinforced concrete structures and have exactly the same dimension. One has the ordinary foundation and the other has the base-isolated foundation. Fig. 1 shows the plan and elevation of the test buildings. The base-isolated building has 6-rubber bearings and 12-oil dampers.

3.2 Earthquake observation

An earthquake observation array system is installed to record the earthquake response of the building and the ground. In this array, there are 2 observation points on the ordinary building, 5 observation points on the base-isolated building and 4 points in the surrounding ground. Each of 11-points has 2 or 3 components, (X, Y) or (X, Y, Z), with a total of 31 observed components. Observation points in the array are also shown in Fig. 1. Location of the test building and the observation point of ground are shown in Fig. 2.

Forty-nine earthquakes are observed by the array from 1986/June/24 to 1990/July/28. The characteristics of these records are shown in Figs. 3 and 4.

3.3 Statistic analysis result

With the above records, five kinds of seismic indices are studied by the regression analysis. These indices include maximum acceleration (a_{max}), square root of total energy ($\sqrt{E_T}$), response spectrum (S_A), response spectrum normalized by the square root of its total energy ($S_A/\sqrt{E_T}$) and the response spectrum normalized by its maximum acceleration (S_A/a_{max}). Basic form of the regression formula is as follows.

$$\log_{10}A = aM + b\log_{10}X + c \quad (3)$$

where
A : the earthquake data
M : the magnitude of the earthquake
X : the focal distance of the earthquake
a, b, c : the constant coefficients obtained from regression analysis

According to the ref. [2], we introduced the segmentation method by which a direct S-wave segment of the record is identified, instead of employing the usual method which takes the whole duration of a ground motion into account. The duration time of the direct S-wave (T_D) is obtained from the following formula.

$$\log_{10}T_D = 0.31M - 0.774 \quad (4)$$

The regression formula for a_{max} is

$$\log_{10}a_{max}=0.571M-1.61\log_{10}X+1.057 \quad (5)$$

and the βa_{max} is 0.553. EW component and NS component of each record are treated as independent data in this analysis.

The regression formula for $\sqrt{E_T}$ is

$$\log_{10}\sqrt{E_T}=0.686M-1.58\log_{10}X+0.472 \quad (6)$$

and the βE_T is 0.516.

The coefficients (a, b, c) and the βr of the regression formula for S_A, $S_A/\sqrt{E_T}$ and S_A/a_{max} are shown in Figs. 5, 6 and 7. EW component and NS component of each record are treated as independent data. Following the discussion in the previous section, the βE_T (=0.516) is used as the variation of the attenuation formula. The variation of the ground response (β_R) is obtained from the formula (2) in Fig. 8. This figure also shows the β_R evaluating in ref. [4]. The quantity of both variation is almost same. This means that the resultant variation of formula (2) can be treated as β_R.

4. Hazard analysis

Hazard analysis is performed both by the previous method and the modified method. The evaluating site is Sendai in Japan where the earthquake observation used in this report is conducted. The seismic source model is shown in Fig. 9 [3].

βa_{max} and $\beta S_A/a_{max}$ are used in the previous PRA method. Fig. 10(a) is the hazard curve and the expected spectral shape according to the method. The expected spectral shape is obtained using the regression formula for S_A/a_{max}.

As the modified PRA method, βE_T is used as the variation of the attenuation formula. β_R is used as the βr of the response. The hazard curves for the expected maximum ground motion (a_{max}, Fig. 10(b)) , $\sqrt{E_T}$ (Fig. 10(e)) and the expected response of the oscillator with the dominant period of the structures (S_A, Fig. 10(c), (d)) are evaluated. The dominant period of the ordinary building is 0.3 sec. and that of the base-isolated building is 1.6 sec.

5. Fragility analysis

Fragility analysis is performed to the test buildings of the observation system. Flexural crack of any column is regarded as the failure mode.

5.1 Expected response

The analysis model of two buildings and specifications of the model are shown in Fig. 11. The variation of the soil property is accounted for in the previous PRA method. From our recent study [4], it is suggested that the variation of the expected spectral shape includes that of the soil property. To clarify the difference of hazard evaluation method, we do not consider the variation of the soil property in this report.

Tables 1 and 2 compare the expected response shear force of each story.

The previous method and the modified method evaluate similar median response value. The βr of the response calculated based on S_A is as same as that based on a_{max} in the case of the ordinary building. This is because the variation of the both spectra is the same. Even though the variation of spectral shape at 1.60 sec. is different between the expected spectral shape for the response of the oscillator and the expected spectral shape for the maximum ground motion (0.20 and 0.30 respectively), the βr of the response of the base-isolated building has not changed very much between the two. This is because the transfer function of the base-isolated building is wide bound as shown in Fig. 12.

5.2 Expected capacity

The expected capacity of each story on both buildings is evaluated on the shear force of columns. The evaluation error of this formula is assumed as log-normal standard deviation.

The material properties are shown on Table 3. The section of the column and evaluation formula of flexural crack moment are shown in Fig. 13. Relationship between story maximum moment and the spring force at each story is calculated from the static analysis against the lateral rectangular load on each story. The inflection point is assumed to be in the middle of the story. The expected capacity of each building is summarized in Table 4.

5.3 Fragility Curve

The expected ground acceleration capacity by both the previous method and the modified method is shown in Table 5 for each story. The expected response capacity for each dominant period and the expected total energy capacity is also shown in Table 5. System analysis is performed to obtain the fragility curve for the failure probability of the buildings. The failure of each story is assumed to be perfectly correlated. Figs. 14 and 15 show the fragility curve of the ordinary building and the base-isolated building. Each figure compares the previous method and the modified method. Both methods give the same result for the ordinary building. The previous method, however, gives larger randomness to the fragility of the base-isolated test building than to the ordinary building.

6. Failure probability

Fig. 16 shows the expected annual frequency of flexural cracking of the building. For the ordinary building, all cases show a good agreement. This building is very stiff and the maximum ground acceleration can represent the effective input force to the building. On the other hand, the annual failure frequency derived by the previous method is much larger than that modified procedure. The difference of seismic index, such as a_{max}, $\sqrt{E_T}$ or S_A, does not have large effect on the annual failure frequency. The variation of the expected spectral shape for the oscillator with the dominant period of 1.60 sec. was very different from other cases. This difference, however, does not cause any effect on the annual failure frequency.

7. Conclusion

The modified PRA method is proposed considering the correlation between the input motion and dynamic characteristics, and the correlation between the maximum ground acceleration and the spectral shape. The previous method and the modified method show little difference in the evaluation of the annual flexural cracking frequency of the ordinary building. The previous method gives an irregularly larger result than the modified method in the evaluation of the annual flexural cracking frequency of the base-isolated building, .

Acknowledgments

The earthquake observation referred to in this study was carried out through the joint research of Tohoku University and Shimizu Corporation. The authors would like to express their appreciation to all members of the project team. We also would like to express our gratitude to Mr. M. Mizutani, TEPSCO; M. Saruta, SRI; and Mr. T. Okumura, Mr. H. Katukura, ORI; for their valuable comments and suggestions.

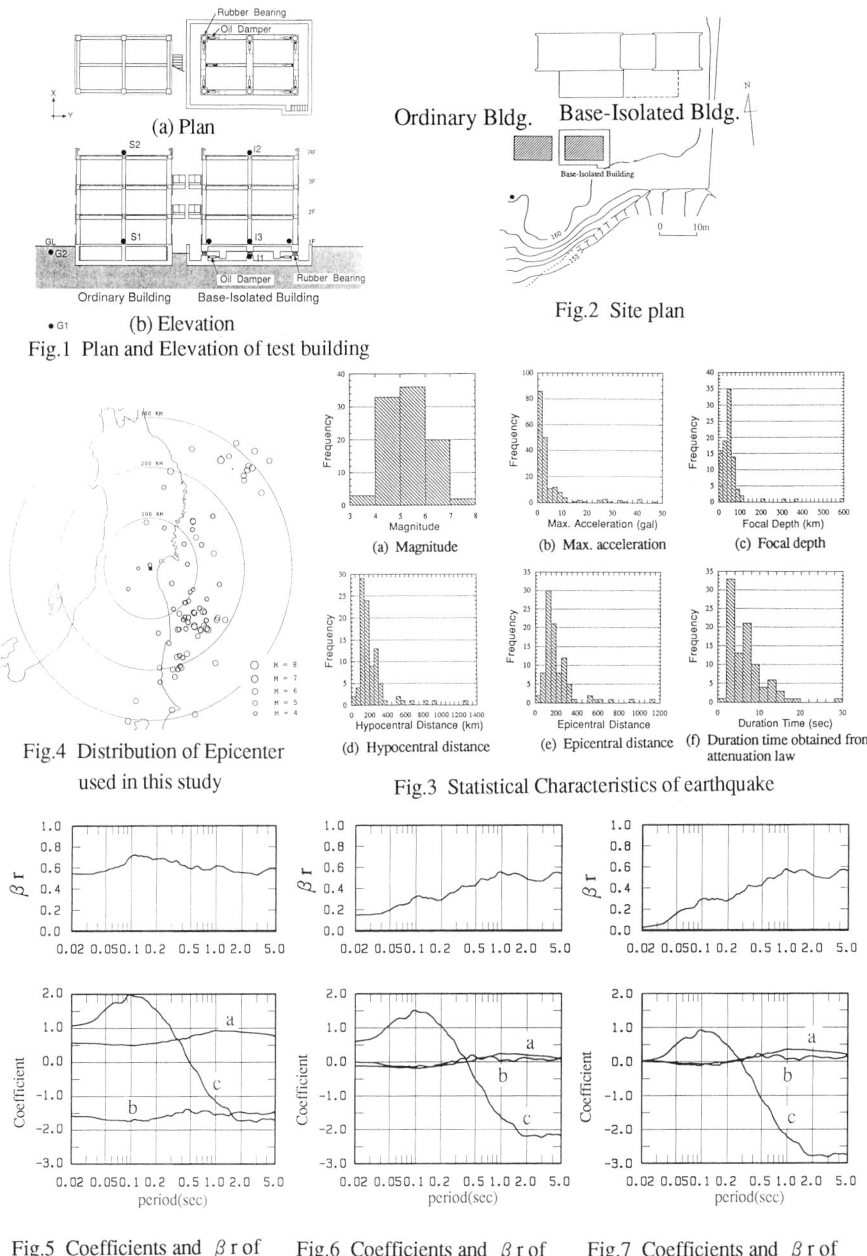

Fig.1 Plan and Elevation of test building

Fig.2 Site plan

Fig.3 Statistical Characteristics of earthquake

Fig.4 Distribution of Epicenter used in this study

Fig.5 Coefficients and β r of Regression Formula(S_A)

Fig.6 Coefficients and β r of Regression Formula($S_A / \sqrt{E_T}$)

Fig.7 Coefficients and β r of Regression Formula(S_A / a_{max})

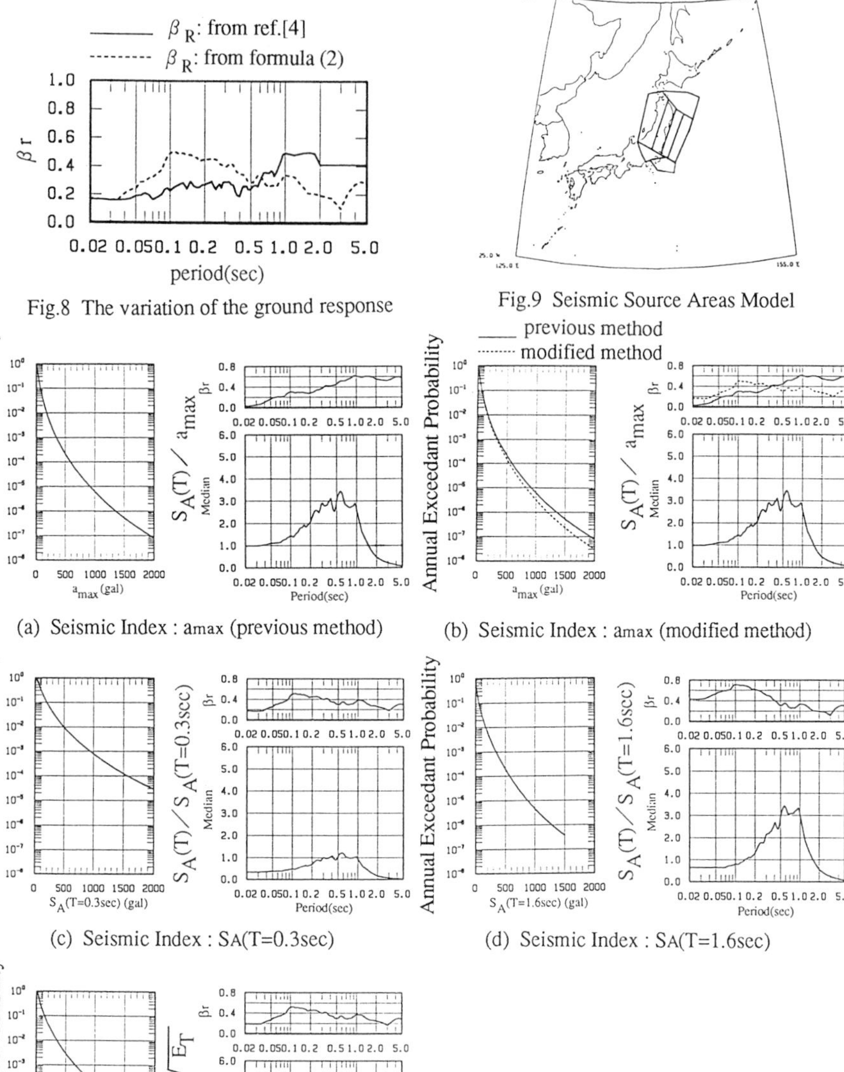

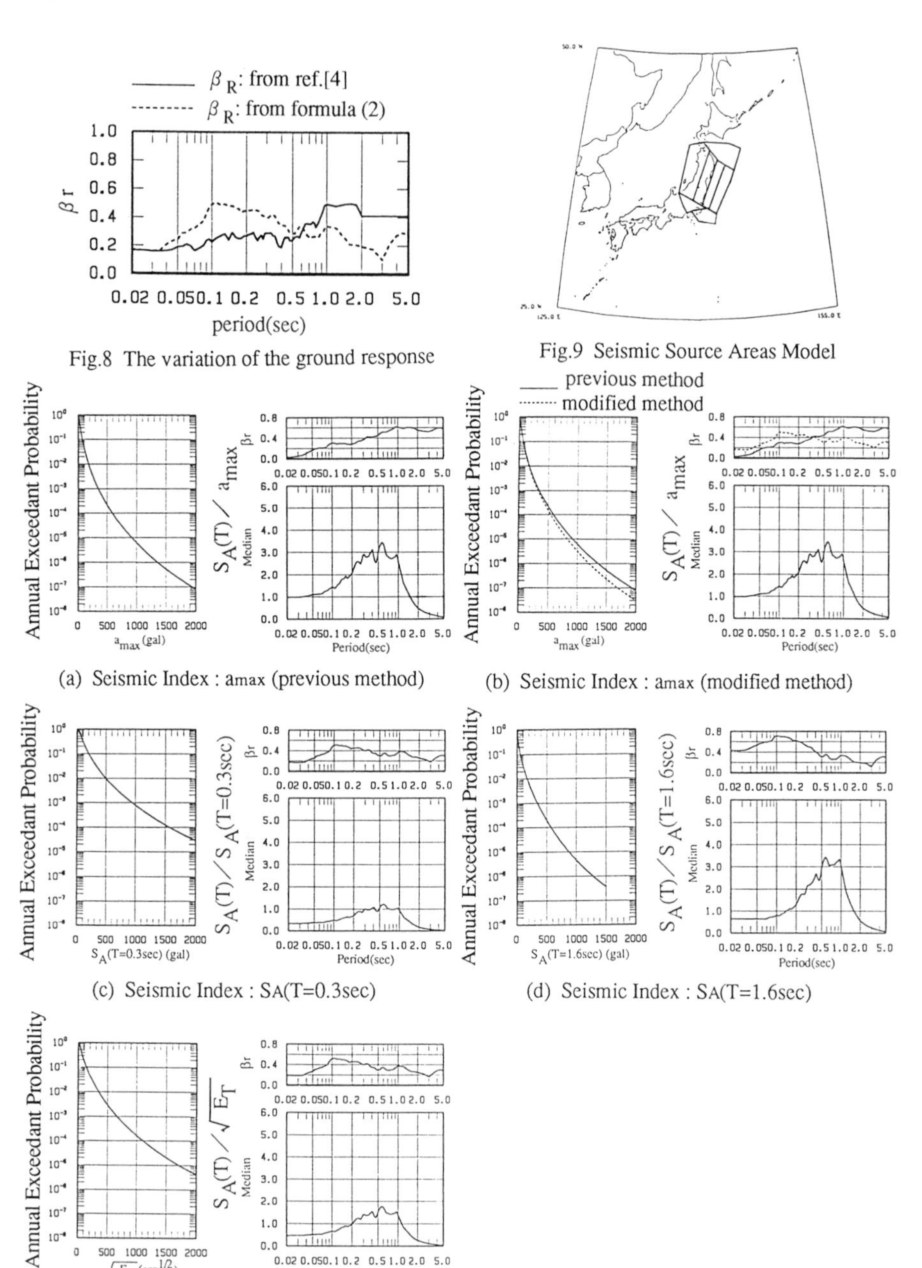

Fig.8 The variation of the ground response

Fig.9 Seismic Source Areas Model

(a) Seismic Index : amax (previous method)

(b) Seismic Index : amax (modified method)

(c) Seismic Index : SA(T=0.3sec)

(d) Seismic Index : SA(T=1.6sec)

(e) Seismic Index : $\sqrt{E_T}$

Fig.10 Hazard Curve and Expected Spectral Shape

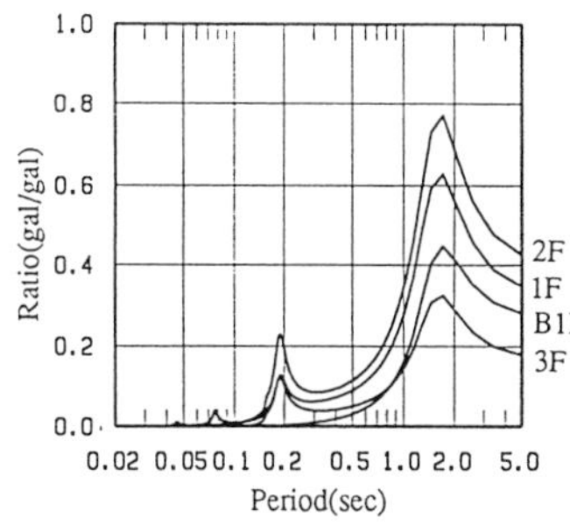

Fig.12 Transfer Function of Shear Force of Base-Isolated Building

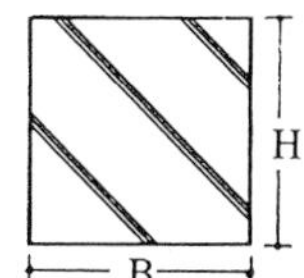

Q=2Mc/h

- Q : flexural crack shear strength
- h : story height
- Mc : flexural crack moment
  $Mc=(BH^2/6)\times(cft+\sigma 0)$
- B : width of column(=500mm)
- H : depth of column(=500mm)
- cft : tensile strength of concrete
- $\sigma 0$: axial stress of column

Fig.13 Section of Column

Table 1 Expected Shear Response of Ordinary Building

	amax(previous)			amax(modified)			SA(T=0.30)			$\sqrt{E_T}$		
	Median (t)	βr	βu	Median (t)	βr	βu	Median (t)	βr	βu	Median (t)	βr	βu
3F	0.281	0.34	0.34	0.281	0.41	0.34	0.096	0.41	0.34	0.135	0.41	0.34
2F	0.366	0.35	0.34	0.367	0.41	0.34	0.126	0.41	0.34	0.177	0.41	0.34
1F	0.368	0.36	0.34	0.368	0.41	0.34	0.127	0.41	0.34	0.178	0.41	0.34

Table 2 Expected Shear Response of Base-Isolated Building

	amax(previous)			amax(modified)			SA(T=1.60)			$\sqrt{E_T}$		
	Median (t)	βr	βu	Median (t)	βr	βu	Median (t)	βr	βu	Median (t)	βr	βu
3F	0.049	0.56	0.34	0.050	0.34	0.34	0.052	0.31	0.34	0.026	0.32	0.34
2F	0.079	0.56	0.34	0.080	0.34	0.34	0.084	0.30	0.34	0.041	0.31	0.34
1F	0.091	0.57	0.34	0.092	0.33	0.34	0.098	0.29	0.34	0.048	0.31	0.34
Is. Fl.	0.118	0.60	0.34	0.120	0.32	0.34	0.131	0.27	0.34	0.062	0.29	0.34

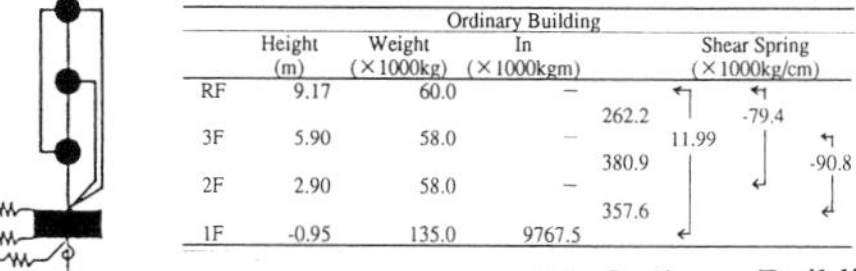

Ordinary Building						
	Height (m)	Weight (×1000kg)	In (×1000kgm)	Shear Spring (×1000kg/cm)		
RF	9.17	60.0	–			
				262.2		-79.4
3F	5.90	58.0	–		11.99	
				380.9		-90.8
2F	2.90	58.0	–			
				357.6		
1F	-0.95	135.0	9767.5			

Ordinary building			
		impedance	damping
Side Spring	Sway	2.11×10^5kg/cm	1.50×10^3kg/cm/sec
	Sway	3.32×10^5kg/cm	1.86×10^3kg/cm/sec
Bottom Spring	Sway	3.34×10^6kg/cm	2.96×10^3kg/cm/sec
	Rocking	3.30×10^{11}kgcm/rad	7.56×10^7kgcm/rad/sec

(a) Ordinary Building

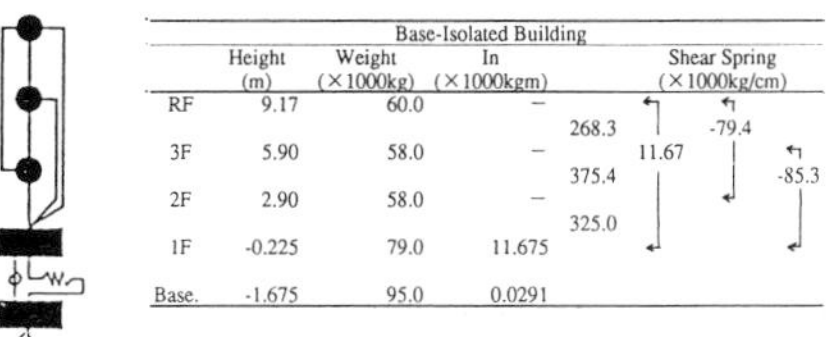

Base-Isolated Building						
	Height (m)	Weight (×1000kg)	In (×1000kgm)	Shear Spring (×1000kg/cm)		
RF	9.17	60.0	–			
				268.3		-79.4
3F	5.90	58.0	–		11.67	
				375.4		-85.3
2F	2.90	58.0	–			
				325.0		
1F	-0.225	79.0	11.675			
Base.	-1.675	95.0	0.0291			

Base-Isolated building			
		impedance	damping
Side Spring	Sway	2.11×10^5kg/cm	3.28×10^3kg/cm/sec
	Sway	3.32×10^5kg/cm	4.32×10^3kg/cm/sec
Bottom Spring	Sway	4.61×10^6kg/cm	5.62×10^3kg/cm/sec
	Rocking	9.86×10^{11}kgcm/rad	3.51×10^7kgcm/rad/sec

Specification of isolation device model

Horizontal	$Kh=4.48\times10^3$kg/cm	C=680kg/cm/sec
Rotational	$Kr=2.65\times10^{11}$kgcm/rad	h=0.02

(b) Base-Isolated Building

Fig.11 Response Model

Table 5 Expected Flexural Crack Capacity

(a) Ordinary Building

	amax(previous)			amax(modified)			SA(T=0.30)			$\sqrt{E_T}$		
	Median (gal)	βr	βu	Median (gal)	βr	βu	Median (gal)	βr	βu	Median ($\sqrt{erg}$)	βr	βu
3F	106.8	0.35	0.35	106.8	0.42	0.35	311.6	0.42	0.35	221.5	0.42	0.35
2F	95.1	0.36	0.35	95.1	0.42	0.35	277.3	0.42	0.35	197.0	0.42	0.35
1F	82.3	0.36	0.35	82.3	0.41	0.35	240.1	0.41	0.35	170.5	0.41	0.35

(b) Base-Isolated Building

	amax(previous)			amax(modified)			SA(T=1.60)			$\sqrt{E_T}$		
	Median (gal)	βr	βu	Median (gal)	βr	βu	Median (gal)	βr	βu	Median ($\sqrt{erg}$)	βr	βu
3F	574.3	0.56	0.35	567.4	0.35	0.35	550.8	0.32	0.35	1106.4	0.33	0.35
2F	421.4	0.57	0.35	416.5	0.34	0.35	398.9	0.31	0.35	810.5	0.32	0.35
1F	385.1	0.58	0.35	380.2	0.34	0.35	358.7	0.30	0.35	737.0	0.32	0.35
Is. Fl.	758.5	0.60	0.34	746.8	0.32	0.34	685.0	0.27	0.34	1439.6	0.29	0.34

Table 3 Properties of Materials

Fc (kg/cm2)	
Median	Lognormal Standard Deviation
317.10	0.14

Table 4 Expected Flexural Crack Shear Strength

	Ordinary Building			Base-Isolated Building		
	S (t)	βr	βu	S (t)	βr	βu
3F	29.94	0.07	0.10	28.38	0.07	0.10
2F	34.96	0.07	0.10	33.32	0.07	0.10
1F	30.38	0.07	0.10	35.04	0.07	0.10
Is. Fl.	–	–	–	89.60	0.00	0.00

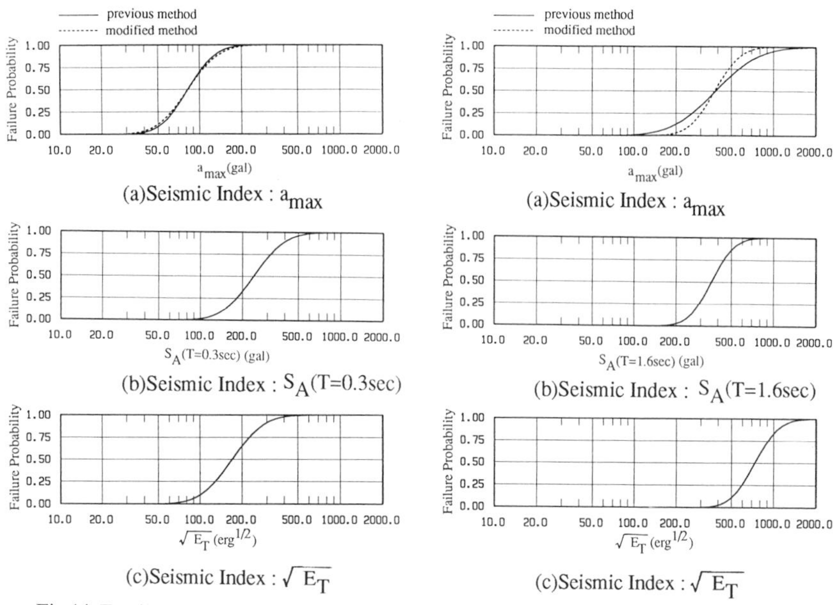

(a)Seismic Index : a_{max} (b)Seismic Index : S_A(T=0.3sec) (c)Seismic Index : $\sqrt{E_T}$

(a)Seismic Index : a_{max} (b)Seismic Index : S_A(T=1.6sec) (c)Seismic Index : $\sqrt{E_T}$

Fig.14 Fragility Curve of Ordinary building Fig.15 Fragility Curve of Base-Isolated building

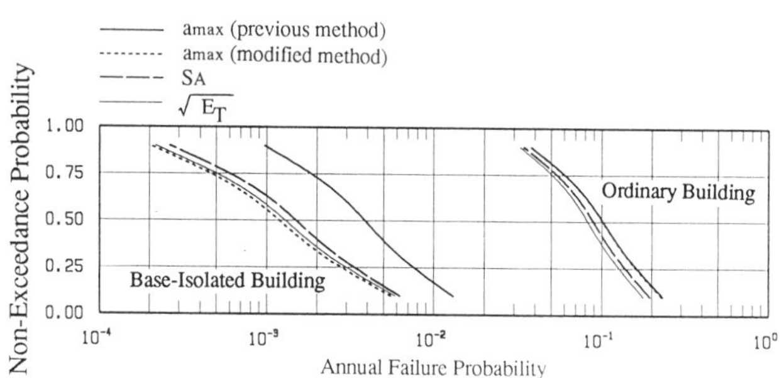

Fig.16 Expected Annual Frequency of Flexural Cracking of Building

References

[1] M.Takemura et al., A Seismic PRA Procedure in Japan and its Application to a Building Performance Safety Estimation Part 1 Seismic Hazard Analysis, ICOSSAR, Vol. 1 pp621-628, 1989

[2] M.Takeda, Y.Kai, M.Mizutani, A Seismic PRA Procedure in Japan and its Application to a Building Performance Safety Estimation Part 2 Fragility Analysis, ICOSSAR, Vol. 1 pp62-636, 1989

[3] M.Shinozuka, M.Mizutani, M.Takeda, Y.Kai, A Seismic PRA Procedure in Japan and its Application to a Building Performance Safety Estimation Part 3 Estimation of Building and Equipment Performance Safety, ICOSSAR, Vol. 1 pp637-644, 1989

[4] M.Matsubara, M.Takeda, Y.Kai, Study on the method of seismic PRA model by utilization of observed earthquake data, 10th WCEE, Vol. IV, pp521-527, 1992

[5] M.Mizutani, S.Fukushima, Y.Akao, H.Katukura, Modeling of Seismic Hazards for Dynamic Reliability Analysis, 12th SMiRT, 1993

[6] T.Itoh, K.Ishi, Y.Ishikawa, T.Okumura, Development of seismic hazard analysis in Japan, 9th SMiRT

Reliability and Optimization of Structural Systems, V (B-12)
P. Thoft-Christensen and H. Ishikawa (Editors)
Elsevier Science Publishers B.V. (North-Holland)

Basic Study on Development of Efficient Integration Procedure and Its Application to First-Passage Problem

S. Matsuho, W. Shiraki

Department of Civil Engineering, Tottori University
Tottori-shi, 680, Japan

Abstract

In this study, two types of efficient procedures for numerical integration are developed. One is based on efficient Monte-Carlo simulation technique and the other on lattice discretization method. Fathermore, first-passage problem of the stochastic process is considered as an application of these procedures to illustrate the applicability to various engineering problems.

Keyword Codes: I.1.2; I.1.4; I.6.3
Keywords: Algebraic Manipulation, Algorithms; Applications;
Simulation and Modeling, Applications

1. Introduction

In general, almost models can be described by differential equation, which can be solved by integration. And if we use precise mathematical model in analysis of the problem, such model is made up of many parameters and has necessary of multiple integral to solve the problem. Therefore, an efficient calculation procedure of multiple integral is needed. In this study, two efficient procedures of numerical integration are developed. One is based on efficient Monte-Carlo simulation technique and the other on lattice discretization method.

The former method is hybrid one making use of the Haber's method and the method separating principal part from integrand. Integral of the separated principal part is achieved analytically. The latter method is based on Chebyshev interpolatory quadrature. In calculation, first, each parameter of the integrand is discretized. Second, integrand between values of discretized integrand is interpolated by Chebyshev formula. Last, integral is carried out by expectation operation of integrand at the points sampled from all the discretized points of each parameter. Efficiency of these methods is proven by numerical examples.

In the latter half of our paper, the first-passage problem of stochastic process is considered as an application of the above-mentioned efficient procedures. First step of calculation of the probability that response process never exceeds a given level is to discretize the parameter of response process. Second step is to decide the joint probability density function of the magnitudes of the response process at all the discretized points. Last step is to calculate multiple integral of the joint probability density function over safety region. As a simple numerical example, the first-passage problem of response stochastic process of a single degree of freedom (SDOF) elastic system is considered. This SDOF system is assumed to be subjected to a ground acceleration by earthquake that is modeled by the Kanai-Tajimi spectral formula. And response process of this SDOF system is also assumed to be stationary for simplicity. Random response analysis can be easily gotten by using the correlation theory. Therefore, the calculation of the first-excursion probability by the preceding three steps can be carried out using the proposed multiple integration methods. Effectiveness of this method is shown by making a comparison between the result and analytical solution.

2. Hybrid Method of Separating Method of Integrand from Principal Part and Haber's Method

In this paper, efficient calculation means reducing variance of integral estimation. In this chapter, several general methods [1,2] of efficient integral calculation are introduced briefly. Furthermore, more efficient integration method is presented using these methods.

2.1. Stratified Sampling Method

In this method, domain of integration [a,b] is divided into m small domains $[a_1,b_1]$, $[a_2,b_2],\cdots,[a_m,b_m]$ (where $a=a_1<b_1=a_2<b_2=\cdots=a_m<b_m=b$). In each small domain, integral is carried out by using crude Monte-Carlo simulation technique. Then, integral I of integrand f(x) expressed by Eq.(1) is approximately provided as Eq.(2).

$$I=\int_a^b f(x)\,dx \qquad (b>a) \tag{1}$$

$$I=\sum_{j=1}^{m}\int_{a_j}^{b_j} f(x)\,dx \fallingdotseq I_1=\sum_{j=1}^{m}\frac{b_j-a_j}{N_j}\sum_{i=1}^{N_j} f(\xi_i^{(j)}) \tag{2}$$

In Eq.(2), $\xi_i^{(j)}$ $(i=1,\cdots,N_j)$ is random number distributed uniformly within j-th small domain $[a_j,b_j]$ and N_j is number of random numbers generated in j-th domain $[a_j,b_j]$. Each small domain of integration $[a_j,b_j]$ is selected so that change of integrand f(x) in $[a_j,b_j]$ may be small and $N_j \geqq (b_j-a_j)N/(b-a)$ $(N=N_1+\cdots+N_m)$.

2.2. Method of Antithetic Variates

A certain random variable X which has expected value I is considered. If another random variable X' has same expected value I and has strong antithetic correlation with X, expected value of new variable (X+X')/2 is also same to expected X (or expected X'). But variance of estimation for expected value of (X+X')/2 can be reduced more than in the case of X (or X'). In evaluating Eq.(1) whose domain of integration is [a,b]=[0,1], expected value of $\zeta=[f(\xi)+f(1-\xi)]/2$ is $E[\zeta]=I$, if integrand f(x) is monotone function of x. Therefore, approximate value I_2 of Eq.(1) is given as follows:

$$I=\int_0^1 f(x)\,dx \fallingdotseq I_2=\frac{1}{2N}\sum_{i=1}^{N}[f(\xi_i)+f(1-\xi_i)] \tag{3}$$

2.3. Haber's Method

This method is hybrid method of the above-mentioned methods (see sections 2.1 and 2.2). In this section, multiple integral of function of vector $\vec{x}$ in the s-dimensional space is considered as shown in Eq.(4) for general discussion.

$$I=\int_0^1\cdots\int_0^1 f(\vec{x})\,d\vec{x} \tag{4}$$

In this method, domain of integration, that is unit length s-dimensional hyper-cube $G_s=\{x \mid 0\leqq x_i\leqq 1,\ i=1,\cdots,n\}$, is divided into $N=K^s$ small domains which are expressed as $A_1,\cdots,A_N$. To generate vector point $\vec{x}_r$ $(r=1,\cdots,N)$ in each small domain of integration, approximate value I_3 of integral I is provided as Eq.(5).

$$I_3 = \frac{1}{N}\sum_{r=1}^{N}\frac{f(\overrightarrow{x_r})+f(\overrightarrow{x_r}')}{2} \tag{5}$$

where $\vec{x}_r' = 2\vec{c}_r - \vec{x}_r$ is symmetric point to centre point $\vec{c}_r$ of small domain A_r. Number of evaluating integrand in each A_r is 2. So, this method is efficient.

2.4. Development of More Efficient Integration Method [3,4]

In this section, we propose hybrid method using Haber's method and the method separating principal part from integrand. Principal part of integrand is one which is able to make integral analytically. For simplicity of illustration, integrand f(x) in the case of single dimension is shown in Fig.1. In this study, shade part expressed by slanted lines in Fig.1 is considered as principal part g(x) of integrand. Division of integration domain [a,b] is performed in accordance with division method in Haber's method. Integral I_4' of f(x)−g(x) is estimated by Haber's method. The sum of Integral I_4' and integral of principal part give precise estimation I_4 of integral I of Eq.(1).

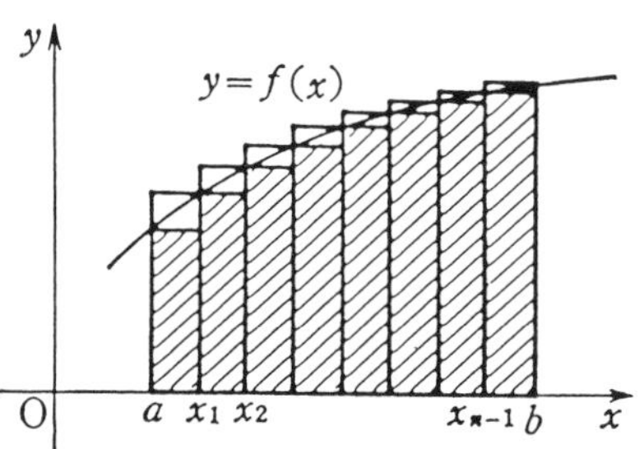

Fig.1 Principal Parts of Integrand and Its Termwise Integration

3. Efficient Method Based on Lattice Discretization Method

In this chapter, lattice discretization method which is one of efficient integration methods is illustrated briefly. Next, modified method based on this method is proposed.

3.1. Lattice Discretization Method [1,4]

In this method, coordinate along each axis in integration domain is divided into n intervals and is discretized into n+1 points. Then, coordinate in k-dimensional integration domain D is discretized into $(n+1)^k$ points, which is called lattice points. Lattice points are indicated by the notation $(x_1^{(i1)}, x_2^{(i2)}, \cdots, x_k^{(ik)})$ where each superscript (i_r) is lattice point number along r-th coordinate axis. Lattice points are also written in the form of Eq.(6).

$$\vec{i} = (i_1, i_2, \cdots, i_k), \quad i_r = 0, 1, \cdots, n \quad (r=1,2,\cdots,k) \tag{6}$$

Assuming that integrand in k-dimensional space can be approximated by arbitrary interpolatory formula as shown in Eq.(7), estimation formula is given by Eq.(8).

$$f(\vec{x}) \fallingdotseq \sum_{\vec{i}} f(x_1^{(i1)}, \cdots, x_k^{(ik)}) L(\vec{i}; \vec{x}) \tag{7}$$

$$I \fallingdotseq \sum_{\vec{i}} f(x_1^{(i1)}, \cdots, x_k^{(ik)}) C(\vec{i}), \quad C(\vec{i}) = \int_D L(\vec{i}; \vec{x}) dx \tag{8}$$

If coordinate of each variable is scaled appropriately, integration domain D can be considered as D=[−1,1]×⋯×[−1,1]. Then, using Chebyshev formula (9) as interpolatory formula of Eq.(7), Eq.(8) is written in the form of Eq.(10).

$$f(x_1, \cdots, x_k) \fallingdotseq \sum_{i1=0}^{n} \cdots \sum_{ik=0}^{n} f(x_1^{(i1)}, \cdots, x_k^{(ik)}) \prod_{r=1}^{k} \prod_{\substack{j=0 \\ \neq ir}}^{n} \frac{x_r - x_r^{(j)}}{x_r^{(ir)} - x_r^{(j)}} \tag{9}$$

$$I \fallingdotseq \sum_{\vec{i}} f(\vec{i}) \prod_{r=1}^{k} \int_{-1}^{1} \prod_{\substack{j=0 \\ \neq ir}}^{n} \frac{x_r - x_r^{(j)}}{x_r^{(ir)} - x_r^{(j)}} \equiv \sum_{\vec{i}} f(\vec{i}) C(\vec{i}) \tag{10}$$

Lattice points $x_r^{(j)}$ in Eq.(10) are given by zero points (see Eq.(12)) of Chebyshev polynomial (Eq.(11)). Furthermore, coefficients $C(\vec{i})$ in Eq.(10) are provided by Eq.(13).

$$T_{n+1}(x) = 2^{-n} \cos[(n+1)\cos^{-1}x] \equiv (x - x^{(0)}) \cdots (x - x^{(n)}) \tag{11}$$

$$x_r^{(j)} = \cos\left(\frac{2j+1}{n+1}\frac{\pi}{2}\right), \qquad (r=1,\cdots,k;\ j=0,\cdots,n) \tag{12}$$

$$C(\vec{i}) = c(i_1) \cdots c(i_k), \quad c(i_r) \equiv \int_{-1}^{1} \prod_{\substack{j=0 \\ \neq ir}}^{n} \frac{x_r - x_r^{(j)}}{x_r^{(ir)} - x_r^{(j)}} \tag{13}$$

Considering the case of $f(x) \equiv 1$ in Eq.(10) yields relationship of Eq.(14).

$$\sum_{\vec{i}} C(\vec{i}) = 2^k, \quad \sum_{i=0}^{n} c(i) = 2 \tag{14}$$

Then, assuming Eq.(15), Eq.(10) is written in the form of Eq.(16).

$$c'(i) = c(i)/2 \ (i=0,1,\cdots,n), \ C'(\vec{i}) = C(\vec{i})/2^k \tag{15}$$

$$I = 2^k \sum_{\vec{i}} C'(\vec{i}) f(\vec{i}), \qquad C'(\vec{i}) = c'(i_1) c'(i_2) \cdots c'(i_k) \tag{16}$$

C' and c' in Eq.(15) satisfy the relationship of Eq.(17).

$$\sum_{\vec{i}} C'(\vec{i}) = 1, \ 0 \leqq C'(\vec{i}) \leqq 1 \ ; \ \sum_{i=0}^{n} c'(i) = 1, \ 0 \leqq c'(i) \leqq 1 \tag{17}$$

Eq.(17) shows that C' can be interpreted as probability which lattice point $(i_1,\cdots,i_k)$ in k-dimensional space takes place. Each coordinate i_r of lattice point $(i_1,\cdots,i_k)$ can be obtained by sampling in accordance with the probability c' independently each other.

3.2. Sampling of lattice points

In the lattice discretization method, integrand f is sampled with probability C' (see Eq.(16)). But, in this procedure, precision of integral estimation is determined by the precision of generated random number with probability C'. In this study, using the idea of the importance sampling, Eq.(16) is written in the form of Eq.(18) where N is number of sampling.

$$I = 2^k \sum_{\vec{i}} N C'(\vec{i}) f(\vec{i}) \cdot \frac{1}{N} \tag{18}$$

Eq.(18) shows that NC'f may be sampled using uniformly distributed random numbers.

4. Numerical Examples of Efficient Integration Methods

Several numerical examples are achieved for checking effectiveness of the above-mentioned methods [3,4]. In this chapter, one of them is introduced. Table 1 shows estimation results of definite integral given by Eq.(19) using various estimation methods.

$$\int_0^1 \cdots \int_0^1 (e^{x_1} e^{x_2} e^{x_3} e^{x_4} - 1)\, dx_1 dx_2 dx_3 dx_4 = 0.0693976 \tag{19}$$

Table 1 Calculation Results of Integral by Various Methods

Crude	Antithetic	Haber	Hybrid	Lattice
.0695476	.0698262	.0693878	.0693993	.0693975

In Table 1, "Crude", "Antithetic" and "Haber" mean estimation results by simulation techniques of crude Monte-Carlo, antithetic variates and Haber, respectively. "Hybrid" and "Lattice" indicate methods proposed in this study. In Haber's and Hybrid methods, unit length hyper-cube of 4-dimensional integration domain is divided into 14641(=11^4) small domains, namely, integration interval of each variable is divided into 11 intervals. In lattice discretization method, each variable is divided into 5 intervals. Therefore, integral is estimated using 1296=$(5+1)^4$ lattice points. It takes about 10 sec to calculate Eq.(19) by Haber's and Hybrid methods using personal computer (NEC 9801RA, CPU i386™DX, 16MHz). In the case of using lattice discretization method, it takes about 1 sec under same condition. These results suggest that the proposed methods are very powerful numerical integration.

5. Application of Proposed Efficient Method to First-Passage Problem

The first-passage problem of stochastic process (stochastic wave or stochastic field) with continuous argument(s) is generally difficult to treat. But, this problem can be solved simply when the parameters are adopted as discretizations and random series model. Moreover, one considers that such the method will have the advantages of extensive applicability and simplicity of formulation, if keeping in one's mind that the excitation records can be obtained in discrete form, or the extensively used numerical analysis method, such as FEM etc., be treated as discretization method, or the analysis of continuous values can also be discretely calculated by using digital computer. In this chapter, the integration methods proposed in chapters 2 and 3 are applied to the first-excursion problem of the discretized response processes. This method can be considered to be applicable to the first-excursion problem of not only stochastic process but also stochastic wave or stochastic field.

5.1. General Formulation of Failure Probability

To calculate the failure probability P_f, the limit state, which decides whether the system is safety or not, should be determined at first. In the case of static problem, the limit state can be given by using the estimation function $g(\vec{x})$ of the system as shown in Eq.(20).

$$g(\vec{x})>0 \quad \text{Safety Region,} \qquad g(\vec{x})\leqq 0 \quad \text{Failure Region} \tag{20}$$

The estimation function $g(\vec{x})$, which is also called limit state function in the reliability theory, is the function of vector $\vec{x}$, whose elements $x_1,\cdots,x_n$ are parameters to control the system. The similar function of limit state can also be considered in the case of dynamic problem as shown in Eq.(21).

$$g(\vec{x},T)>0 \quad \text{Safety Region,} \qquad g(\vec{x},T)\leqq 0 \quad \text{Failure Region} \tag{21}$$

Eq.(21) shows that the system is considered to be safe when the first equation is satisfied during the whole duration T. However, if the limit state function satisfies the second equation of Eq.(21) even once, it may be considered to be dangerous in the sense of the first-excursion. When limit state of the system is defined (see Eqs.(20) and (21)), the failure probability of the system P_f (which is called the probability of the first-excursion (or first-passage) failure in the dynamic case) can be formalized by using the joint probability density function $f_{x1 \cdots xn}(x_1,\cdots,x_n)$ of parameters $x_1,\cdots,x_n$ as shown in Eq.(22) [5].

$$P_f = \int_{g(\vec{x},T) \leq 0} f_{x1,\cdots,xn}(\vec{x})\, d\vec{x} \tag{22}$$

In the dynamic case, Eq.(22) shows that the response process a(t) is discretized in time axis (as shown in Fig. 2) and the joint probability density function $f_{x1 \cdots xn}(x_1,\cdots,x_n)$ of the response values (random variables) at all the discretized time points is integrated over failure domain. Eq.(22) is a strict estimation function of the failure probability and may be extensively used because it may not be influenced by the stationary or non-stationary, linear or non-linear as well as the probability distribution of the problem. However, the problem, that the multiple integral of probability density function of large dimension may be required, are arisen. Traditionally, such kind of analyses may be impossible. However, such problems modeled by Eq.(22) can be solved by using the proposed efficient integration methods.

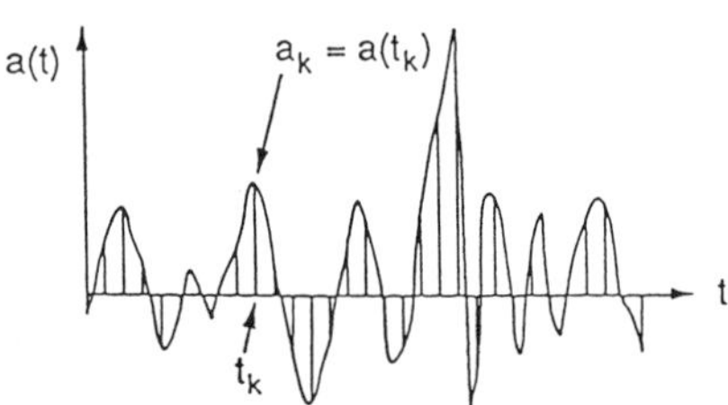

Fig.2 Discretization of A Continuous Process [5]

5.2. Numerical Example of First-Passage Problem

To examine the above-mentioned method, a simple problem is considered. As an example, a linear SDOF (Single Degree of Freedom) system subjected to earthquake excitation is adopted as shown in Eq.(23)

$$\ddot{x}+2\zeta_s\omega_s\dot{x}+\omega_s^2 x=\xi(t) \tag{23}$$

where ζ_s and ω_s are the damping factor and undamped natural circular frequency of the system, respectively. ξ(t) is the excitation of the seismic acceleration which is modeled as white noise process with the intensity of 2πS. However, the real seismic acceleration is non-white noise process in fact. Its (both-sided) spectrum density function, for example, can be given by the Kanai-Tajimi expression of earthquake acceleration process model as shown in Eq.(24).

$$Sa(\omega)=\frac{\left[1+4\zeta_g^2\left(\frac{\omega}{\omega_g}\right)^2\right]S_0}{\left[1-\left(\frac{\omega}{\omega_g}\right)^2\right]^2+4\zeta_g^2\left(\frac{\omega}{\omega_g}\right)^2} \tag{24}$$

So, the spectrum density function S of ξ(t) is taken as S = $S_a(\omega_s)$ in this study. This means that the system of Eq.(23) with small damping factor may absorb the spectrum density of the excitation which located at the domain near the natural circular frequency ω_s because the system is a narrow-band filter. In Eq.(24), S_0 is the spectrum density of the seismic excitation of the basement when it is modeled as white noise. ω_g is the natural circular frequency of the basement. ζ_g is the damping factor of the basement. Based on the above-mentioned conditions, the analysis of random response are achieved using the correlation theory. In the analysis, the displacement response x(t) of the system are assumed to be a Gaussian stationary process with mean zero. The covariance function $K_x(\tau)$ of the response process and the variance D_v of the derivative process v(t)=dx(t)/dt are given by Eqs.(25) and (26), respectively.

$$K_x(\tau)=D_x \exp(-\zeta_s \omega_s |\tau|)(\cos\omega_d \tau + \frac{\zeta_s \omega_s}{\omega_d} \sin\omega_d |\tau|)\ ,\quad \omega_d=\omega_s\sqrt{1-\zeta_s^2} \tag{25}$$

$$D_v = D_{\dot{x}} = K_{\dot{x}}(\tau)\big|_{\tau=0} = -\frac{K_x(\tau)}{d\tau}\bigg|_{\tau=0} = (\omega_d^2+\zeta_s^2\omega_s^2)D_x \tag{26}$$

In Eqs.(25) and (26), τ is a time lag and D_x is a variance of the response process which is given by Eq.(27).

$$D_x=\frac{\pi S_a(\omega_s)}{2\zeta_s \omega_s^3} \tag{27}$$

Using the preceding equations, the probability P_f of the threshold-crossing failure can be calculated. The displacement response spectrum u_s=4.0 (cm) is considered as a level of threshold-crossing. The parameters used in this paper are as follows: S_0=37.801 (cm^2/sec^3), duration of the earthquake T=9.294 (sec), ω_g =20.256(rad/sec), ζ_g =0.321, ζ_s=0.05, ω_s=2π/0.5=12.566 (rad/sec). The calculation results are shown in Table 2. It is assumed

Table 2 Calculation Results of Eq.(22) [Calculation Results for Whole T]

Numb. of Discretization	4	5	6	7	…	Analysis
Failure Probability P_f	.001723	.002248	.003524	.005158	…	.004075

that the first-passage failure takes place both in the positive direction and in the negative direction. In the table, "Analysis" indicates the analytical solution of the probability that extreme value of the response process exceeds a given level (the displacement response spectrum u_s in this case). On the assumption that extreme value of the response process is Rayleigh distributed, this analytical solution can be calculated as shown in Eq.(28).

$$P_f = \int_{g\leq 0} f_R(a)\,da\ , \qquad f_R(a)=\frac{a}{D_x}\exp\left(-\frac{a^2}{2D_x}\right) \tag{28}$$

In Eq.(28), $f_R(a)$ is probability density function of Rayleigh distribution and the excursions of the positive and negative response extremes are considered. Table 2 shows that the estimated values of Eq.(22) increase on the increase of dimensional size of multiple integral (i.e. number of the discretization of T (sec)), or on the decrease of the divided time interval Δt (sec). In this table, calculation results up to 7-dimensional multiple integral are given, because calculation of multiple integral in very large dimensional space is a waste of time. So, twice of calculation result for half the whole duration T(=9.294 (sec)), which is much long than correlation time τ_c (=3.256 (sec) in this example), is adopted as estimation of Eq.(22) in this study and its result is shown in Table 3. Estimation of Eq.(22) in Table 3 looks like to converge to P_{f2}≒0.008 with the increase of number of discretization. Three times of calculation result for duration T/3 (=3.098 (sec)) is P_{f3}=0.009513, whose dimensional size of the multiple integral is 8, for reference.

Table 3 Calculation Results of Eq.(22) [Twice of Calculation Results for T/2]

Numb. of Discretization	5	6	7	8	...
Failure Probability P_f	.004482	.006517	.007773	.007956	...

6. Summary and Conclusions

In analyzing engineering problems, we frequently face the problem that the multiple integral are required. In this paper, an efficient integration methods were developed for solving such problem. Its efficiency and effectiveness were shown through the numerical examples. The results showed that, based on the proposed efficient integration methods, the calculation of multiple integral can be achieved efficiently no matter what kinds of the integrands may be. Moreover, the conventional simulation method can solve the multi-variable problems effectively no matter whether linear or non-linear of the problems are. The proposed efficient integration method can be combined with the conventional simulation techniques well and its applicability to other numerical methods was shown by the numerical example of the first-excursion problem of stochastic process. In the first-excursion problem, it is a key to consider the correlation between values of response process (field or wave) at the discretized points. In developing more efficient method of analyzing the first-excursion problem, reliability theory of structural system such as PNET (Probabilistic Network Evaluation Technique) may be introduced. Furthermore, one considers that the proposed methods can be adopted to solve much more complex problems as long as the stochastic characteristics of the discretized points are gotten by using stochastic finite element method or recently developed simulation methods of stochastic wave (or field). This study was performed by a Grant-in-Aid for Scientific Research provided by the Japanese Ministry of Education in 1992 (Grant No.04855094).

References

1. T.Tsuda: Numerical Analysis of Multi-Variable Problem by Computer, Science Company, 1976. (in Japanese)
2. T.Tsuda: Monte-Carlo method and Simulation, Baifukan Co. Ltd., 1987. (in Japanese)
3. S.Matsuho et al.: Basic Study on Efficient Calculation Method of Riemann Integral and Its Application to First-Excursion Problem, Reports on the Information System presented at the 17th Symposium on Civil Engineering Information Processing System, pp.143-146, 1992-10. (in Japanese)
4. S.Matsuho et al.: Basic Study on Efficient Calculation of Riemann Integral, Reports of the Faculty of Engineering, Tottori University, Vol.23, No.1, 1992-11. (in Japanese)
5. G.I.Schuëller and C.G.Bucher: Computational stochastic structural analysis - a contribution to the software development for the reliability assessment of structures under dynamic loading, Probabilistic Engineering Mechanics, Vol.6, Nos.3/4, 1991-9/10, pp.134-138.

Reliability and Optimization of Structural Systems, V (B-12)
P. Thoft-Christensen and H. Ishikawa (Editors)
Elsevier Science Publishers B.V. (North-Holland)

Studies on Assessment of Structural Reliability by Response Surface Method and Neural Network

Y. Murotsu*, S. Shao*, N. Chiku*, K. Fujita**, Y. Shinohara**

* University of Osaka Prefecture, Sakai, Osaka 593, Japan

** Mitsubishi Heavy Industries, Ltd., Takasago, Hyogo 676, Japan

Abstract

This paper is concerned with assessment of structural reliability by a response surface method and a neural network when a limit state function is not given in an explicit form of basic variables. The response surface method approximates the limit state function in a polynomial form of the basic variables by using the response points while the neural network does in the sum of sigmoid functions. The two methods are applied to reliability assessment of stochastic structural systems.

Keyword Codes: J.6
Keywords: Computer-Aided Engineering

1 Introduction

In estimating reliability or failure probability of a structural system, the failure criterion for the system should be expressed as a so-called limit state function with respect to random design variables involved in the system. In general, a limit state function is defined in the following form:

$$g(x)\begin{cases} >0: & safety \\ \leq 0: & failure \end{cases} \tag{1}$$

where $x = (x_1, x_2, \cdots, x_n)^T$ represents a random vector. $g(x)$ is actually a structural response corresponding to x, and it is not easy to be represented explicitly. When dealing with complex structural systems or the structures under a dynamic load, a complicated structural analysis or a simulation of the system behavior should be carried out to obtain $g(x)$. Since the estimation of structural failure probability requires repeated calls for the limit state function $g(x)$, the computation is very expensive.

For the above reason, predicting techniques for a limit state function are needed to get a functional form relating the structural responses to random design parameters. This paper presents some results in developing an approximating limit state function (response surface) with the Taylor series expansion and a neural network. The procedures are applied to the reliability assessment of a three-dimensional piping system where the lower bound of the natural frequency of first mode is taken as a design criterion.

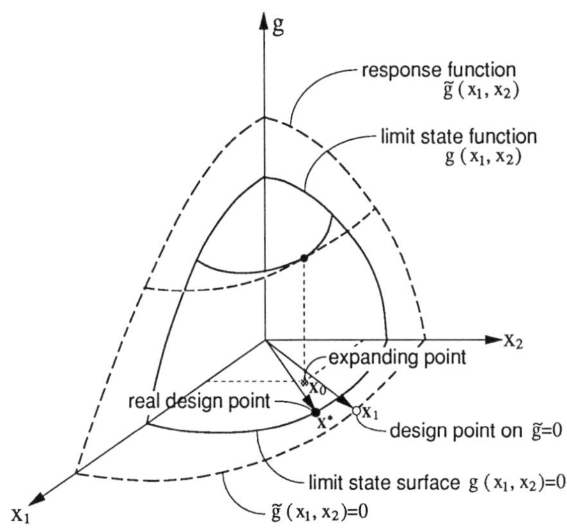

Figure 1. Response Surface by Taylor Series Expansion

2 Estimation of Limit State Function

In order to predict the limit state function $g(\boldsymbol{x})$, the structural responses are at first calculated in some points of $\boldsymbol{x}$, i.e., $g(\boldsymbol{x}_1)$, $g(\boldsymbol{x}_2)$, $g(\boldsymbol{x}_3)$, $\cdots$ corresponding to $\boldsymbol{x}_1$, $\boldsymbol{x}_2$, $\boldsymbol{x}_3$, $\cdots$. Then, an appropriate function $\tilde{g}(\boldsymbol{x})$ is determined to fit the real limit state function $g(\boldsymbol{x})$. $\tilde{g}(\boldsymbol{x})$ is also called a response surface. In pioneer researches, a regression analysis[1~3] or an interpolation[4] had been used in fitting such a response surface, and a second order polynomial form is recommended. Since a real limit state function $g(\boldsymbol{x})$ may have an arbitrary shape, it is in fact difficult to find an ideal $\tilde{g}(\boldsymbol{x})$ fitting $g(\boldsymbol{x})$ very well throughout the whole area. However, if only the most important region around the design point is considered, even a first order form of Taylor series expansion at the design point can be sufficient. In the following, two different types of approximating functions, i.e., Taylor series expansions and a neural network, are developed to respectively satisfy the simple case with only one design point and the case where $g(\boldsymbol{x})$ is complicated and a whole-area fitting is required.

2.1 Response Surface by Taylor Series Expansion

Consider a limit state function $g(\boldsymbol{x})$ as shown in Fig. 1. Select an initial point $\boldsymbol{x}_0$, e.g., the means of $\boldsymbol{x}$. The second order form of Taylor series expansion around $\boldsymbol{x}_0$ is:

$$\tilde{g}(\boldsymbol{x}) \;=\; g(\boldsymbol{x}_0) + \sum_{i=1}^{n} \frac{\partial g(\boldsymbol{x})}{\partial x_i}(x_i - x_{i0}) + \frac{1}{2}\sum_{i=1}^{n}\sum_{j=1}^{n} \frac{\partial^2 g(\boldsymbol{x})}{\partial x_i \partial x_j}(x_i - x_{i0})(x_j - x_{j0}) \tag{2}$$

where the partial derivatives of the random variables $\boldsymbol{x}$ can be calculated by a perturbation method. Response function $\tilde{g}(\boldsymbol{x})$ fits the original limit state function $g(\boldsymbol{x})$ well around the

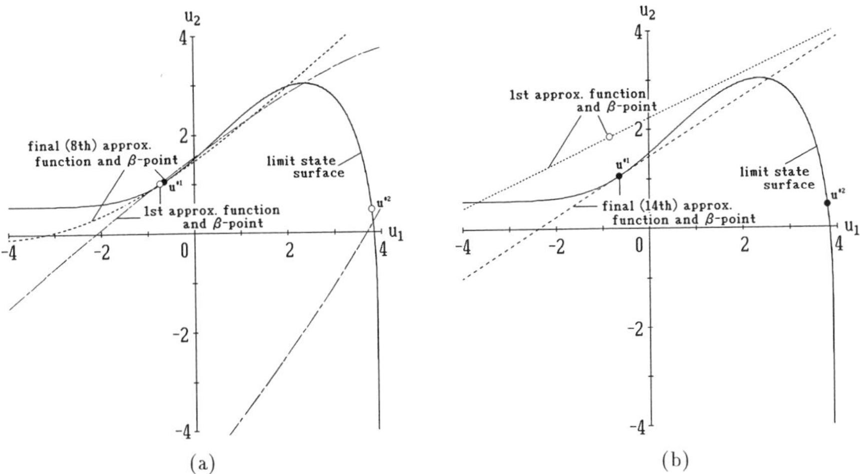

Figure 2. First and Second Order Response Functions

expanding point x_0, while it doesn't ensure the confidence on other parts. The design point x_1 on $\tilde{g}(x) = 0$ is searched and in the next step x_1 is used as a new expanding point to develop another response function. In this way, the expanding point is gradually moved to the real design point x^* and the Taylor series expansion around x^* yields a good approximating function $\tilde{g}(x)$ to replace $g(x)$ in estimating the failure probability of the structure.

A simple example is used here to illustrate the method mentioned above. The real limit state function $g(x_1, x_2)$ and the probability density function $f_x(x_1, x_2)$ is given as:

$$g(x_1, x_2) = 5 - x_1 - x_2 \qquad (x_1, x_2 > 0) \tag{3}$$

$$f_x(x_1, x_2) = (2x_1 + 0.5x_2 + x_1x_2) \times exp(-2x_1 - 0.5x_2 - x_1x_2) \tag{4}$$

Fig. 2 shows the limit state surface transformed into the independent standard normal variable space, i.e., $\boldsymbol{u}$-space. There are two local minimum β-points, i.e., $(u_1^{*1}, u_2^{*1}) = (-0.641, 1.054)$ and $(u_1^{*2}, u_2^{*2}) = (3.799, 0.456)$ with $\beta_1 = 1.234$ and $\beta_2 = 3.826$, respectively. A first order and a second order Taylor series expansion forms are used as response functions with an initial expanding point $(u_{10}, u_{20}) = (0, 0)$. By changing the expanding points, both of them converge to the real design point $\boldsymbol{u}^{*1}$ as shown in Fig. 2(a) and (b). Although the first order response function yields a large error in the beginning and its re-expanding times are much more than that of the second order response function, the total computation time of the former is only half of the latter. This is because the partial derivative information needed by the first order approximation is less than the second order approximation.

2.2 Neural Network for Prediction of Limit State Function

For a complex limit state function such as a multi-mode function, there may exist more than one critical failure region. A simple first or second order polynomial can not reflect the problem of this type. In this section, a neural network is applied to approximate the limit state function. The neural network consists of many sigmoid functions which could be adaptively adjusted to constitute various functional or non-functional relations.

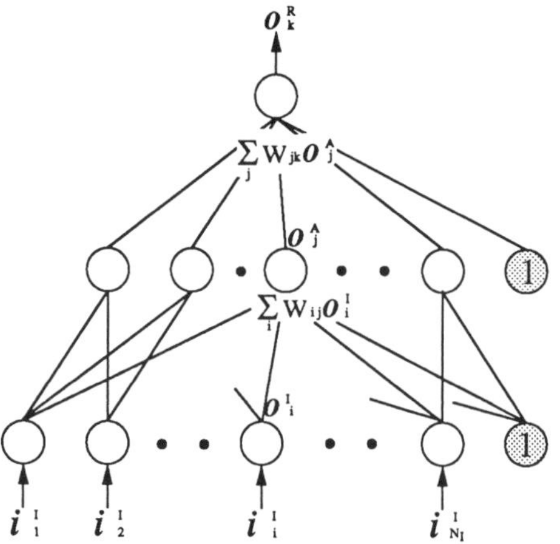

Figure 3. A Multi-Layer Neural Network

Fig. 3 illustrates a three-layer model of neural network. The numbers of units involved in the input layer, the hidden layer and the output layer are N_I, N_A, $N_R(N_R = 1)$, respectively. The input-output relation on each unit in the hidden layer and the output layer is represented by a sigmoid function defined by Eq. (5), while for the input layer Eq. (6) is used.

$$f(x) = \frac{1}{1 + \exp(-x)} \tag{5}$$

$$f(x) = x \tag{6}$$

Further, a unit with a constant output 1 is combined in the input layer and in the hidden layer. Let $\boldsymbol{i}^I_i$, $\boldsymbol{i}^A_j$, $\boldsymbol{i}^R_k$ denote the inputs and $\boldsymbol{o}^I_i$, $\boldsymbol{o}^A_j$, $\boldsymbol{o}^R_k$ the outputs, corresponding to the units in the input layer, the hidden layer and the output layer. Then, the following relations hold:

$$\boldsymbol{o}^I_i = \boldsymbol{i}^I_i \tag{7}$$

$$\boldsymbol{o}^A_j = f(\boldsymbol{i}^A_j) = f(\sum_{i=1}^{N_I+1} w_{ij}\boldsymbol{o}^I_i) \tag{8}$$

$$\boldsymbol{o}^R_k = f(\boldsymbol{i}^R_k) = f(\sum_{j=1}^{N_A+1} w_{jk}\boldsymbol{o}^A_j) \tag{9}$$

where w_{ij}, w_{jk} are weighting coefficients which can be adjusted to generate an arbitrary function between $\boldsymbol{o}^R_k$ and $\boldsymbol{i}^I_i$, $i = 1, 2, \cdots, N_I$. When considering a limit state function $g(\boldsymbol{x})$ of a structure, $\boldsymbol{o}^R_k$ represents the response $g(\boldsymbol{x})$ and $\boldsymbol{i}^I_i$ represents the random variables x_i, $i = 1, 2, \cdots, n = N_I$.

In the beginning, the network should have a "learning process", where some actual data, i.e., $\boldsymbol{x}_1$, $\boldsymbol{x}_2$, $\boldsymbol{x}_3$, $\cdots$ and the responses $g(\boldsymbol{x}_1)$, $g(\boldsymbol{x}_2)$, $g(\boldsymbol{x}_3)$, $\cdots$ are provided, and the error between the output $\boldsymbol{o}^R_k$ of the network and every response is calculated. The squared error E is written as a function of the weighting coefficients w_{ij} and w_{jk}. Using a so called back-propagating-error algorithm[5] to minimize E, the weighting coefficients are modified step by step, following

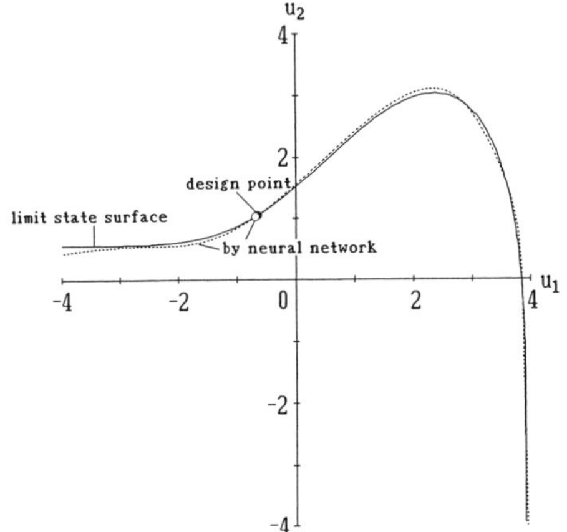

Figure 4. Response Surface by a Neural Network

Table 1. Design Point from a Neural Network

	Neural network	Real value
Design point (u_1^*, u_2^*)	$(-0.687, 1.022)$	$(-0.641, 1.054)$
β	1.232	1.234
$\Phi(-\beta)$	1.090×10^{-1}	1.085×10^{-1}

226195 learning times
Calculation time: $1907s$ (CPU : Sun 4/75)

Eq. (10):

$$\Delta w^{t+1} = -\varepsilon \frac{\partial E}{\partial w} + \alpha \Delta w^t \tag{10}$$

where ε and α are constants, t denotes a step number and Δw^t means the change of the weighting coefficient w in step t. The minimization of the squared error E is continued until E converges to 0 for all test points.

The neural network mentioned above is used to simulate the limit state function of Eqs. (3) and (4). The number of the units in the hidden layer, i.e., N_A, is selected as 4 times of N_I. The constants ε and α in Eq. (10) is set to be 0.1 and 0.9 based on experience. In this example, test points used in "learning process" are randomly generated from the region $-4 \leq u_1, u_2 \leq 4$. Table 1 and Fig. 4 show the results of the simulation. It is seen that the neural network approximates the limit state function very well throughout the whole area.

3 Application Problem — A Piping System

In order to demonstrate the effectiveness of the proposed procedures, the reliability estimation of an actual piping system is carried out in this section using predicted limit state functions.

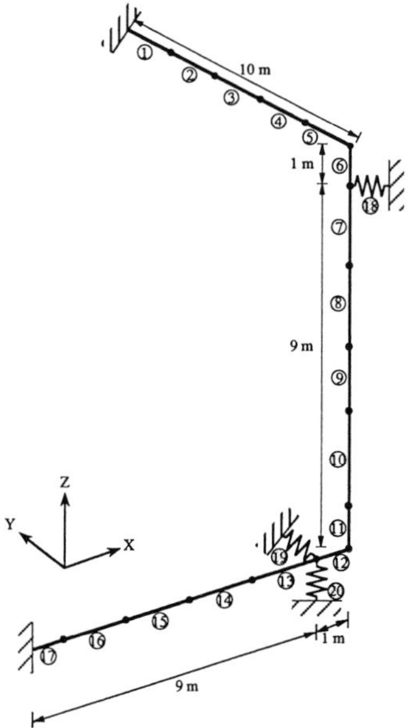

Figure 5. A Three-Dimensional Piping System

Consider a three-dimensional piping system model as shown in Fig. 5. The outside diameter and the thickness of the pipe are $D = 0.4064m$ and $t = 21.4mm$. The density, Young's modulus and Poisson's ratio are $\rho = 1.1784 \times 10^4 kg/m^3$, $E = 2.1 \times 10^{10} kgf/m^2$ and $\nu = 0.3$. The supports are modeled as columns with the length $l = 1m$. Their cross-sectional areas and materials are assumed to be the same as those of the pipe. As shown in Fig. 5, the system is divided into 20 elements. In this case, the natural frequency of the first mode of the system is $f_1 = 12.854\,(Hz)$. Then, in the following, the Young' modulus, densities, pipe thicknesses and the lengths of the supports are assumed as normal random variables with the same coefficient of variation $CV = 0.1$. Consequently, the system is characterized by 60 basic random variables. The lower bound for the natural frequency of first mode is set to be $f^* = 10.854\,(Hz)$ which is considered as a failure criterion.

Before the calculation with the 20-element model, a 8-element model for the same system is used to compare the two procedures for the prediction of limit state functions. Table 2 lists the results from a first order Taylor series expansion and a neural network. They are found to be very close. In the following, only the approximation by a first order Taylor series expansion is used in the detailed analysis.

Table 3 shows the design point and the failure probability calculated with the 20-element model. In order to investigate the influences of individual random variables, the sensitivity factors $\frac{u_i^*}{\beta}$[6] of the important elements are given in Table 4. It is seen that the elements 8 and 9 have the most significant influence on the system reliability.

Table 2. Results from Taylor Series Expansion Polynomial and Neural Network

	Reliability index β	Failure probability P_f
Taylor series expansion	5.0889	1.8007×10^{-7}
Neural network	4.9631	3.4693×10^{-7}

Table 3. Design Point and Failure Probability

Element No.	Young's modulus ($\times 10^{10}\ kgf/m^2$)	Thickness (mm)	Density ($\times 10^4\ kgf/m^3$)	Length (m)
1	1.912	19.82	1.180	–
2	2.068	21.61	1.205	–
3	2.000	22.12	1.263	–
4	1.968	22.34	1.287	–
5	2.053	22.19	1.246	–
6	2.097	21.56	1.189	–
7	2.057	21.99	1.233	–
8	1.681	23.09	1.406	–
9	1.402	23.29	1.460	–
10	2.024	22.84	1.293	–
11	2.071	21.32	1.188	–
12	1.938	20.05	1.180	–
13	2.013	20.73	1.183	–
14	2.038	21.12	1.193	–
15	2.078	21.37	1.188	–
16	2.084	21.28	1.180	–
17	2.071	21.14	1.178	–
18	1.912	–	1.180	1.134
19	1.985	–	1.178	1.030
20	2.064	–	1.178	1.045
	$\beta = 6.0587$		$P_f = 6.8614 \times 10^{-10}$	

$CV_{Ei} = 0.1$, $CV_{ti} = 0.1$, $CV_{\rho i} = 0.1$, $CV_{li} = 0.1$

Calculation time: 11200.9s, (CPU : Sun 4/75)

Table 4. Sensitivity Factors of Important Elements

No.	Young's modulus	Thickness	Density	Length
1	$-0.5484^{(9)}$	$0.1455^{(9)}$	$0.3949^{(9)}$	$0.2214^{(18^*)}$
2	$-0.3296^{(8)}$	$0.1304^{(8)}$	$0.3195^{(8)}$	$0.0740^{(20^*)}$
3	$-0.1482^{(1)}$	$-0.1220^{(1)}$	$0.1607^{(10)}$	$0.0494^{(19^*)}$
4	$-0.1481^{(18^*)}$	$0.1108^{(10)}$	$0.1516^{(4)}$	–
5	$-0.1274^{(12)}$	$-0.1045^{(12)}$	$0.1184^{(3)}$	–
6	$-0.1041^{(4)}$	$0.0721^{(4)}$	$0.0941^{(5)}$	–

( ): Element no.

4 Conclusions

The reliability estimation with a predicted limit state function, i.e., a response surface has been discussed. For a problem with only one critical failure domain, a low order Taylor series expansion form is applicable by modifying the expanding point step by step. From the numerical results it is seen that a first order form shows high efficiency and good convergence property in the movement to the critical domain. To simulate complex limit state functions, a neural network is applied to make a whole-area approximation. For this case, the "learning process" costs some of time, but a good fitness to an actual limit state function could be expected. The proposed procedures have been applied to the reliability assessment of a piping system to illustrate the effectiveness.

References

[1] Box, G. P., and Draper, N. R., Empirical Model Building and Response Surfaces, John Wiley and Sons, New York, 1987.

[2] Collins, J. D., Evensen, D. A., and Heubach, W. F., Failure Prediction for Non-Linear Structural Systems, Proc. of ICOSSAR '89, Vol. 3, pp. 839-846, 1989.

[3] Engelund, S., and Rackwitz, R., Experiences with Experimental Design Schemes for Failures Surface Estimation and Reliability, Proc. of ASCE 6th Spec. Conf. on Probabilistic Mechanics and Structural and Geotechnical Reliability, pp. 252-255, Denver, July 8-10, 1992.

[4] Schuëller, G. I., Bucher, C. G., Bourgund, U., and Ouypornprasert, W., On Efficient Computational Schemes to Calculate Structural Failure Probabilities, Probabilistic Engineering Mechanics, Vol. 4, No. 1, pp. 10-18, 1989.

[5] Aso, H, Information Processing by Neural Network (in Japanese), Sangyotosho, 1990.

[6] Madsen, H. O., Krenk, S. & Lind, N. C., Methods of Structural Safety, Prentice Hall, 1986.

Reliability and Optimization of Structural Systems, V (B-12)
P. Thoft-Christensen and H. Ishikawa (Editors)
Elsevier Science Publishers B.V. (North-Holland)

Dynamic Reliability Assessment of a Nuclear Reactor Building

N. Nakamura*, T. Takada**, S. Ogawa**, M. Mizutani***

* Tokyo Electric Power Co., Tokyo, 100, Japan

** Ohsaki Research Institute, Tokyo, 100, Japan

*** Tokyo Electric Power Service Co., Tokyo, 100, Japan

ABSTRACT

In the present paper, dynamic reliability assessment of an imaginary nuclear reactor building designed according to Japan seismic code requirement is demonstrated to study the spatial imbalance of local structural safety within the building subject to seismic hazard. The seismic hazard may be expressed in terms of a hazard curve and a non-white stationary random vector process. The building is modeled into finite elements and a linear random vibration analysis as well as structural reliability theory are fully utilized to compute the failure probability associated with the elastic limit states defined in each finite element. Besides the above study, sensitivity analyses are also performed to know the influence of some key parameters and of modelings on the limit state probability (LSP).

1. OBJECTIVE

Most advanced and complex structural systems such as nuclear power plants require well-sophisticated methods to ensure structural integrity. For the above purpose, both analytical and experimental efforts have been intensively made for last decades in Japan. However, considering the uncertainty inherent to real data to be used in analyses, and physical limitations of experiments such as the number of experiments, specimens, conditions under consideration, etc., the structural safety of buildings has to be studied more from different aspects.

On the other hand, a great attention has been paid so far on structural reliability evaluation methods which are fully based on the structural reliability theory. These methods have been successfully established especially for elastic linear structures even in dynamic problems. Many practical applications of these methods have been reported in various engineering fields, and the obtained results do enable us to quantitatively evaluate the structural integrity and give us very useful insight.

Although importance, effectiveness and rationality of the reliability theory has been recognized for a couple of decades by researchers and practitioners, people are still reluctant to implement the theory to codification. One of the major obstacles to the theory might be small numbers of applications of the theory, that might not have people get aware of the merits of the theory. For the above reason, this study was done and the major motivation behind this study is to establish a reliability-based evaluation method towards a future probability-based design method for nuclear plant facilities. It should be noted that this study is basically different from the line of probabilistic safety assessment (PSA) studies having been conducted before for a number of nuclear plants in the US and Japan as well.

2. EXTERNAL LOADS AND MODELING

External loads to be taken into account in the evaluation are the following most influential loads: seismic load, dead and live loads, and thermal load in normal operation of the plant. Modeling of these loads are briefly described below [1,2].

2.1. Seismic Load

Seismic load to a reactor building will be determined similarly to the way of the Japanese design practice. Figure 1 shows how this load is evaluated in this study. Seismic loads are modeled as a zero-mean, stationary Gaussian vector process with a Kanai-Tajimi (KT) power spectral density function. Also, the variation of the intensity of the process is taken into consideration in a seismic hazard analysis. The vector process has two components: horizontal and vertical seismic components acting from the bottom of the reactor building.

The seismic load is firstly defined on the base rock and its intensity is determined from the seismic hazard analysis, namely, the determination of so-called hazard curves. The hazard curves associated with a peak ground acceleration (PGA) are assumed to have the following type II asymptotic distribution:

$$F_{A_{max}}(a_{max}) = 1 - \exp\left[-\left(\frac{\alpha}{a_{max}}\right)^{\beta}\right] \tag{1}$$

On the other hand, to characterize the probabilistic nature of the process, a cross-power spectrum matrix is defined, based on the results of the detailed spectral analysis of seismic wave observed at a particular building site. The resulting cross-power spectrum matrix denoted by $\mathbf{S}_{HV}(\omega)$ is formally written as follows:

$$\mathbf{S}_{HV}(\omega) = \begin{bmatrix} S_{HH}(\omega) & S_{HV}(\omega) \\ S_{VH}(\omega) & S_{VV}(\omega) \end{bmatrix} \tag{2}$$

where both $S_{HV}(\omega)$ and $S_{VH}(\omega)$ are practically zero, as its coherence function of an actual observation is shown in Fig.2. This obviously implies that the horizontal component is statistically independent of the vertical one.

According to Fig.1, the amplification effect of soil ground between the bed rock surface and the building is evaluated using 1-D wave propagation theory. The input ground motion to the building, which is the product of the transfer function of the soil and the power spectra at the bed rock surface, is modeled into Kanai-Tajimi (KT) power spectra, as shown in Figs. 3. The assumption of the KT spectra was selected so that the ensuing random vibration analysis can be performed easily although the KT spectra do not always reflect the actual one.

Before closing this sub-section, it is noted that the relationship between the power spectrum and the PGA in the seismic hazard curve is derived from the following practical equation.

$$a_{max} \approx \gamma\sqrt{\pi\omega_g\left(\frac{1}{2\zeta_g} + 2\zeta_g\right)S_0} \tag{3}$$

where γ is a peak factor which practically takes 3 from past experience, and ω_g, ζ_g and S_0 are parameters which determines the KT spectrum, and the above root stands for the root mean square value of the process with the KT spectrum.

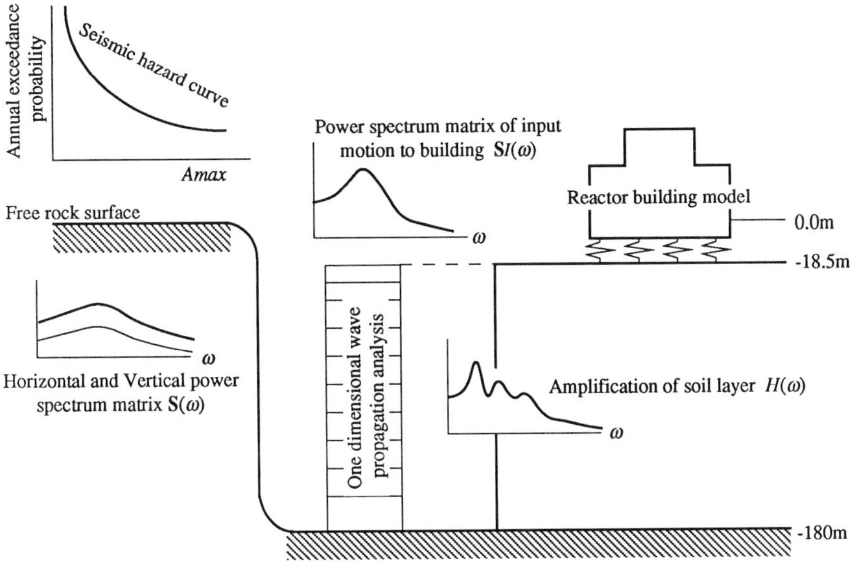

Figure 1 Procedure of evaluation of the seismic load to the building

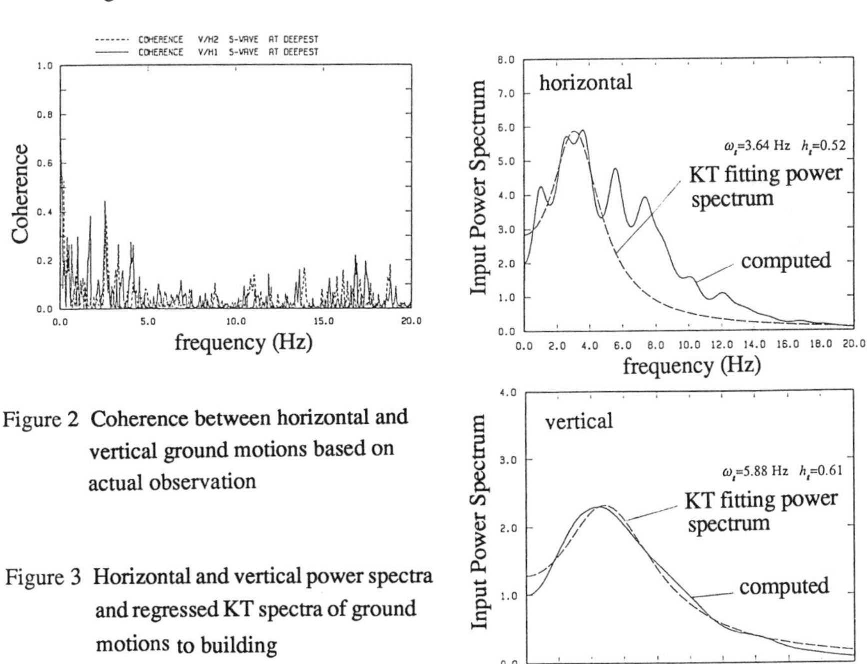

Figure 2 Coherence between horizontal and vertical ground motions based on actual observation

Figure 3 Horizontal and vertical power spectra and regressed KT spectra of ground motions to building

2.2. Other Loads

Dead and live loads and thermal load in normal operation are taken into account in the following evaluation. The dead and live loads are assumed to be deterministic since the variation of the loads can give less influence on the structural safety. Since the thermal load varies annually, the variation of the load is considered in an approximate manner as will be stated later.

3. STRUCTURAL MODEL AND LIMIT STATES OF INTEREST

3.1. Structural Model

The building to be analyzed is a 1100 KW BWR type nuclear reactor building not embedded into the soil layer, and its finite element model is shown in Fig.4. The building is assumed to be designed accordingly to the Japanese design criteria. The model composes of more than two hundred finite elements: mainly shell elements and beam elements. To properly evaluate the soil-structure interaction effect, Winkler type soil springs are attached below the model in three directions. Main walls: inner walls (I/W) and outer walls (O/W), slabs, mat foundation (MAT) and steel roof trusses (RT) are properly modeled into finite elements.

3.2. Determination and Modeling of Limit States

Limit states to be considered in this study are the events that sectional stresses of each finite element exceed the elastic limit which is determined from the information of concrete section and reinforcements. The elastic limit of the stress is defined in such a way that the stress reaches either two-third of the real concrete strength or the yielding strength of steel bars.

When a finite element is a reinforced concrete shell element, which is mainly the case, the limit state function is a nonlinear function with respect to the sectional stresses, which are defined in multi-stress space, as illustrated in Fig. 5. Therefore, for simplicity of the analysis, the nonlinear function in an implicit form is approximated by fifty six hyperplanes so that the ensuing dynamic reliability analysis can be performed with great ease. The limit state functions thus approximated $\mathbf{g}_k$ can be written as follows.

$$\mathbf{g}_k = \mathbf{f}_k - \mathbf{G}_k \mathbf{s}_k(t) = \mathbf{0} \tag{4}$$

where $\mathbf{s}_k(t)$ is the element stress vector of the k-th finite element, which denotes both in-plane sectional stresses such as N_x, N_y and N_{xy}, and out-of-plane stresses such as M_x, M_y and M_{xy}.

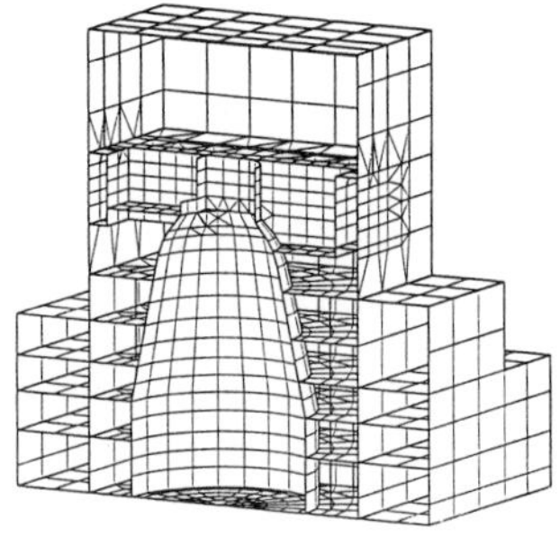

Figure 4 Finite element model

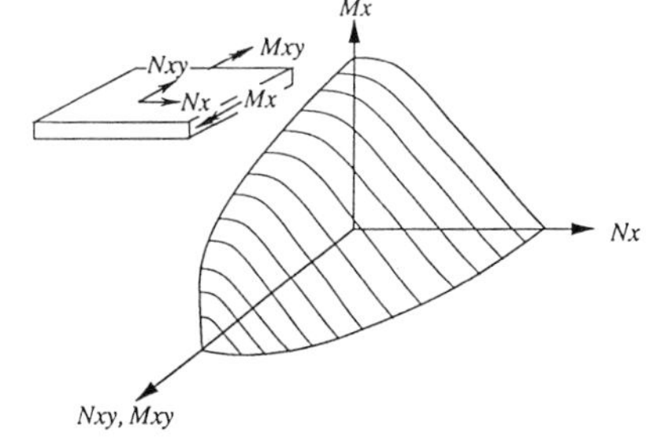

Figure 5 Limit state functions of RC shell elements

4. METHOD OF DYNAMIC RELIABILITY ESTIMATION

Assuming the linearity and stationarity of structural response, a random vibration analysis of the finite element system based on the modal analysis can be employed [1]. Figures 6 and Table 1 show the results of the modal analysis. The first two modes are identified as rocking-sway modes of the building, the third and fourth modes are the vertical vibrational modes. To perform the response analysis, modal damping ratios are determined from the references [4].

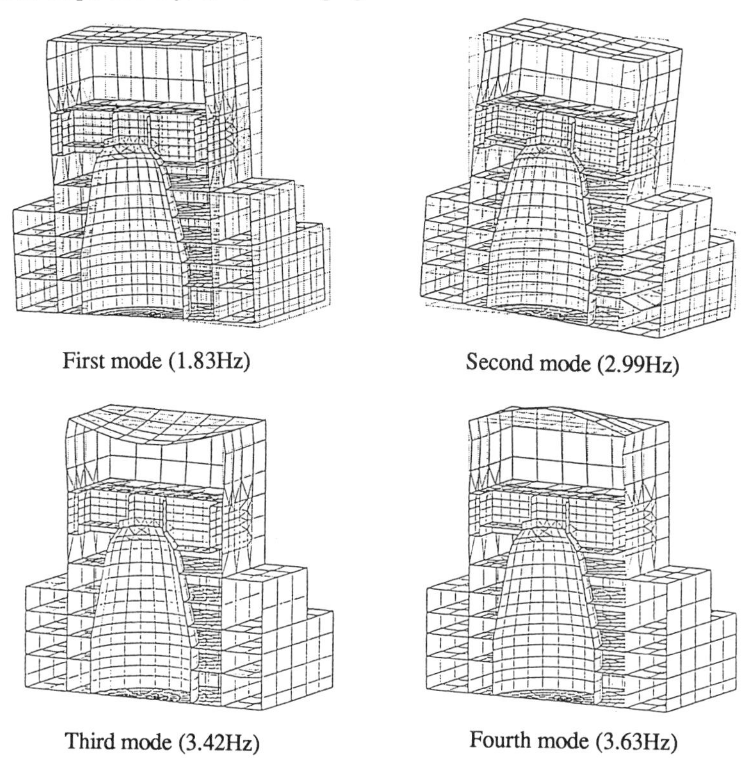

First mode (1.83Hz) Second mode (2.99Hz)

Third mode (3.42Hz) Fourth mode (3.63Hz)

Figure 6 Major mode shapes

Table 1 Results of eigen value analysis and modal damping ratios

	Freq.(Hz)	period(sec)	participation factor. (H)	participation factor. (V)	modal damping
1	1.83	0.546	-119.6	0.5	0.10
2	2.99	0.334	-55.1	-3.9	0.10
3	3.42	0.293	1.1	-111.6	0.50
4	3.63	0.275	0.5	-69.2	0.02
5	4.29	0.233	1.0	5.6	0.05

After the analysis, the statistics of the element stress under the prespecified seismic loading can be computed, and a first excursion probability that the element stress exceeds the above-mentioned hyperplanes in the time interval T_d as follows:

$$\mathbf{P}_k(A_{max}) = \text{Prob}\left[\min_{0 \le t \le T_d}\{\mathbf{g}_k(A_{max}, t)\} < \mathbf{0}\right] \tag{5}$$

is computed as a conditional probability under the seismic intensity A_{max}. Since the limit state functions of k-th element are modeled into many hyperplanes, the following Cornell's upper and lower bounds of the LSP are used.

$$\max_{j=1,56}\{\text{Prob}\{g_{jk} < 0\}\} < P_k(A_{max}) = \Pi_{j=1}^{56}(\cup\text{Prob}\{g_{jk} < 0\}) < \sum_{j=1}^{56}\text{Prob}\{g_{jk} < 0\} \tag{6}$$

Finally, the LSP thus computed is integrated with respect to the intensity of the seismic load A_{max} to obtain the final LSP.

To take into consideration of the variation of the annual thermal stresses and material strength, a Latin hypercube sampling method with a sample size two is used.

5. SPATIAL DISTRIBUTION OF LIMIT STATE PROBABILITY

The contour plots shown in Fig.7 clearly shows that the LSP of I/W is relatively large, compared with other portions. It can be guessed from this figure that I/W are subjected mostly to horizontal seismic force. From detail examination of the failure modes of each portion, the most likely failure mode occurring in I/W is mainly in-plane shear failure although the failure occurs in multi-stress state mode, while those of the upper part of I/W perpendicular to the horizontal seismic loading is an out-of-plane flexible failure due to the excessive vibration of the roof subjected to vertical components of the seismic loading. The LSP of the center of RT

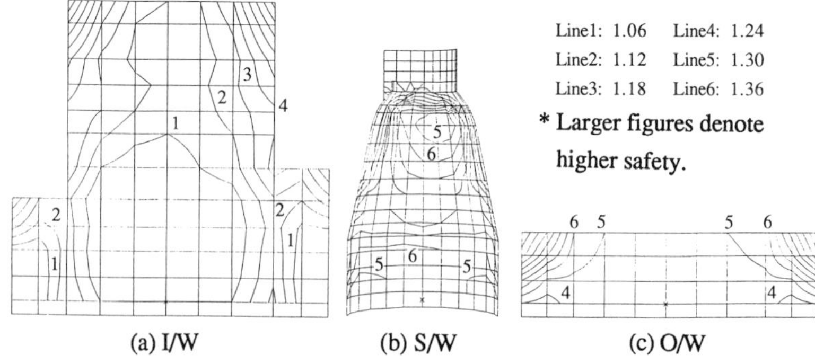

Figure 7 A contour plot of limit state probability

is largest due to the fundamental mode of the roof truss subjected to vertical seismic load. The other portions such as a pool, S/W and MAT are still safer. This can be considered that an accidental load including thermal and pressure loads is most significant in these portions, and that thermal stress in normal operation contributes most to the design of MAT, thereby these parts are considered to have larger margin against the seismic load.

6. SENSITIVITY ANALYSES

The following sensitivity to the LSP will be examined: (1) the effect of material strength variation, (2) the effect of vertical components of seismic loading and (3) the effect of modeling of seismic load.

6.1. Effect of Material Strength Variation

The material strength variation of concrete and rebar may influence the LSP more or less. In this sub-section, it is investigated how large this influence is. To do so, the results of the case in which whole material variation are taken into account will be compared with the case in which the material strength variation take their mean values. From the comparison of the results, it is found that there is no discrepancy between the results from the two cases, therefore it is concluded that the material strength variation does not influence on the LSP at all. This conclusion, however, may change when the limit state under consideration is defined as the ones beyond the elastic limit such as structurally ultimate states.

6.2. Effect of Vertical Seismic Motion

The portions which are affected most by whether or not the vertical ground motion is simultaneously acting are the RT and walls supporting the RT. Since the vibrational mode associated with the roof itself is induced so much from the result of the modal analysis, it simultaneously generates the out-of-plane bending moment at the upper edge of the walls.

6.3. Effect of Seismic Load Modeling

The seismic load has been modeled as a random process which enables us to use the random vibration theory. Here, the following model is considered: the seismic load action is an equivalent static load which can be given by the deterministic time history response analysis with a lumped mass-stick model, which is usually used in an actual design practice, with some deterministic artificial ground time histories based on the site geology and seismicity. Using this design model, the distribution of seismic external force can be determined, and then this force will be applied horizontally and statically to each floor levels of the FEM model. Table 2 shows the results of the two cases, in both of which the vertical seismic load is disregarded for comparison.

It can be observed from the table that the static case tends to generate more conservative result and the difference between the two cases is most significant in the parts of MAT and POOL. It can be guessed from the results that local vibration modes which the lumped mass-stick model fails to evaluate affect much on the LSP there. In view of the fact that the similar result are obtained in I/W from both cases, the adequacy of the modeling of the seismically resistant elements such as I/W, O/W and S/W into the stick model is emphasized even if the equivalent static load is assumed.

Table 2 Effect of seismic load modeling

	S/W	I/W	O/W	MAT	POOL
Dynamic load modeling	1.37	1.00	1.37	6.40	6.40
Equivalent static load modeling	1.30	0.87	1.24	2.10	1.94

These figures in the table are normalized by the maximum LSP within the building after taking the logarithm of the LSP.

7. CONCLUSIONS

From the results obtained, the followings are concluded:

(1) When only a horizontal component of seismic load exists, the LSP of I/W is one-order higher than those of other portions. The LSP of MAT and POOL are relatively low.

Through the sensitivity analyses, the following findings are obtained:

(1) Material strength variation do not contribute to the LSP at all as far as the LS is defined as an elastic limit.

(2) When a vertical components of seismic load is simultaneously applied, the LSP of the RT is the largest, and those of the wall edge supporting the RT is also large.

(3) The static seismic force equivalent to the dynamic one yields the results similar to that from the dynamic load modeling.

For further study, a reliability-based design method following the procedure mentioned in the present paper will have to be established. To do so, the technique of determining thickness of RC members, size and location of rebars is needed, and these study is now under the way.

REFERENCES

[1] Shinozuka, M., Kako, T. and Tsurui, A., Random Vibration Analysis in Finite Element Formulation, Random Vibration, Elesevier Science Publishers, 1986, pp.415-450

[2] Shinozuka, M., Hwang, H.M., H. and Reich, M., Reliability assessment of reinforced concrete containment structures, Nuclear Engineering and Design, Vol.80, pp.247-267, 1984

[3] Lin, Y. K., Probabilistic Theory of Structural Dynamics, Robert E. Krieger Publishing Co. Malabar, FL, 1976

[4] Hirashima, S., Damping Property of Reactor Buildings, Annual meeting of Architectural Institute of Japan, 1980 (in Japanese)

Reliability and Optimization of Structural Systems, V (B-12)
P. Thoft-Christensen and H. Ishikawa (Editors)
Elsevier Science Publishers B.V. (North-Holland)

Uncertainties of Safety Index on Structural Safety

S. Nakanishi, H. Nakayasu

Department of Industrial Management,
Osaka Institute of Technology, Osaka 535, Japan

Abstract

This paper deals with the problem of the uncertainties of safety index on structural safety. Though a second moment safety index is widely used for a probabilistic design code, the index has intrisically statistical uncertainties since the random parameters are estimated only from the obtained field data such as actual strength data of structural members and of the applied loads. In this paper, the new measures of uncertainties of safety index of second moment reliability are proposed as a confidence region is around the design point with the significance level of the sample statistics. These confidence regions are described as an ellipsoid on super plane whose region corresponding to the value of significance level. As the practical value for the probabilistic design code, procedures of deciding the lower and upper confidence level instead of safety index are derived in the paper which are depend on the sample size of the specified sample statistics of random parameters.

Keyword Codes : G.1.2; G.1.4; G.1.6
Keywords : Mathematics of Computing, Approximation;
Quadrature and Numerical Differentiation; Optimization,
Safety Index, Advanced First Order Second Moment, Sample Statistics, Reliability
Confidence Region, Lower and Upper Bound of Safety Index, Statistical Theory

1. INTRODUCTION

Though it is important for structural safety to maintain structural systems reliability[1-2]. It is not easy to compute the exact value of reliability of structural system analytically. Then the structural reliability is estimated from the results of structural members with some physical characters. In particular, the design parameters which construct these characters have statistical uncertainties. So these parameters must be treated as random variables. The characteristic information of design parameters are estimated by these sampling inspections or simulations based on these sample sizes. In considering this things, the random parameters of the design parameters also have statistical uncertainties. Therefore it is necessity that the random parameters are provided with these confidence region by confidence level. These random parameters has been used as fixed values in the traditional design of structural safety.

In this paper, it is reviewed that uncertainties of failure probability which is based on statistical uncertainties of the random parameters are existed, and presented the new measure on uncertainty of failure probability is computed as lower and upper bound of safety index based on confidence level.

2. STRUCTURAL SAFETY AND SAFETY INDEX

In the theory of structural safety[2], the design parameters which have uncertainties are dealed as random variables. If these random variables construct n-dimensional area, it is able to express these the basic random variable vector $\boldsymbol{x}$. That is

$$\boldsymbol{x}=(x_1 \ldots x_n)^T. \tag{1}$$

Then the elements of $\boldsymbol{x}$ are constructed by strength, external load of structural component and so on. On $\boldsymbol{x}$, these failure mode of structural component is able to define failure surface as:

$$g(\boldsymbol{x})=0 \tag{2}$$

where $g(\boldsymbol{x})$ is called that limit state function or performance function whose boundary is on the failure domain $F=\{g(\boldsymbol{x})\leqq 0\}$ or safety domain $S=\{g(\boldsymbol{x})>0\}$.

In consideration of these things, failure probability of structural component is expressed by the equation

$$Pf=Prob\{F\}=\int dF(\boldsymbol{x}) \tag{3}$$

However, it is difficult to calculate Eq.(3). Therefore, it is more effective to use safety index β derived from the second-moment of $\boldsymbol{x}$.

Firstly, it is necessary to transform Eq.(1) into the following equation :

$$\boldsymbol{y}=(y_1 \ldots y_n)^T \tag{4}$$

where $\boldsymbol{y}$ is independent among its element y_i, and normalized variable vector. y_i is conformed standard normal distribution $N(0,1^2)$, and there is no correlation between y_i and y_j. These transformation are represented by the standradizd transformation[3]. Eq.(2) is able to rewrite in the standardized area

$$g(\boldsymbol{y})=0. \tag{5}$$

Secondly, if Eq.(5) is nonlinear function, it is possible to show the first-order approximate equation on the design point $\boldsymbol{y}^*$

$$g(\boldsymbol{y})=\nabla g(\boldsymbol{y}^*)^T(\boldsymbol{y}-\boldsymbol{y}^*)=0 \tag{6}$$

where $\nabla g(\boldsymbol{y}^*)$ is gradient vector of $g(\boldsymbol{y}^*)$

$$\nabla g(\boldsymbol{y}^*)=(\partial g/\partial y_1 \ldots \partial g/\partial y_n)^T. \tag{7}$$

Thus $g(\boldsymbol{y})$ is able to be treated as linear function. Therefore safety index β is expressed as the distances from origin to the design point $\boldsymbol{y}^*$

$$\beta^*=-\nabla g(\boldsymbol{y}^*)^T\boldsymbol{y}^*/\{\nabla g(\boldsymbol{y}^*)^T\nabla g(\boldsymbol{y}^*)\}^{1/2} \tag{8}$$

where $\boldsymbol{y}^*$ is satisfied the convergent judgment : $|\boldsymbol{y}^{i+1}-\boldsymbol{y}^i|<\varepsilon$ and β is calculated by the following iteration :

$$\boldsymbol{y}^{i+1}=\nabla g(\boldsymbol{y}^i)^T\cdot\{\nabla g(\boldsymbol{y}^i)^T\boldsymbol{y}^i-g(\boldsymbol{y}^i)\}/\{\nabla g(\boldsymbol{y}^i)^T\nabla g(\boldsymbol{y}^i)\}. \tag{9}$$

Because $\boldsymbol{y}^*$ is location vector called β points, it is able to define as its norm form [1-2]

$$\beta^*=\|\boldsymbol{y}^*\|. \tag{10}$$

Thus Eqs.(8), (10) yield the unique value. Therefore, β^* by Eqs.(8) or (10) is invariant index, whose probability can be calculated as

$$Pf=\Phi(-\beta^*) \tag{11}$$

where $\Phi(.)$ is the cumulative normal distribution function whose value can be easily obtained by mathematical table. Therefore $\boldsymbol{y}^*$ is upper percent point of the cumulative normal distribution function.

3. STATISTICAL UNCERTAINTIES OF SAFETY INDEX BASED ON SIGNIFICANCE LEVEL

In practical problem, design parameters of each structural members must be maintained high reliability. However, because of economic or time constraint, it is difficult that from material test, load test or computing the characters of design parameters are obtained as the satisfied solutions. Therefore sample size and significance level of x have to be defined as vector $\mathbf{m}$, γ

$$\mathbf{m}=(m_1 \ldots m_n)^T$$
$$\gamma=(\gamma_1 \ldots \gamma_n)^T. \tag{12}$$

For example, when expectation vector μ of x is unknown, and covariance matrix Σ of x is already known on x, the following confidence region of μ exists

$$\bar{x}-\Sigma_m\mathbf{k} \leq \mu \leq \bar{x}+\Sigma_m\mathbf{k} \tag{13}$$

where $\mathbf{k}=(k_1 \ldots k_n)^T$, and Σ_m is covariance matrix of $\bar{x}$ based on the sample size vector $\mathbf{m}$, whose elements are

$$\sigma_{m\,i\,i}=\sigma^2/m_i \qquad (i=j)$$
$$\sigma_{m\,i\,j}=\rho\sigma_i\sigma_j/(m_i m_j)^{1/2} \qquad (i \neq j) \tag{14}$$

and $\mathbf{k}$ is both lower and upper limit point vector with significance level γ.

In the standardized space, $\bar{y} \sim N(0, \Sigma_{Ym})$, since Σ_m is transformed into Σ_{Ym} by standardized transformation[3] as well as from Σ to unit matrix I. In considering these things, if y^*_o is the unique value derived from Eqs.(8) or (10) which is the distribution parameter of ramdom variable y^* with sample size $\mathbf{m}$, the confidence reigion of y^*_o with significance level γ in correnpondence to Eq.(13) is given by

$$y^*-\Sigma_{Ym}\mathbf{k} \leq y^*_o \leq y^*+\Sigma_{Ym}\mathbf{k} \tag{15}$$

where y^* is β point, and if Q is decided as the inverse matrix of $K=diag[ki]$. Eq.(15) is rewritten as the following formation

$$(y^*-y^*_o)^T Q^2 (y^*-y^*_o)-1 \leq 0 \tag{16}$$

which shows the confidence region of safety index called ***Confidence Hyperellipsoid***, since the equation expresses the region within an ellipsoid in multidimensional area. There is the real safety index β in the region of confidence hyperellipsoid. Therefore as represented in the Figure 1 and 2, it is able to decide that the lower safety index β_L as a minimum distance from origin to confidence hyperellipsoid and the upper safety index β_U as a maximum distance.

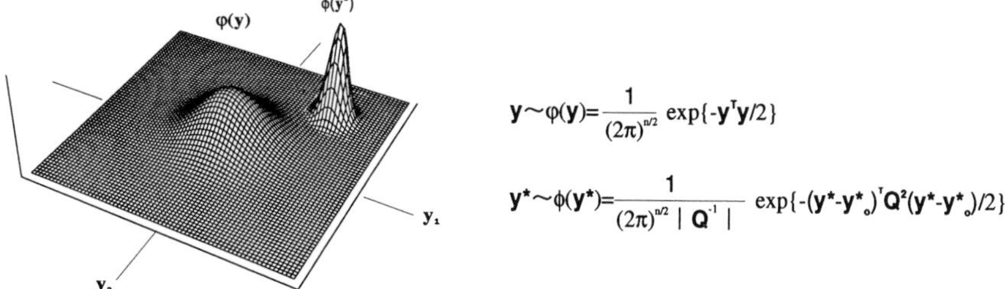

Figure 1. Joint density distribution of y and y^*.

Therefore, Lower and upper limit values of β on the same axis in the Figure 2 are computed the following equation

$$\beta^*_L=\beta^*\{1-1/(y^*Q^2y^*)^{1/2}\},$$
$$\beta^*_U=\beta^*\{1+1/(y^*Q^2y^*)^{1/2}\}. \tag{17}$$

If sample size and confidence level on each axis are equal, β^*_L is β_L and β^*_U is β_U. However, as represented the Figure 2., β^*_L and β^*_U are not the exact lower and upper bounds of β is practical. In the following section 4, the algorithmic techniques are proposed for the computation of the lower and upper bounds of β.

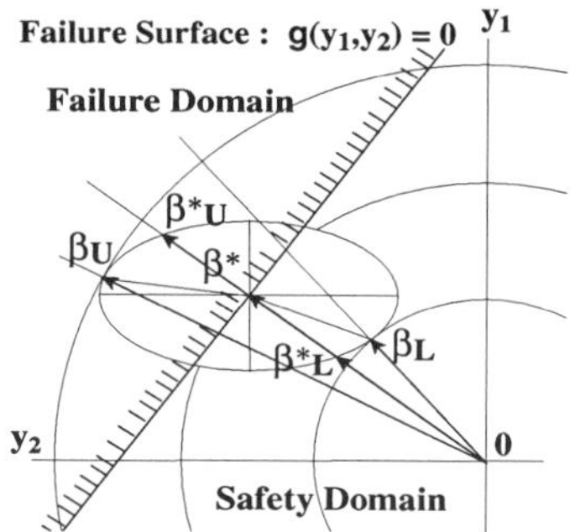

Figure 2. Confidence hyperellipsoid and upper and lower bound of safety index β.

4. SOLUTION FOR LOWER AND UPPER BOUNDOF SAFETY INDEX

4.1 Analytical method

To obtain the lower and upper bound β, it is able to formulate the mathematical programming problem with the objctive function $\beta(y)$ and constraint function

$$\beta(y^*_o)=\|y^*_o\|$$

$$\text{s.t.} \quad (y^*-y^*_o)^T Q^2 (y^*-y^*_o)-1=0. \tag{18}$$

Eq. (18) is the problem of the conditional extreme value. Therefore, this problem is able to solve with the method of using Lagrangian multiplier method. However, there are some difficulty in this solution. Firstly, it is not easy to decide the first values of the design parameters for calculation. Secondly, it is difficult that the constraint function is exactly satisfied in its process.

Then we introduce the following conversational equation.

$$y^*_o(\theta)=y^*+Kc$$

$$\text{where} \quad \theta=(\theta_2\ \theta_3 \ldots \theta_n)^T$$

$$c=(c_1 \ldots c_n)^T \tag{19}$$

θ_k $(k=2,3,\ldots,n)$ is the angle between axis of subscript k and the join axis constructed from the subscript number less than k. c is the vector of which the elements conform to the following equation.

$$c_1 = \cos\theta_n \cos\theta_{n-1} \cdots \cos\theta_3 \cos\theta_2$$

$$c_2 = \cos\theta_n \cos\theta_{n-1} \cdots \cos\theta_3 \sin\theta_2$$

$$c_3 = \cos\theta_n \cos\theta_{n-1} \cdots \sin\theta_3$$

$$\cdots$$

$$c_{n-1} = \cos\theta_n \sin\theta_{n-1}$$

$$c_n = \sin\theta_n \tag{20}$$

According to use Eq. (19), this problem is represented as the following formulation.

$$\beta(\theta)=\|y^*+Kc\| \tag{21}$$

On this formulation, there are not the above mentioned difficulties. Therefore it is able to compute effectively.

4.2 Solution by mathematical programming

From Eq.(18), it is possible to formulate the problem of mathematical programming for β_L and β_U. That is to say, lower bound of β is minimal solution under the equality constraints. And upper bound β is optimum solution on the equality constraints.

minimization of β : $min\ \beta(y_o^*)=\|y_o^*\|$

$$\text{s.t.}\ (y^*-y_o^*)^T Q^2 (y^*-y_o^*)-1=0 \tag{22}$$

maximization of β : $max\ \beta(y_o^*)=\|y_o^*\|$

$$\text{s.t.}\ (y^*-y_o^*)^T Q^2 (y^*-y_o^*)-1=0 \tag{23}$$

Each computing the Eq.(22), (23), it is possible to obtain the solutions of β_L and β_U. However there are some difficulty about Eq.(18) in this method. Because Eq.(21) is the equations without constraints, it is easy to computing its optimal solution. That is to say

$$\beta_L = min\ \|y^*+Kc\|$$

$$\beta_U = max\ \|y^*+Kc\| \tag{24}$$

When number of deign parameter increases, it is effective that this method is used to obtain rather than analytical solution.

5. NUMERICAL ANALYSIS

5.1. 2-Dimensional case

Supporse the stress-strength model. In this case, let x_1 is strength and x_2 is load. Each parameter is conforms normal distribution, and it is independent in the relation both x_1 and x_2. i.e. We consider the following case based on Eq.(2).

$$g(x)=x_1-x_2=0$$

where $\quad x_1 \sim N(10.0,\ 2.0^2),\ x_2 \sim N(1.25,\ 0.65^2) \qquad (25)$

According to Eq.(8) or Eq.(10), β^* that based on Eq.(25) was calculated $\beta^*=4.16$, $(y_1=-3.95703,\ y_2=1.28603)$.

On the other hand, for the random parameter of design parameters has uncertainties, β^* has also uncertainty. Therefore it is necessity that this uncertainty are constrained by hyperellipsoid. The condition of bound, based on confidence level are given by Table 1.

Table 1 Measure of radius on axis based on confidence level : $\gamma=0.95$

Comdition	Both side percent point : t	Sample size : m	Radius of axis : $k=t/\sqrt{m}$
1	1.95996	50	0.277180
2	1.95996	100	0.195996
3	1.95996	500	0.087652
4	1.95996	1000	0.061979

For Example, when x_1 was condition 2 and x_2 condition 3, confidence ellipsoid is expressed as the following equation in the normalized area. i.e.

$$\frac{(y^*_1+3.95703)^2}{0.195996^2}+\frac{(y^*_2+1.28603)^2}{0.087652^2}=1 \tag{26}$$

Since confidence ellipsoid is Eq. (26), it is able to computed β^*_L and β^*_U by Eq. (17).
$\beta^*_L=3.99$, $(y_1^*=-3.79847,\ y_2^*=1.23449)$, $\beta^*_U=4.32$, $(y_1^*=-4.11557,\ y_2^*=1.33755)$
Figure 3 shows the behavior of β^*_L and β^*_U under any sample size m_2.

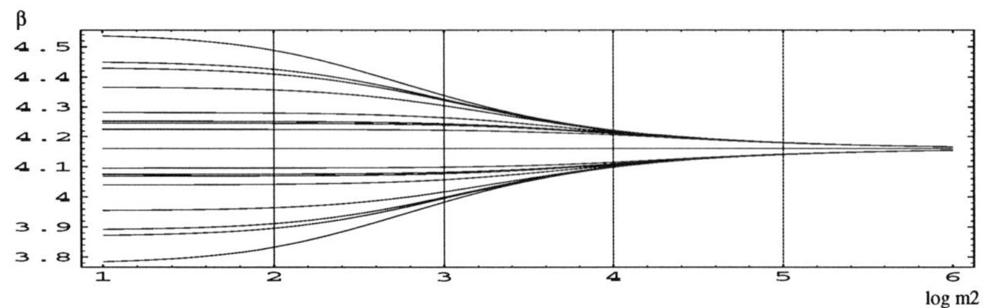

Figure 3 The same axial lower & upper limit of safety index : β^*_L, β^*_U.

5.1.1. Computing analytical solution

To obtain the extreme values based on Eq. (21), the following equation is given from Eq. (19).

$$\beta(\theta)=\sqrt{(-3.95703+0.195996\ \cos\theta)^2+(1.28603+0.087652\ \sin\theta)^2} \tag{27}$$

Further differentiations of this equation about θ is

$$d\beta(\theta)/d\theta=\{-0.195996(-3.95703+0.195996\ \cos\theta)\sin\theta$$
$$+0.087652(1.28603+0.087652\ \sin\theta)\cos\theta\}$$
$$/\{(-3.95703+0.195996\ \cos\theta)^2+(1.28603+0.087652\ \sin\theta)^2\}^{1/2}. \tag{28}$$

From the extreme values calculated by Eq. (27), (28), the following solution are obtained.
$\beta_L=3.97$, $(y_1^*=-3.76324,\ y_2^*=1.27292)$, $\beta_U=4.34$, $(y_1^*=-4.15114,\ y_2^*=1.29817)$

5.1.2. Solution by mathematical programming

In the section 5.1.2., β_L and β_U are shown by Eq. (27), (28). These solutions are computed as the optimum solutions in this section. i.e.

$$\beta(\theta)=\min\ \{(-3.95703+0.195996\ \cos\theta)^2+(1.28603+0.087652\ \sin\theta)^2\}^{1/2}$$
$$\beta(\theta)=\max\ \{(-3.95703+0.195996\ \cos\theta)^2+(1.28603+0.087652\ \sin\theta)^2\}^{1/2} \tag{29}$$

From Eq. (29), these optimum solutions are
$\beta_L=3.97$, $(y_1^*=-3.76324,\ y_2^*=1.27292)$, $\beta_U=4.34$, $(y_1^*=-4.15114,\ y_2^*=1.29817)$

These solutions coincide with the extreme solutions in the section 5.1.1. It is clear that β_L is lower bound of β^* and β_U is upper bound of β^*. β_L and β_U that based on the other case are computed as these values presented by Table.2. We are able to understand these bound when these values are represented in Figure 4.

Table 2.
Lower and upper bound of β^* and the same axial lower and upper limit of β^*

m_2	β_L	β^*_L	β^*_U	β_U
50	3.9559	3.9599	4.3616	4.3662
100	3.9647	3.9647	4.3567	4.3567
500	3.9726	3.9940	4.3274	4.3493
1000	3.9737	4.0170	4.3044	4.3484

β^*=4.1607, Siginificance level 95%, m_1=100

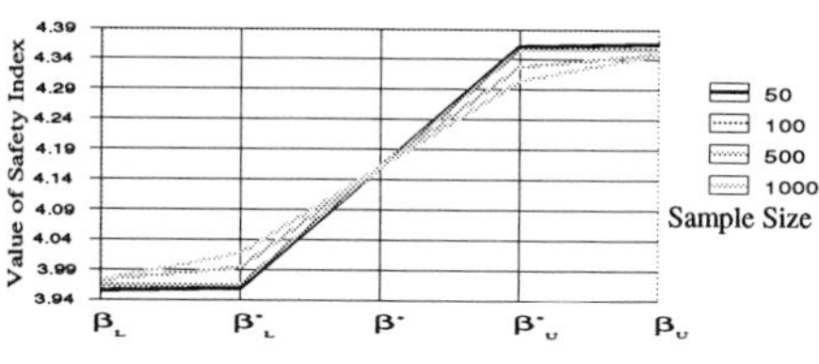

Figure 4. Effect of sample size for β_L β^*_L β^* β^*_U β_U.

5.2. Multidimensional case

In this case, it is valid to obtain optimum solution rather than analytical methods. Consider the problem which composed by 6 parameters is presented by Figure 5.

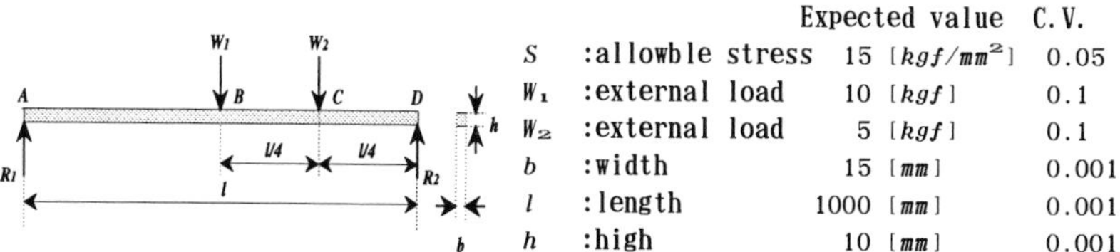

		Expected value	C.V.
S	:allowble stress	15 [kgf/mm^2]	0.05
W_1	:external load	10 [kgf]	0.1
W_2	:external load	5 [kgf]	0.1
b	:width	15 [mm]	0.001
l	:length	1000 [mm]	0.001
h	:high	10 [mm]	0.001

Figure 5. Simply supported beam.

Limit state function is formulated by allowable stress design.

$$S-3W_1l/2bh^2-3W_2l/4bh^2=0 \tag{30}$$

Where each parameters are distributed as normal distribution. And each correlation coefficients are 0. Therefore, the relation of both different parameters is each independent. Then β^* based on Eq.(8) or (10) is obtained.

β^*=4.05585, (y_1^*=-2.384664, y_2^*=3.180515, y_3^*=0.795129,
y_4^*=-0.050507, y_5^*=0.050502, y_6^*=-0.101019)

Statistical uncertainties based on confidence level are presented as hyperellipsoid.

$$\begin{aligned}\beta_L=min\{&(-2.384664-0.19996\cos\theta_2\cos\theta_3\cos\theta_4\cos\theta_5\cos\theta_6)^2\\&+(3.180515-0.19996\sin\theta_2\cos\theta_3\cos\theta_4\cos\theta_5\cos\theta_6)^2\\&+(0.795129-0.19996\sin\theta_3\cos\theta_4\cos\theta_5\cos\theta_6)^2\\&+(-0.050507-0.19996\sin\theta_4\cos\theta_5\cos\theta_6)^2\\&+(0.050502-0.19996\sin\theta_5\cos\theta_6)^2\\&+(-0.101019-0.19996\sin\theta_6)^2\}^{1/2}\\\beta_U=max\{&(-2.384664-0.19996\cos\theta_2\cos\theta_3\cos\theta_4\cos\theta_5\cos\theta_6)^2\\&+(3.180515-0.19996\sin\theta_2\cos\theta_3\cos\theta_4\cos\theta_5\cos\theta_6)^2\\&+(0.795129-0.19996\sin\theta_3\cos\theta_4\cos\theta_5\cos\theta_6)^2\\&+(-0.050507-0.19996\sin\theta_4\cos\theta_5\cos\theta_6)^2\\&+(0.050502-0.19996\sin\theta_5\cos\theta_6)^2\\&+(-0.101019-0.19996\sin\theta_6)^2\}^{1/2}\end{aligned} \tag{31}$$

When it is computed by Eq.(31), β_L and β_U are obtained in the table 3. These values of Table 3 are represented as Figure 7. Where it is known that the less the sample size is the larger the interval both β_L and β_U is.

Table 3.
Lower and upper bound of β^* and the same axial lower and upper limit of β^*

m_2	β_L	β^*_L	β^*_U	β_U
50	3.5701	3.7625	4.3491	4.5695
100	3.8598	3.8598	4.2518	4.2518
500	3.9269	3.9791	4.1325	4.1888
1000	3.9361	4.0309	4.0807	4.1811

β^*=4.0558, Confidence level 95%,
$m_1=m_2=m_4=m_5=m_6=100$

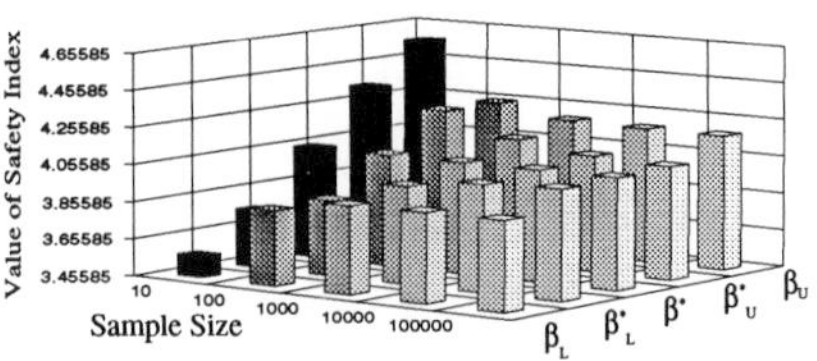

Figure 6.
Relation between sample size and lower and upper bound safety index.

6. CONCLUSIONS

In this paper, it is remarkable that about the statistical uncertainties of safety index β must be taken into account. β^* which is presented by Hasofer and Lind [1] is used as the exact and invariant measure. However, when the design parameters have to be estimated by sample sizes and siginificance levels, we should recognize these random parameter have statistical uncertainties. These uncertainties are shown by the region within the hyperellipsoid proposed by the method in this paper.

ACKNOWLEDGMENTS

This work was party supported by the Science Research Fund of the Ministry Education, Science and Culture of Japan, and the Science Research Promotion Fund of Japan, and the Science Research Promotion Fund of Japan Private School Promotion Foundation.

REFERENCES

1. A. M. Hasofer and N. C. Lind, Proc. ASCE, J. Eng. Mech. Div., 100-EM1, (1974) 111.
2. M. Sinozuka, Proc. ASCE, J.Str. Eng., 109-SE (1983) 721.
3. M. Rosenblatt, Annals Math. Stat., 23. (1959) 470.

Reliability and Optimization of Structural Systems, V (B-12)
P. Thoft-Christensen and H. Ishikawa (Editors)
Elsevier Science Publishers B.V. (North-Holland)

Probabilistic Validation of EC 2 Partial Safety Factors using Full Distribution Reliability Methods

A. Neuenhofer, K. Zilch

Dept. of Civ. Eng., Technical University Aachen, 5100 Aachen, Germany

Abstract

Several parametric studies are carried out in order to examine the stochastic mechanical behavior of reinforced concrete structural members under combined axial force-bending moment interaction. The emphasis is on pointing out different influences of both the two materials concrete and reinforcing steel and the two load components. Current code design regulations are investigated in terms of the validity of the set of applied partial safety factors. Topics include reliability analyses of reinforced concrete cross sections as well as EC 2 designed model columns accounting for the effect of geometric nonlinearity. A probabilistic design with respect to a target reliability index is carried out to yield an adequate set of partial safety factors for each design situation which may be compared with the respective values of a deterministic code format.

1 Introduction

In the process of further realizing the European Single Market that was officially initiated with the beginning of this year, common and uniform design rules for structures (Eurocodes) have to be developed among the members of the European Community. Eurocode No. 2 [1] covers the design of concrete structures. Compared to the former German regulation DIN 1045 [2] EC 2 incorporates basically two entirely new concepts, namely the partial safety factor approach and the application of nonlinear methods to determine the internal forces in the structure. This presentation concentrates on the former. Design of cross sections of reinforced concrete structures is among other key assumptions usually based on stress-strain-relationships for concrete and reinforcing steel, respectively with corresponding values for stress and strain in the ultimate limit state. The combined behavior of the two components is complex in terms of their contribution to a generalized resistance quantity. Consequently a fixed set of partial safety factors is not always capable of adequately reflecting the inherent variability of the design parameters.

2 Basis of Design of Eurocode 2

The limit state design of modern structural codes is based on fundamental principles of reliability and probability theory. In structural design, limit states are usually classified into ultimate limit states and serviceability limit states. The investigations in this study only refer to the ultimate limit state and checking may require consideration of loss of equilibrium or stability as well as rupture or failure due to excessive deformations. In general the condition for non-exceedance of a limit state of a structural component can be written as $g(x_1, \ldots, x_n) \geq 0$ where $\mathbf{x} = (x_1 \ldots x_n)^T$ is a set of basic random variables that define the state of the component and $g(\mathbf{x})$ denotes the limit state or failure function. The variables $\mathbf{x}$ represent basic uncertain quantities, e.g. action variables, material properties and geometrical parameters. The formulation of the before mentioned limit state criterion in terms of a deterministic code format can then be expressed as $g(x_{d_1}, \ldots, x_{d_n}) \geq 0$ where x_{d_i} is a deterministic design value of the respective variable x_i.

3 Determination of Partial Safety Factors Using FORM

It is well known that the first-order estimate

$$p_{f1} = \int\limits_{\nabla G(\mathbf{y}^*)(\mathbf{y}-\mathbf{y}^*)\leq 0} \varphi_{\mathbf{Y}}(\mathbf{y})\mathrm{d}\mathbf{y} = \Phi(-\beta) \tag{1}$$

for the probability of failure p_f and the reliability index β provides an acceptable approximation for the exact value of the probability integral

$$p_f = \int_{G(\mathbf{y})\leq 0} \varphi_{\mathbf{Y}}(\mathbf{y})d\mathbf{y} \tag{2}$$

as well as for the strict reliability index β especially for the vast majority of structural engineering problems where failure probabilities are usually very low. The expressions in Eqs. 1 and 2 are formulated in the transformed space of independent standard normal variables $\mathbf{y}$ where $\varphi_{\mathbf{Y}}(\mathbf{y})$ and $\Phi(.)$ denote the multidimensional standard normal PDF of $\mathbf{Y}$ and the standard normal CDF, respectively. In Eq. 1 the limit state surface $G(\mathbf{y})$ in the standard normal space is replaced by an approximating tangent hyperplane at the design point $\mathbf{y}^*$, the point with minimum distance β to the origin of space $\mathbf{y}$ such that $\beta = \sqrt{\mathbf{y}^{*T}\mathbf{y}^*}$. The transformed variables $\mathbf{y}$ are obtained from the space $\mathbf{x}$ of basic random variables through the probability transformation $\mathbf{y} = \mathbf{T}(\mathbf{x})$. Using the inverse mapping the design point $\mathbf{x}^*$ in the original space is obtained by $\mathbf{x}^* = \mathbf{T}^{-1}(\mathbf{y}^*)$. It can easily be seen that a structure designed using the values x_i^* of the probabilistic design point $\mathbf{x}^*$ for the deterministic design values x_{d_i} has the reliability index β with respect to a specified limit state function $g(\mathbf{x})$. Hence, if the reliability index is acceptable for certain design conditions, adequate partial safety coefficients $\gamma_i = x_i^*/x_{k_i}$ for action variables are obtained by normalizing the coordinates x_i^* of the design point with respect to the characteristic values x_{k_i} of the respective quantity. For resistance quantities the reciprocal representation $\gamma_i = x_{k_i}/x_i^*$ is usually used.

4 Reliability Analyses and Probabilistic Design of Reinforced Concrete Structural Components Subjected to M-P Interaction

4.1 Introductory Remarks

Topics of the present section are parametric first-order reliability analyses of both symmetrically reinforced rectangular cross sections and isolated columns subjected to bending moment-axial force interaction. It focusses on two main aspects:

- Reliability level of reinforced concrete cross sections with dimensions b and h and so-called model columns designed according to Eurocode 2 design rules (see Fig. 1)
- Effects of a probabilistic design with respect to a target reliability index on deterministic design parameters like reinforcement ratio and partial safety factors.

We discuss those aspects within four different parametric studies: (1) Reliability of Eurocode 2 designed reinforced concrete cross sections; (2) Reliability of Eurocode 2 designed model columns (accounting for second order effects); (3) Probabilistic design of cross sections; (4) Probabilistic design of isolated columns (accounting for second order effects). In each of the four studies we analyze the limit state function (see Fig. 1)

$$g = M_u(N) - M \tag{3}$$

where M_u and M denote the ultimate bending capacity of the bottom cross section and the acting bending moment, respectively and N denotes the respective axial force. The vector of basic random variables is assumed to be

$$\mathbf{x}^T = (\; P \quad H \quad f_c \quad f_{y1} \quad E_{s1} \quad f_{y2} \quad E_{s2} \;) \tag{4}$$

where

P = Vertical load
H = Horizontal load
f_c = Concrete compressive strength
f_{y1} = Reinforcing steel strength at the tension side of the cross section
E_{s1} = Young's modulus of elasticity of the reinforcing steel at the tension side of the cross section
f_{y2} = Reinforcing steel strength at the compression side of the cross section
E_{s2} = Young's modulus of elasticity of the reinforcing steel at the compression side of the cross section

All the analyses in this study, design as well as first-order reliability calculations are based on the stress strain relationships for the two components concrete and reinforcing steel that are shown in Fig. 3, together with the deterministic values of ultimate strains that are needed for the evaluation of the resistance term M_u in Eq. 3. They are shown in Fig. 4. The statistical assumptions in terms of the load and resistance properties as well as the set of partial safety factors on which Eurocode 2 design is based are listed in Table 1. The values of f_y and E_s refer

Table 1: Summary of Load and Resistance Properties

Variable	P	H	f_y	E_s	f_c
Dimension	MN	MN	MN/m^2	GN/m^2	MN/m^2
Mean Value μ	μ_P	μ_H	560	200	28
Standard Deviation σ	$0.1\,\mu_P$	$0.1\,\mu_H$	33.6	10	5
Characteristic Value	μ_P	μ_H	500	200	20
Partial Safety Factor γ	1.35	1.35	1.15	1.00	1.50
Deterministic Design Value	$\gamma \cdot \mu_p$	$\gamma \cdot \mu_H$	435	200	13.3

to both steel layers (see Fig. 1). The vertical load P and the horizontal load H are assumed to be normally distributed with a coefficient of variation of $v = 0.10$, for the steel strength f_y and Young's modulus of elasticity E_s, as well as for the concrete compressive strength f_c, we choose a lognormal distribution. All basic random variables are considered to be mutually statistically independent.

Probabilistic Design

In the parametric studies 3 and 4 mentioned above, the reinforcement A_s is optimized with respect to a target reliability index of $\beta^* = 4.7$. Such a level of safety is recommended e.g. by the Nordic Committee on Building Regulations [3] for very serious failure consequences under ductile failure with no reserve capacity or serious failure consequences under brittle failure and instability. The reinforcement A_s leading to the given target reliability index β^* is obtained by applying the standard NEWTON-algorithm

$$A_{s,i+1} = A_{s,i} + \frac{\partial A_s}{\partial \beta} \,|_{\beta_i}\, (\beta^* - \beta_i) \tag{5}$$

The necessary derivative $\partial\beta/\partial A_s$ is computed according to a conventional *FORM* parametric sensitivity analysis [4] to yield

$$\frac{\partial \beta}{\partial A_s} = \frac{1}{\mid \nabla G_{\mathbf{y}}(\mathbf{y}^*, A_s) \mid} \frac{\partial G(\mathbf{y}^*, A_s)}{\partial A_s} \tag{6}$$

where $\nabla G_{\mathbf{y}}$ is the gradient of the limit-state function in the standard normal space with respect to the standard variates $\mathbf{y}$ evaluated at the design point $\mathbf{y}^*$. The application of partial safety factors evaluated according to Section 3 (with $\mathbf{x}^*$ being the final design point after convergence of Eq. 5) would lead to a uniform safety level of the deterministic code format.

Differences Between Cross Section and Column Analyses

As already mentioned, we perform analyses on both cross sections and isolated columns. In the case of the latter, not only physical but also geometrical nonlinearities (second order effects) are included in the analyses.

Cross Sections

A conventional reinforced concrete design is performed with design values $N_d = -P_k \cdot \gamma_P$ and $M_d = H_k \cdot \gamma_H \cdot L$ with the characteristic values P_k and H_k and the partial safety factors γ_P and γ_H for the actions P and H, respectively shown in Table 1. The design values on the material side are also listed in Table 1. With such a design a reliability analysis is performed with the internal forces $N = -P$ and $M = H \cdot L$ acting on the cross section (see Fig. 1), where P and H and the resistance parameters f_c, f_y and E_s are now random variables with statistical properties listed in Table 1.

Columns
With the total design eccentricity $e_{tot} = e_a + e_2$ the second order analysis is transformed to a cross section design with $N_d = -P_k \cdot \gamma_P$ and $M_d = H_k \cdot \gamma_H \cdot L + P_k \cdot \gamma_P \cdot e_{tot}$ in the framework of the Eurocode 2 model column method (see Fig. 2 and Table 1). The eccentricity e_a accounts for the possible effects of imperfections by assuming that the structure is inclined at a certain angle to the vertical, e_2 represents the eccentricity due to second order effects evaluated by approximate methods. More details on this issue are beyond the scope of this paper. On the inclined column we perform a reliability analysis considering a column slenderness of $\lambda = 70$ and we have $N = -P$ and $M = H \cdot L + P \cdot e$ where the eccentricity $e = f(P, H, f_c, f_{y1}, E_{s1}, f_{y2}, E_{s2})$ is now a derived random variable that in general depends on all basic random variables involved in the analyses. As opposed to the cross section analyses, the material parameters do not only influence the resistance side of Eq. 3 but also effect the action side as a result of second order effects. Hence, one has to model the material properties random field along the column. Our investigations, as well as those of other authors, have shown that different correlation models for these random fields have only a minor influence on the reliability as far as internal force or stress limit states are concerned [5], even in the presence of significant second order effects. This phenomenon is due to the predominant influence of the material properties on the resistance side. For this reason, a simplified model for the statistical distribution of the concrete compressive strength f_c along the column is selected. We assume perfect correlation between f_c in the critical section at the base of the column (resistance side) and the f_c within the lowest quarter of the column (action side). The strength f_c of the remaining part of the column is taken as an additional random variable statistically independent from the former. The steel properties f_y and E_s of one steel layer are assumed to be perfectly correlated over the column, the quantities in the two different layers at the tension and compression side of the cross section, respectively, are assumed to be statistically independent.

4.2 Reliability Analyses of Cross Sections and Model Columns

In this parametric study, first-order reliability indices are computed with the general purpose reliability program CALREL [6] for varying levels of the vertical load P and the horizontal load H. In Fig. 5 and Fig. 6 contour lines of the required geometrical reinforcement ratio ρ (Fig. 5a and 6a) and of the reliability index β (Fig. 5b and 6b) are plotted versus the normalized mean axial force and the normalized first order bending moment at the base of the column $\mu_n = -\mu_P/bh\mu_{f_c}$ and $\mu_m = \mu_H/Lbh^2\mu_{f_c}$, respectively, where Fig. 5 and Fig. 6 refer to the cross section and model column analysis, respectively. Reinforcement ratios that are beyond the recommended values of EC 2 are added for completeness (shown in dashed lines). The results obtained from FORM indicate a rather non-uniform safety level. Starting from pure bending ($\mu_n = 0$) with reliability indices at $\beta \approx 5$ that lie pretty much within bounds, we observe a steady increase of β with decreasing normalized eccentricity $\mu_m / \mid \mu_n \mid$ until a maximum of $\beta \approx 10$ occurs for a reinforcement ratio of $\approx 2\%$. We note that the maximum of β approximately coincides with a state of strain in the cross section of $\varepsilon_c = \varepsilon_{cu}$ and $\varepsilon_s = \varepsilon_{su}$ at the probabilistic design point thus representing the transition from steel failure to concrete failure (see Fig. 4). This phenomenon is due to the fact that in this range of loading the two partial safety factors γ_s and γ_c for reinforcing steel and concrete, respectively, have maximum effect and add their influence to yield this high level of reliability. It is important to mention the decreasing reliability index in the range of small reinforcement ratios and small axial forces. This behavior results from the favorable effect of a compression force of small magnitude. This resistance effect of the axial force P can also be observed in the deterministic interaction diagram in Fig. 5a where an increasing P leads to a smaller reinforcement ratio ρ in the lower part of the figure. From a deterministic point of view, this effect arises up to a certain compressive force, independent of the reinforcement ratio. A reliability analysis, however, leads to significant reductions of β only in a very small range of m-n-combinations associated with low reinforcement ratios. Starting from identical results in Fig. 6 compared to Fig. 5 for pure bending (no second order effects) we notice a slightly steeper increase of the reliability index in the case of the model column investigation. This behavior indicates that the EC 2 estimate of the eccentricity e_2 (see Fig. 2) due to second order effects is clearly on the safe side because the results show an increased reliability index compared to the cross section analysis even for rather large eccentricities of $\mu_m / |\mu_n| > 1$. Although we again notice a small range of favorable compressive forces in the lower part of the deterministic m-n-interaction diagram in Fig. 6a (column study) this effect does not lead to a decreasing reliability index as opposed to the observation in Fig. 5b. Maximum values of β are found for almost the same m-n combinations compared to Fig. 5b but we note a broader plateau of large reliability indices e.g. $\beta > 8$.

4.3 Probabilistic Design

As mentioned earlier in this paper, the parametric studies 3 and 4 cover a probabilistic design of reinforced concrete structural components with a target reliability index of $\beta^* = 4.7$. The resulting probabilistic interaction diagrams are shown in Fig. 7a and 8a for the cross section and column analyses, respectively. Qualitatively they coincide with the corresponding deterministic results in Fig. 5a and 6a. A more quantitative comparison between the deterministic and the probabilistic design concept is provided in Fig. 5b and 6b where the normalized difference of reinforcement $\Delta A_s = (A_{s,\mathrm{EC2}} - A_{s,\mathrm{PB}})/A_{s,\mathrm{EC2}}$ between a code design ($A_{s,\mathrm{EC2}}$) and a probabilistic design ($A_{s,\mathrm{PB}}$) is illustrated. In accordance with Fig. 5b and 6b, where a design based on the application of constant partial safety factors leads to rather strongly varying reliability indices, we obtain significant differences in the reinforcement ratio when the design is based on a constant reliability index. Fig. 9 and 10 show contour lines of the partial safety factors that correspond to the probabilistic design. For the sake of brevity, the representation is restricted to the values γ_P and γ_c for the longitudinal force and the concrete compressive strength, respectively. In terms of the concrete compressive strength it should be noted again that the partial safety factor is equal to the normalized design point coordinate with respect to the specified characteristic value which in case of material properties are lower fractiles (see Tab. 1). This is the reason for getting partial safety factors smaller that unity for resistance variables in some situations (see Fig. 9b) which at first glance may appear somewhat surprising. We notice quite a good agreement between the maximal values of the partial safety factors max γ_P and max γ_c in Fig. 9 and 10 and the corresponding values $\gamma_G = 1.35$ for permanent loads and $\gamma_G = 1.50$ for the compressive strength of concrete of Eurocode 2. Surprisingly, we observe γ_P values far below unity for the first order cross section analysis, indicating that in a small range of m-n-combinations the main influence of P as a resistance variable and its main influence as an action variable is of approximately the same magnitude. This behavior results in the significant increase in reinforcement of a probabilistic design compared to Eurocode 2 design (see Fig. 7). In the second order column investigation this phenomenon is largely reduced because of the fact that an increase in the longitudinal load P automatically increases the bending moment. As expected, the partial safety factor for P increases with decreasing eccentricity of the loading, maximal values are obtained for almost axial loading ($m/n \approx 0$) and high reinforcement ratios. Contour lines for the partial safety factor γ_c show a strong dependence on the reinforcement ratio ρ in the range of concrete failure ($\gamma_c > 1.2$) due to different load sharing of the two components with varying reinforcement ratio. Closely spaced contour lines for both γ_P and γ_c in the range of $-n \approx 0.25$ indicate a rapid change from steel failure ($-n \leq 0.25$) to concrete failure (see Fig. 4). Significant reductions in the necessary amount of reinforcement that can be observed for a wide range of m-n combinations are predominantly due to the fact that the maximum values for the partial safety factors occur in quite different regions of the m-n-plane.

The sole second order effects, isolated from the influence of eccentricity and reinforcement ratio appear to be negligible (see Fig. 9 and 10). The extreme values of the partial safety coefficients are not effected, nor do we find significantly different values of γ when fixing the reinforcement ratio and the load eccentricity, with the exception of observing an extreme favorable effect of the vertical load in the first order study that doesn't exist in the column investigation for reasons already mentioned.

The basic computational steps that have been performed in the probabilistic design are summarized in the flow diagram presented in Fig. 11. Particularly, a probabilistic design that involves a nonlinear finite element analysis is computationally extremely costly due to three nested iterative procedures. Despite an effective approach for determining the gradients [5] the necessary computer time on a 486 personal computer amounts to 220 hours for analyzing 50 x 50 m-n combinations.

5 Summary and Conclusions

A comprehensive numerical study is carried out on the reliability of reinforced structural components subjected to plane bending moment-axial force interaction. Both simple rectangular cross sections and isolated columns with geometrically nonlinear behavior are considered. Results show a substantially varying reliability index depending on the load eccentricity and the reinforcement ratio. Through a probabilistic design of the same structural members, insight is gained into the performance of Eurocode 2 partial safety factors in terms of a safe and economic design.

References

1. DIN 1045 (1988). *Beton und Stahlbeton, Bemessung und Ausführung.* Beuth Verlag GmbH, Berlin, Germany.
2. Eurocode 2 (1992). *Design of concrete structures.* Part 1: General Rules and Rules for Buildings. Commission of the European Communities, Revised Final Draft.
3. Nordic Committee on Building Regulations (1978). Recommendation for Loading and Safety Regulations for Structural Design. *NKB-Report* No. 36.
4. Madsen, H. O., Krenk, S., Lind, N. C. (1986). *Methods of structural safety.* Prentice-Hall, Englewood Cliffs, New Jersey, USA.
5. Liu, P. L., Der Kiureghian, A. (1991). Finite element reliability of geometrically nonlinear uncertain structures. *Journal of Engineering Mechanics*, ASCE, 117, 1806-1825.
6. Liu, P. L., Lin, H. Z., Der Kiureghian, A. (1989). CALREL User Manual. *Report No. UCB/SEMM-89/18.* Department of Civil Engineering, University of California, Berkeley, California, USA.

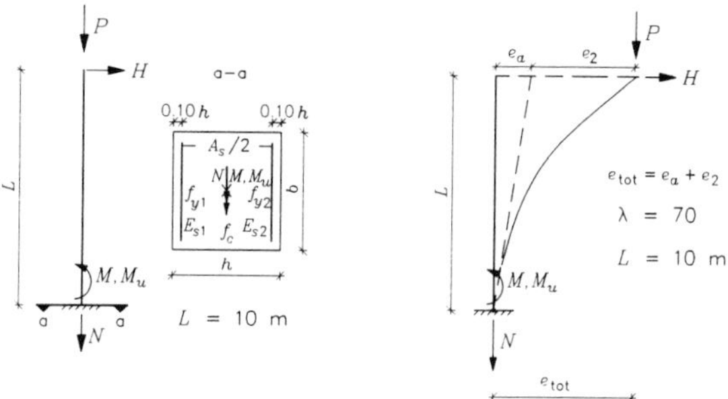

Fig. 1. Basic Random Variables and Example Structure   Fig. 2. EC 2 Model Column

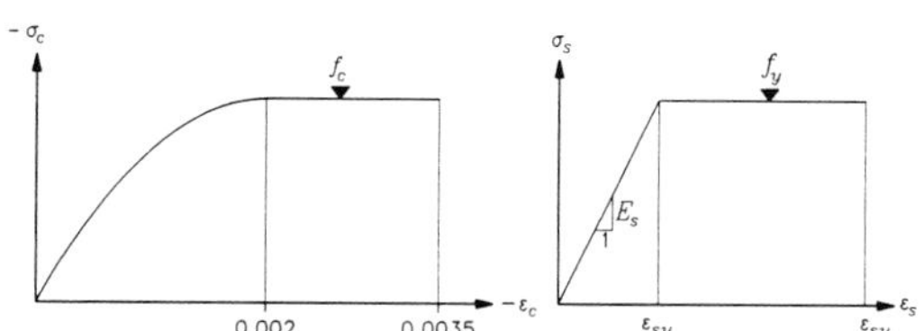

Fig. 3. Stress-Strain-Diagrams for Concrete and Reinforcing Steel

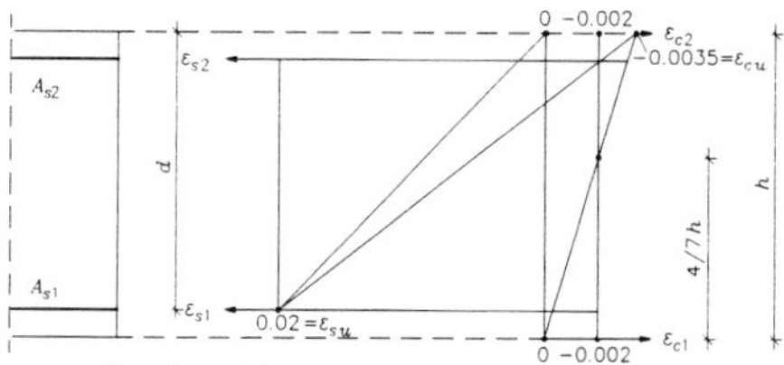

Fig. 4. Strain Diagram for the Ultimate Limit State According to Eurocode 2

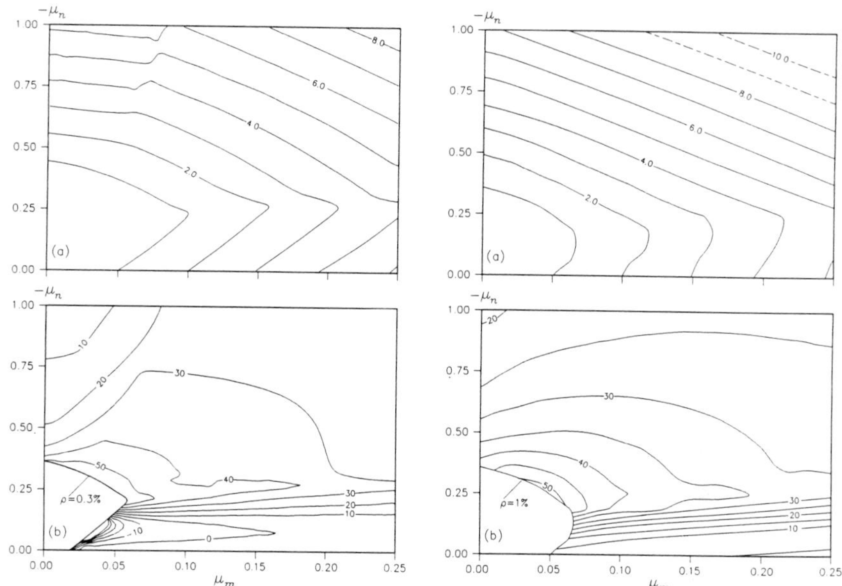

Fig. 5. Contour Lines of the Reinforcement Ratio ρ (a) in % and the Reliability Index β (b) of EC 2 Designed Cross Sections

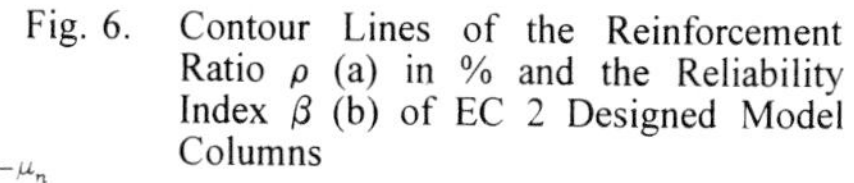

Fig. 6. Contour Lines of the Reinforcement Ratio ρ (a) in % and the Reliability Index β (b) of EC 2 Designed Model Columns

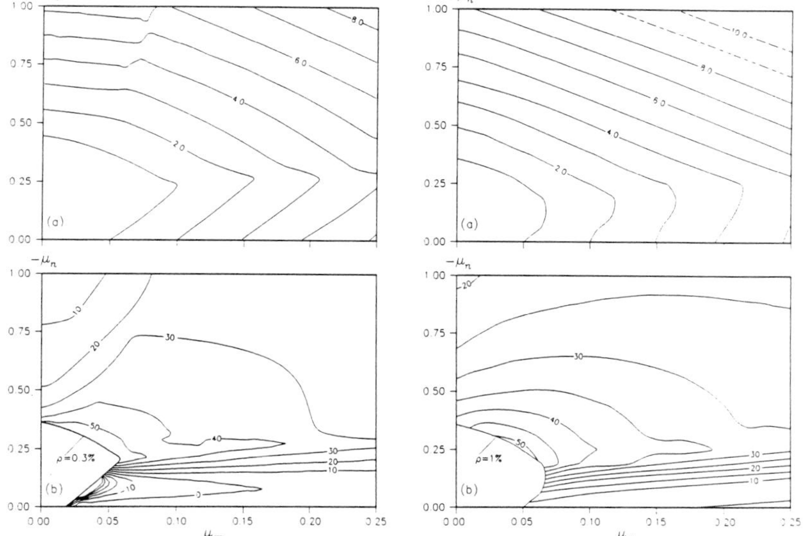

Fig. 7. Probabilistic Interaction Diagram (Contour Lines of ρ) (a) and Contour Lines of Normalized ΔA_s Between EC 2 Code Design and Probabilistic Design (b) in % (Cross Sections)

Fig. 8. Probabilistic Interaction Diagram (Contour Lines of ρ) (a) and Contour Lines of Normalized ΔA_s Between EC 2 Code Design and Probabilistic Design (b) in % (Columns)

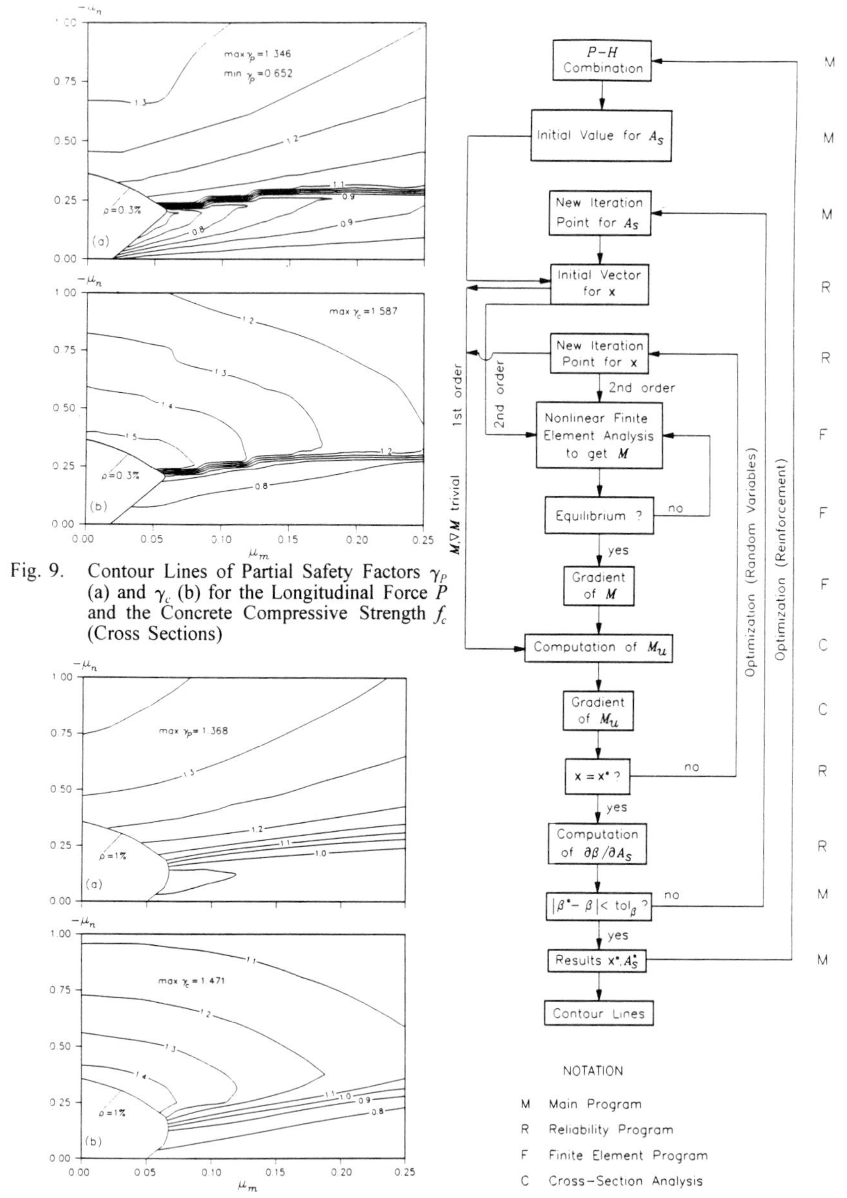

Fig. 9. Contour Lines of Partial Safety Factors γ_P (a) and γ_c (b) for the Longitudinal Force P and the Concrete Compressive Strength f_c (Cross Sections)

Fig. 10. Contour Lines of Partial Safety Factors γ_P (a) and γ_c (b) for the Longitudinal Force P and the Concrete Compressive Strength f_c (Isolated Columns)

Fig. 11. Flow Diagram for Probabilistic Design of Reinforced Concrete Cross Sections and Isolated Columns

Reliability and Optimization of Structural Systems, V (B-12)
P. Thoft-Christensen and H. Ishikawa (Editors)
Elsevier Science Publishers B.V. (North-Holland)

Fundamental Study on Application of AI Techniques to Reliability-Based Design

Y. Okada, S. Ogawa

Nuclear Power Department, Shimizu Corporation,
Seavans South, Shibaura 1-chome, Tokyo 105-07, Japan

Design is to choose a set of optimum members. Artificial Intelligence techniques, e.g. genetic algorithms and neural network, have been recently developed to be applied to optimization problems.

In the present paper, we conduct a fundamental research toward reliability-based design to apply genetic algorithms and neural network to designing steel frames subjected to seismic load. In optimizing the steel members, we take the safety index as an objective function and structural cost, etc. as constraint conditions.

As the result, we found that a set of optimum members can be readily selected by using genetic algorithms. Therefore, we can see the possibility that genetic algorithms are combined with reliability assessment to develop a Level 2 reliability-based design.

Keyword codes:G.1.6; G.3; I.2.m
Keywords: Numerical Analysis, Optimization; Probability and Statistics;
Artificial Intelligence, Miscellaneous

1. Introduction

Recently, Load and Resistance Factor Design has been adopted for designing ordinary buildings in quite a few countries. It is a kind of reliability-based design, so called Level 1 design. One of the merits can be that it is practical since a designer is not required to possess a good knowledge of probability theory in designing structures. However, Level 1 design is a structural design procedure for a lot of similar structures which are planned and built repeatedly. Furthermore, the safety of the structures designed by Level 1 may scatter because of its simplicity.

For a very important structure, e.g. a nuclear power plant, Level 2 and 3 designs can be more suitable than Level 1. However, not only their detailed procedures but their design concept have not yet been established. The final aim of the present paper is to show an attempt to develop Level 2 and 3 reliability-based design.

We try to perform an optimum design such that selected members should be the best set or, say within the best five choices, among all possible combinations satisfied given conditions. One calls it combinatorial optimization. To solve it, an objective function and constraint conditions are to be selected. We may take the safety index as an objective function and structural cost, etc. as constraint conditions. To the optimization, new Artificial Intelligence (AI) techniques, i.e. genetic algorithms (GA) and neural network (NN), are applied. These techniques have already been applied to the field of structural combinatorial optimization.

The aim of this fundamental study is to examine whether GA and NN are applicable to reliability-based design. We take steel frames as example structures and check whether a set of selected members is the best or reasonably good choice.

We first explain briefly GA and NN when they are applied to structural optimization problems. Second, we take a steel portal frame to examine and compare these applicability. Finally, for a more realistic structure, a 6-story steel frame, we confirm whether GA is an effective sampling technique.

2. Outline of GA and NN

In optimizing structural members, we adopt two recent AI techniques. One is GA, which simulates the mechanics of natural selection and is basically a Generate-and-Test type algorithm. The other is NN, which simulates the activity of neuron and is based on the calculus method. The outlines of these methods applied to optimal design in this study are described in the present section.

2. 1 Genetic Algorithms

A simple GA (SGA) consists of three genetic operators, that is, reproduction(or selection), crossover and mutation.[1)] In SGA, we first assign randomly an integer for each string simulating chromosome, which corresponds to a certain combination of design parameters. Then according to the fitness value of the string calculated using design parameters, the probability of selection is determined and mating strings are selected through roulette wheel selection. After selecting mating strings, crossover is performed at a high probability, where a cross site of the strings is determined randomly and two offsprings, which are going to constitute next generation, are generated by combining separated strings of the parents. After crossover, mutation is performed at a low probability on each child string to extend the search range and help improvement of fitness values.

In applying GA, we define the following evaluation function F for the fitness value.

$$F=1.0/\Psi \tag{2–1}$$

$$\Psi = Z+\sum_{j=1}^{n} r_j \langle g_j \rangle^2 \tag{2–2}$$

Here, Ψ is so called exterior penalty function.[2)] In Eq. (2–2), an objective function is represented by Z and constraint conditions are '$g_j \le 0$' (j=1,2,...,n). The operator <> means,

$$\langle g_j \rangle = \begin{cases} g_j & (g_j > 0) \\ 0 & (g_j \le 0) \end{cases} \tag{2–3}$$

The parameter r_j, which is called penalty parameter, has a positive constant value and should be selected appropriately as to acquire optimum parameters. In the final generation, we can find the optimal combination of design parameters in the string which has the largest fitness value.

2. 2 Neural Network

Neural network proposed by Hopfield[3)] can solve combinatorial problems by minimizing its energy function base on the gradient method. In case of dealing with optimizing problems including objective functions and constraint conditions expressed by quadratic functions of parameters, we can easily transform the problem into Hopfield network and find optimum parameters. But in case of non-linear optimizing problems, which consist of a little complicated functions, this method cannot be applied directly. For such problems, we should apply only the essence of the neural network, that is, the steepest descent method and output function of neurons.[4)] The concept of this method is as follows.

First, the following system equations at the time of t are induced.

$$V_i(t) = \Phi[U_i(t)] \tag{2–4}$$

$$\frac{\mathrm{d}U_i}{\mathrm{d}t}=-\frac{\partial E}{\partial V_i} \tag{2–5}$$

In Eq. (2–4), V_i and U_i represent output and input of neuron i respectively. Any monotone increasing function is available to $\Phi[]$, for which the following Sigmoid function is usually used.

$$\Phi[u]=\frac{1}{1+\exp\left(-\frac{u}{T}\right)} \qquad (T\ ;\ \text{positive constant value}) \tag{2-6}$$

In using Sigmoid function, V_i takes values of 0-1.0. In Eq. (2–5), E represents the energy of the network. According to Eqs.(2-4) and (2-5), the time differential of the energy can be induced as follows.

$$\frac{\mathrm{d}E}{\mathrm{d}t}=\sum_i\left(\frac{\partial E}{\partial V_i}\right)\left(\frac{\mathrm{d}V_i}{\mathrm{d}t}\right)=\sum_i\left(\frac{\partial E}{\partial V_i}\right)\left(\frac{\mathrm{d}V_i}{\mathrm{d}U_i}\right)\left(\frac{\mathrm{d}U_i}{\mathrm{d}t}\right)=-\sum_i\left(\frac{\partial E}{\partial V_i}\right)^2\frac{\mathrm{d}\Phi}{\mathrm{d}U_i}\leq 0 \tag{2–7}$$

Thus, the energy of the network decreases as time passes by.

In applying this NN to optimal design, we assign the penalty function Ψ defined in Eq. (2-2) to the energy of the network E and attempt to minimize it. When dealing with continuous design parameters, we relate design parameter X_i to V_i with linear function.

$$X_i=X_0\cdot V_i \qquad (X_0\ ;\ \text{constant value}) \tag{2–8}$$

Then we can get the modification value of U_i by the following equation.

$$\frac{\mathrm{d}U_i}{\mathrm{d}t}=-X_0\cdot\frac{\partial\Psi}{\partial X_i} \tag{2–9}$$

In case of discrete parameters, in stead of using X_0 , we divide V_i into several levels, the number of which is equal to that of candidates of the parameters, and we assign the discrete values to the parameters according to the level of Vi.

In addition, when we deal with multiple-peak function, it becomes necessary to adopt simulated annealing algorithm, which means scheduled changing of value T, or to add white noise to the right side of Eq. (2-9).

3. Comparison between GA and NN

To evaluate the effectiveness and clarify the difference of GA and NN in optimal problems, we apply them to designing a steel portal frame with discrete design parameters. The analytical model and conditions are shown in Fig.3-1, on which various studies have already been performed.[5)] This structure has 10 different failure modes. Aiming at carrying out a Level 2 reliability-based design of the structure, we evaluated the safety index βu by "Cornell's upper bound", that is,

$$\beta u=-\phi^{-1}\left(\sum_{i=1}^{10}\phi(-\beta_i)\right) \tag{3-1}$$

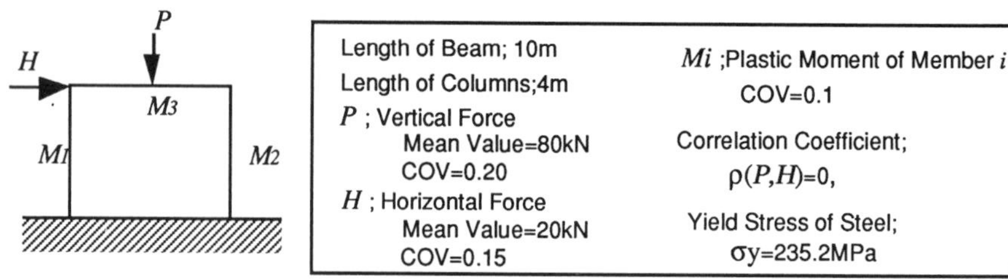

Fig. 3-1 Analytical Model and Conditions

In Eq. (3-1), ϕ and β_i represent the standard normal distribution function and the safety index corresponding to the failure mode i respectively.

In this study, we assign $\phi(-\beta u)$ to the objective function Z and the total weight of the structure (W) to the constraint condition as '$g = W - W_0 \leq 0$' (W_0 ;target weight) in Eq. (2-2).

The design parameters are the mean values of plastic moment for the members. The total weight can be approximately evaluated by W_M below.

$$W_M = \sum_{i=1}^{3} M_i \ell_i \quad (\ell_i;\text{length of member } i) \tag{3-2}$$

Though strictly speaking, W_M does not express the weight itself, we adopted Eq.(3-2) for the first step to simplify the evaluation function by expressing both the objective function and the constraint conditions by M_i on the hypothesis that plastic moments and cross-sectional areas can be related linearly. We also evaluated the real weight of the structure by cross-sectional areas in optimization by GA, later.

As the candidates of design parameters, we selected 32 standard H-shaped steels in Japan, whose properties(cross-sectional area A and plastic section coefficient Zp) are shown in Fig. 3-2. The plastic moment of each section M_i is given by $M_i = Zp_i \sigma y$. Optimization of the design parameters was performed with this model under the constraint of W_0 =5000 kNm^2.

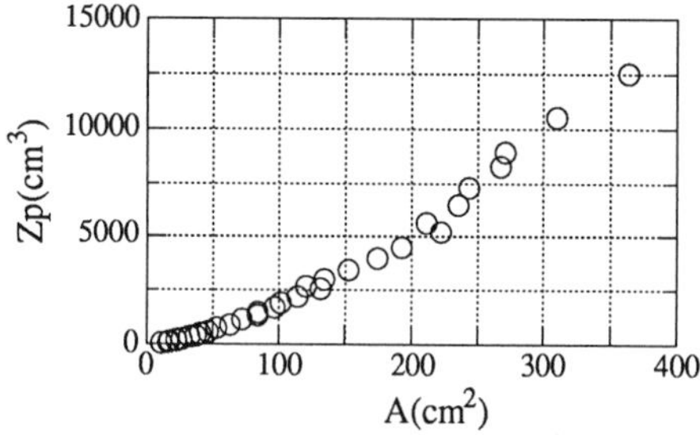

Fig.3-2 Properties of H-shaped steel

In case of GA, various combinations of population sizes, random-number-generation patterns were tried out and some of the results are shown in Fig.3-3 as the relationship between generation and safety index βu. These calculations were conducted with the crossover probability of 0.8, the mutation probability of 0.033 and penalty parameter r_j of

0.01. Generally, it is said that the larger the population size becomes, the best combination appears in the earlier generation and the better combination can be obtained. The result almost accorded with the general trend. In case of P(population size)=100 and 200, it is seen that the objective values are improved by the mutation in the 27th and 13th generation respectively. In case of P=50, a good combination happened to appear in the early generation. The best combination by GA was M_1=128 kNm^2, M_2= 266 kNm^2 ,M_3=341 kNm^2 and βu=5.00.

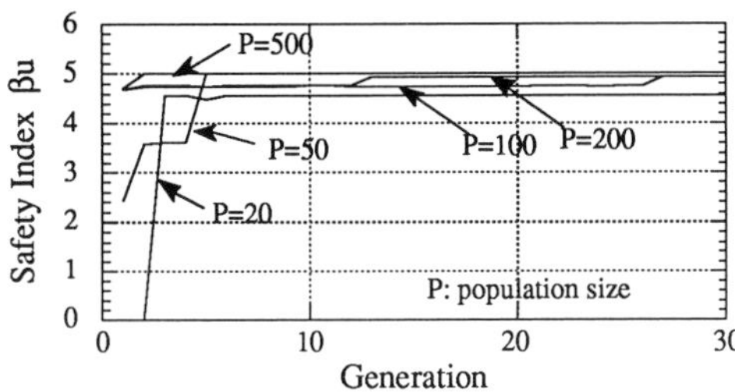

Fig.3-3 History of Optimization by GA

The history of optimization in terms of βu and W_M by NN is shown in Fig.3-4. The first part of the convergence was performed with the continuous method, and the second part with the discrete method. In the continuous part, two discontinuous points are seen, where the penalty parameter is modified. As the initial values for searching discrete optimal values, the continuous optimal values were used, that is, M_1=229 kNm^2, M_2=231 kNm^2 , M_3=316 kNm^2 and βu=5.06. In the second part, a random search is carried out near the continuous optimum point by adding white noise to the right side in Eq.(2-9). The final best combination by NN was M_1=204 kNm^2, M_2=204 kNm^2 , M_3=313 kNm^2 and βu=4.82.

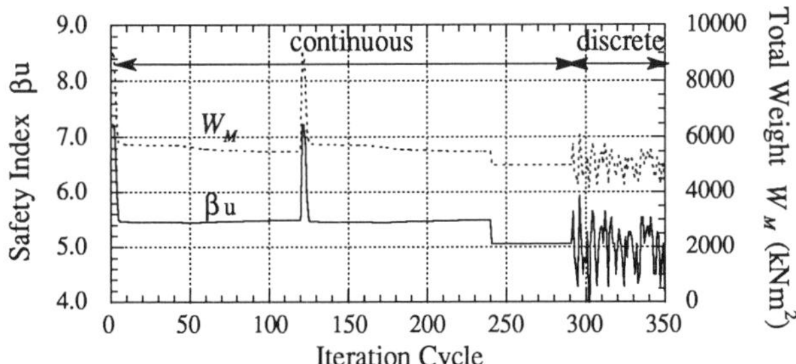

Fig. 3-4 History of Optimization by NN

We confirmed by the round search over the entire combinations that the combination given by GA corresponded to the real optimal one.

Furthermore, we evaluated the constraint by the real weight of the structure W_A by the following equation.

$$W_A = w\sum_{i=1}^{3} A_i \ell_i \qquad (w;\ \text{specific weight of steel=76.93kN/m}^3\text{(7.85tf/m}^3\text{))} \qquad (3\text{-}3)$$

Since A_i in Eq.(3-3) and plastic section coefficient Zp give non-linear relationship including a little randomness as shown in Fig.3-1, the calculus method used in NN cannot deal with such a problem. However, GA can give the optimal combination of M_1=10 kNm^2, M_2=168 kNm^2, M_3=449 kNm^2 and β=5.04 under the constraint of $W_A \leq W_O$=9.8kN(1t f).

Thus, through the comparison between GA and NN, we found that NN is effective for search of continuous optimal parameters or local optimal discrete parameters, but GA is more effective for the optimization with discrete design parameters because of its global search ability and robustness in dealing with non-linear problems.

In the optimization of more realistic structure described in Chapter 4, we adopted GA because of its effectiveness for discrete combinatorial optimization and simplicity of program coding.

4. Numerical Example

Existing structures are not so simple as a portal frame. As the number of members in a structure increases, the number of possible combinations of members goes up on an exponential curve. The increase may lead to difficulties both in assessing the safety index and in optimizing the members. In this chapter, a 6-story steel frame is taken as a numerical example of a more realistic structure to examine.

In Chapter 3, we took the sum of modal safety indexes (the upper bound of the safety) as the objective function. However, when the reviewed structure becomes complicated, assessing the safety index for each failure mode is a formidable task since the number of the failure modes becomes enormous. It indicates that in designing ordinary buildings with optimization techniques, it is necessary to select a suitable objective function and constraint conditions.

In the present study, the safety index for the favorable failure mode specified by a design code is taken as the objective function, and the conditions that make the structure fail with the mode as constraint conditions together with the structural cost. These are realistic and may follow design concept. For the 6-story frame considered, the favorable mode is assumed to be the flexural failure mode with plastic hinges forming 1) at the beam section located at the face of a supporting column, and 2) at the bottom of the basement columns. The condition that brings the frame into the flexural mode is assumed to be as follows.

$$\sum_{\text{all columns}} {}_cM_y \geq 1.2 \times \sum_{\text{all beams}} {}_bM_y \qquad (\text{ at each beam-column joint }) \qquad (4\text{-}1)$$

The analytical model, the favorable mode and analytical conditions are shown in Fig.4-1.

In adopting GA, the fitness function is evaluated basically in the same way as Eq.(2-1). For the objective function, instead of Eq.(3-1), we consider only one failure mode shown in Fig.4-1 and define βu corresponding to it. The condition concerning the total weight of the frame W_A (*i.e.* $W_A \leq W_O$) and the following equations are taken as constraints besides Eq.(4-1).

$$M_i > M_{i+2} \quad (i=1,2,...,10) \qquad (4\text{-}2)$$

Here, M_i is plastic moment of lower column or beam and M_{j+2} is that of upper one. Therefore, 18 constraints including $\beta u \geq 0$ are taken into consideration in all.

In dealing with multiple constraint conditions, each penalty parameter r_j referred in Eq.(2-2) should be tuned well to get the optimum combination efficiently. In this study, the values of r_j were increased systematically with the growth of the generation. Each computation was terminated when the best fitness value stopped improvement and all the constraint conditions were satisfied simultaneously, or the number of processed generations reached 200. The probability of mutation and crossover were taken 0.033 and 0.8.

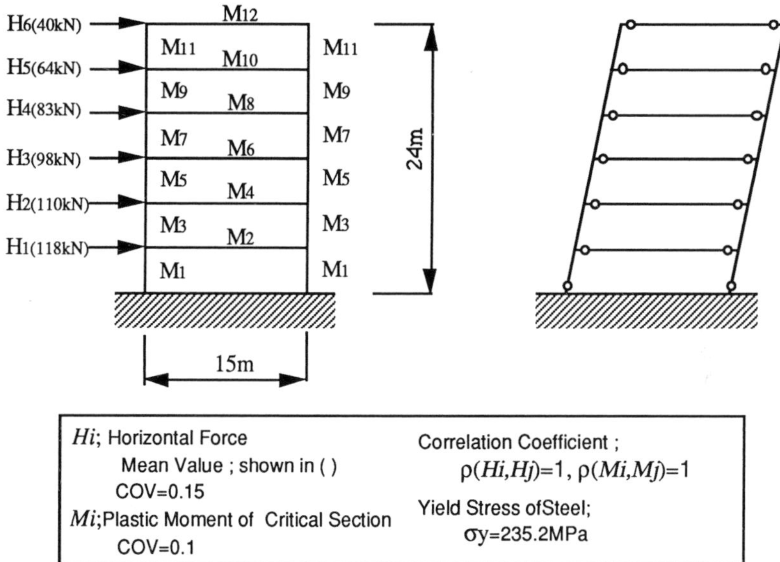

Hi; Horizontal Force Mean Value ; shown in ( ) COV=0.15 *Mi*;Plastic Moment of Critical Section COV=0.1	Correlation Coefficient ; $\rho(Hi,Hj)=1$, $\rho(Mi,Mj)=1$ Yield Stress ofSteel; σy=235.2MPa

Fig.4-1 Analytical Model, Favorable Failure Mode and Analytical Conditions

Some of the optimization results under the condition of $W_A \leq W_O$=98kN(=10tf) are shown in Table 4-1. We took the population sizes as parameters and carried out two runs in each population size to see the effect of starting from different points by choosing different random-number-generator seeds. The best combination was obtained in the largest population size, and the processed generations were almost the same except for case F. The effect of random-number choice was a little sensitive in this optimization.

Table 4-1 Results of Optimization

Case	Population Size	Random Seed	βu	Processed Generation
A	100	1	4.07	66
B	100	1	4.03	56
C	50	1	3.77	66
D	50	1	4.05	53
E	20	1	3.56	51
F	20	1	3.79	200

The order of combinations in this problem reaches 32^{12}, so there exists a little difficulty in finding the optimum combination. Therefore, we generated random numbers and collected

10,000 possible combinations to verify the validity of the acquired combinations by comparing them. Since it's very hard to find combinations satisfying all of the constraint conditions, the constraints by Eq.(4-2) were ignored in Monte Carlo method.

The distribution of safety index βu generated by the Monte Carlo method is shown in Fig.4-2, together with the results by GA.

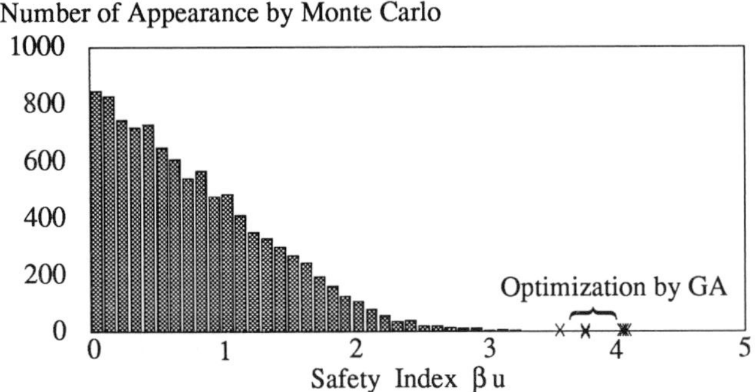

Fig. 4-2 Distribution of βu by Monte Carlo and Results by GA

In Monte Carlo method, we computed on 6,552,557 random combinations and spent 2,591 seconds of execution time on 'SPARC' EWS to find 10,000 possible combinations, but we needed only 62 seconds even in 'case A' in the table with the best combination.

As a result, it was found that GA can be an effective search method in reliability-based combinatorial optimum design of a structure with multiple members and constraint conditions.

5. Conclusions

A fundamental study was conducted on application of AI techniques to Level 2 reliability-based design. We can draw the following conclusions from the present study.

1) Both genetic algorithms and neural network are quite effective to find an optimum solution of structural combinatorial optimization where the safety index is taken as an objective function and structural cost as constraint condition. In this study, the former tends to find better solutions than the latter.
2) Genetic algorithms are effective even when an objective function and constraint conditions are expressed with different parameters in irregular relations.
3) We can see the possibility that genetic algorithms are combined with the assessment of the safety index to develop a Level 2 reliability-based design.

References

1) D. E. Goldberg, Genetic Algorithms in Search, Optimization and Machine Learning, Addison-Wesley, 1989.
2) U. Kirsch (translated into Japanese by Z. Yamada et al), Optimum Structural Design, Maruzen, in Japan, 1983.
3) J. J. Hopfield and D. W. Tank, Neural Computation of Decisions in Optimization Problems Biol. Cybern., Vol. 81(1985).
4) G. Yagawa, Neural Network, Baihu-kan, in Japan, 1992.
5) D. M. Frangopol, A Reliability-Based Optimization Technique for Automatic Plastic Design, Copm. Meth. Appl. Eng.,44(1984).

Reliability and Optimization of Structural Systems, V (B-12)
P. Thoft-Christensen and H. Ishikawa (Editors)
Elsevier Science Publishers B.V. (North-Holland)

A Study on the Second-Order Approximation Method for Engineering Optimization

G.-J. Park, Y.-S. Park, W.-I. Lee

Department of Mechanical Design and Production Engineering,
Hanyang University, 17 Haengdang-Dong, Seongdong-Ku, Seoul, Korea

Optimization has been developed to minimize the cost function while satisfying constraints. The nonlinear programming method is used as a tool for the optimization. Cost and constraint function calculations are normally required in the engineering applications. However, these calculations are extremely expensive. Especially, the function and sensitivity analyses cause a bottleneck in structural optimization which utilizes the Finite Element Method. When the functions are quite noisy, the informations do not carry out a proper role in the optimization process. Recently, the "Second-Order Approximation Method" algorithm has been proposed to overcome the difficulties. The cost and constraint functions are approximated by the second-order Taylor series expansion around a nominal point in the algorithm. An optimal design problem is defined with the approximated functions. The approximated problem is solved by a nonlinear programming algorithm. The solution is included in a candidate point set which is evaluated for a new nominal point. Since the functions are approximated by the function values only, sensitivity informations are not required. The one-dimensional line search is not necessary since the nonlinear algorithm handles the approximated functions. In this research, the method is analyzed and the performance is evaluated. Several mathematical problems are created and some standard engineering problems are selected for the evaluation. Through numerical results, applicabilities of the algorithm to large scale and complex problems are presented.

Keyword Codes: G.1.0; G.1.2; G.1.6
Keywords: Numerical Analysis, General; Approximation; Optimization

1. INTRODUCTION

An optimization problem to be solved is generally stated as

$$\text{Minimize} \quad f(x) \qquad (1)$$

$$\text{Subject to} \quad h_i(x) = 0 \qquad i = 1, m' \qquad (2)$$

$$g_i(x) \le 0 \qquad i = m'+1, m \qquad (3)$$

$$x_i^l \le x_i \le x_i^u \qquad i = 1, p \qquad (4)$$

where $\mathbf{x}$ is a design variable vector, $f(\mathbf{x})$ is a cost function, $h_i(\mathbf{x})$ is an equality constraint function and $g_i(\mathbf{x})$ is an inequality constraint function.

In the conventional optimization technique, function values and their sensitivity informations are required. However, the implementation of the calculations are very costly in engineering application. These calculations may be impossible or inadequate for the quite noisy functions in the optimization process. The function and sensitivity analyses may cause difficulties in the structural optimization which requires the Finite Element Method for calculating the constraint functions. Thus, the application of the structural optimization ends up having many limitations. The exceptions are the well-estabilished problems where F.E.M. informations are easily calculated.[1-4] Efforts have been given to apply the optimization algorithm into Dynamics or Fluid Mechanics. However, the functions and their sensitivity calculations are extremely expensive and sensitivity calculations are impossible in many cases.[1-2]

To overcome these difficulties, an algorithm called "Second-Order Approximation Method" has been recently proposed. The algorithm does not require the sensitivity informations but requires only function calculations to construct the approximated problem in each iteration. It has been developed by Dr. G.N. Vanderplaats a few years ago. However, wide spread applications has not been achieved.[5-7] In this study, the algorithm is analyzed and the results are compared with those of conventional algorithms to evaluate its performance through mathematical and engineering applications.

2. SECOND-ORDER APPROXIMATION ALGORITHM

Overally, all the functions are approximated to quadratic functions and an optimum point is obtained for the approximated problem. The concept of the approximated problem is similar to that of the subproblem of the conventional algorithms. The obtained optimum value is used for the next approximation in the iterative fashion.

2.1 Second-Order Approximation for Given Functions

The Second-Order Approximation algorithm has an arbitrary number of candidate points for the starting point. Among candidate points, a point that minimizes the cost function is selected as a nominal point when constraints are satisfied by the nominal point. If at least a constraint is violated for all the candidate points, a point with the least violation is chosen for the nominal point. The deviations(δx) of design variables from the nominal point and the other deviations(δy) of function values are calculated. The weighted least- square method is performed for a nominal point using δx and δy. It is a curve fitting technique, which constructs the approximated functions through minimizing the summation of the square of the deviations.[8] Cost and constraint functions presented in Eq. (1)-(3) are approximated to explicit functions using the second-order Taylor's series expansion in matrix forms.[5] Through this process, any complicated form of an implicit function can be approximated to an explicit form. The mathematical derivation of the second-order Taylor's series is as follows:

$$f(x^k+\delta x^k) = f(x^k)+\delta x^{kT}\cdot\nabla f(x^k)+ ½\cdot \delta x^{kT}\cdot F\cdot \delta x^k \quad (5)$$

$$h_i(x^k+\delta x^k) = h_i(x^k)+\delta x^{kT}\cdot\nabla h_i(x^k)+½\cdot \delta x^{kT}\cdot H\cdot \delta x^k \quad i=1,m' \quad (6)$$

$$g_i(x^k+\delta x^k) = g_i(x^k)+\delta x^{kT}\cdot\nabla g_i(x^k)+½\cdot \delta x^{kT}\cdot G\cdot \delta x^k \quad i=m'+1,m \quad (7)$$

where x^k is a nominal point in the kth iteration, $\nabla f(x^k)$, $\nabla h_i(x^k)$, $\nabla g_i(x^k)$ are gradient vectors and F,H,G are Hessian matrices of Eq. (1),(2),(3). The values of $f(x^k+\delta x^k)$, $f(x^k)$ and δx^k are known, so the unknowns in Eq. (5) are the components of $\nabla f(x^k)$ and F. As mentioned earlier, the unknowns are calculated by the weighted least-square method using deviations. The more detailed procedure is explained in Reference [8].

2.2 Optimization of the Approximated Problem.

The approximated problem with respect to the design variable δx^k is evaluated by a conventional nonlinear programming algorithm. Since all the functions are approximated to quadratic functions, the function and sensitivity calculations for the approximation problem are quite simple. Therefore, the choice of the algorithm to solve the approximated problem is not essential. If the starting point is taken near a local minimum, the problem can be converged before all unknowns of first order terms are obtained. In this case, the process is iterated until all unknowns of first order term are obtained for the accuracy. However, the process can cease before all the components of second order terms are calculated.

The process is considered to be converged if one of two criterias is satisfied. The first is that the existing nominal point is the minimum point of the approximated problem. The second is the improved point can not be found in four iterations. Occasionally, the final solution may not satisfy all the constraints since any candidate points are made by approximated problems. In this case, an arbitrary point is selected automatically around the nominal point and a certain number of iterations are proceeding. If a feasible solution is not found, the process stops. Those convergence criteria are made based on the heuristic experience of this research. A FORTRAN program has been developed in accordance with the algorithm.[9]

3. APPLICATIONS ON THE SMALL SCALE PROBLEMS

The performance of the Second-Order Approximation algorithm is evaluated by comparisons with some conventional algorithms. Example problems include mathematical problems as well as small and large scale engineering problems. The conventional optimization algorithms used here are Recursive Quadratic Programming(RQP), Cost Function Bounding Method(CFB) and Modified Feasible Direction Method(MFD).[10-11] They are tested under the same condition of Second-Order Approximation Method(SAM). The convergence criterion for the norm of the direction vector is 10^{-4} and that for the maximum constraint violation is 10^{-3}. The utilizes conventional algorithms are well known for their capabilities through extensive optimization researches.[12-14] In this paper, the parameters that are used in the process of each algorithm are set identically for the proper comparison. Example problems include some mathematical and standard structural problems. Since the sensitivity informations are not needed, the

Second-Order Approximation algorithm has been applied for Dynamic problems.[7] However, the conventional algorithms cannot solve these problems easily, the problems that the conventional algorithms can handle are selected and compared with the Second-Order Approximation method.

Throughout the comparisons, the abbreviations are used as follows: They are SPT (starting point), XOV (optimum value), FOV (optimum function value), ITR (No. of Iteration), FEV (No.of Function Evaluation), and GEV (No. of Gradient Evaluation).

3.1 A Problem with Extreme Noise

An ill-conditioned problem is created for this research as shown in Figure 1. This problem has no constraints and has only one design variable. However, the extreme noises are observed near the optimum point in the part A of Figure 1. This condition may appear on a complicated dynamic problems. The practical problems may have many design variables and constraints. The phenomena make the optimization become extremely difficult to obtain the local minimum values. Due to the noise, the sensitivity result on a local point can be different from the overall trend. Although the sensitivity information can be calculated, the convergence of the problem may not be attained due to the inferiorities of the sensitivity analysis result. The mathematical formulation of the problem is represented in reference [9]

RQP and MFD fall into a local minimum near the starting point and do not reach the overall optimum point. The results are shown in Table 1. As specified in Table 1, SAM obtains a same design regardless of the starting point. CFD also obtains the overall optimum point since it iterates without the one-dimensional line search but with bound on the cost function. Therefore, the heuristic algorithms search as SAM and CFB can be more excellent for the problems with extreme noise.

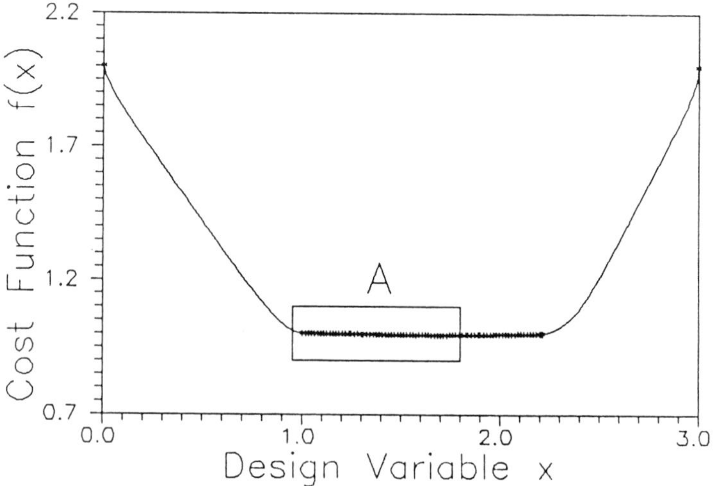

Figure 1. The Entire Graph of a Problem with Extreme Noise

Table 1
Results of the Problem with extreme Noise

	SPT	XOV	FOV	ITR	FEV	SPT	XOV	FOV	ITR	FEV
RQP	1.052	1.4472	0.9990	3	40	1.059	1.5509	0.9937	4	36
CFB	-	1.5432	0.9937	5	43	-	1.1719	0.9974	6	46
MFD	-	1.0487	0.9990	3	11	-	1.7080	0.9934	3	11
SAM	-	1.5202	0.9941	31	32	-	1.5131	0.9940	35	36
RQP	1.085	1.0757	0.9985	3	36	1.06	1.5443	0.9937	3	39
CFB	-	1.5426	0.9937	5	43	-	1.6939	0.9933	11	63
MFD	-	1.0851	0.9987	3	11	-	1.6413	0.9928	4	17
SAM	-	1.6262	0.9932	31	32	-	1.6401	0.9928	6	7

3.2 A Problem not Satisfying Mathematical Conditions

An optimization problem is created to visualize a function which does not have differentiation in a defined region or has deep valleys. The defined optimization problem is shown in Figure 2 and stated in reference [9]. As illustrated in Figure 2, the functions are entangled near the optimum point. The region under the dotted line is the feasible region. When the starting point is from the feasible region (A in Figure 2), only MFD out of conventional algorithms can get the optimum point (* in Figure 2). As shown in Table 2, the solutions are oscillated in RQP or CFB cases. If the starting point is from the feasible region (B in Figure 2) all the conventional algorithms can not obtain a solution. However, SAM makes a quite good solutions although it is not precise optimum values. Therefore, this results indicate that SAM can be applied to the problems which are defined with poor mathematics.

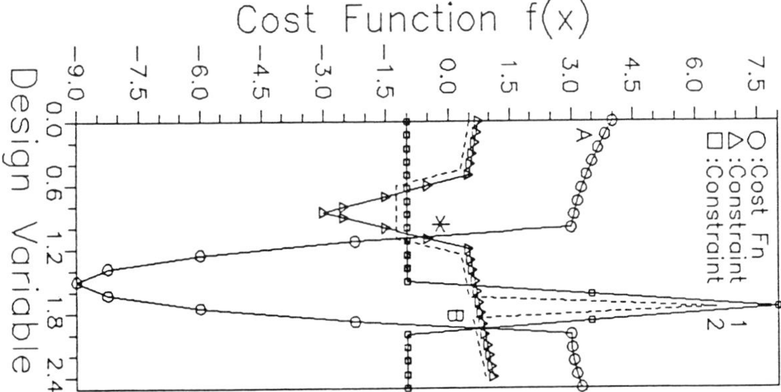

Figure 2. The Graph of the Problem Not Satisfying Mathematical Conditions

Table 2
Results of the Problem Not Satisfying Mathematical Conditions

	SPT	XOV	FOV	ITR	FEV	SPT	XOV	FOV	ITR	FEV
RQP	0.1	1.49	2.875	Oscillate		1.9	1.984	3.000	8	59
CFB	-	1.50	2.877	Oscillate		-	1.894	3.000	10	54
MFD	-	1.15	2.936	5	21	-	1.901	3.011	12	31
SAM	-	0.82	3.106	36	37	-	0.961	3.016	70	72

4. APPLICATIONS ON STRUCTURAL PROBLEMS

SAM is evaluated for the structural optimization. All the conditions are given identically as the previous problems. Finite element method is used for the constraint calculation. In this research, TRUSSOPT program which has been developed for structural optimization is used to calculate the constraint and sensitivities.[15]

4.1 Ten-Bar Truss Problem

The problem formulation is described in detailed in the references [16-17]. The structure is optimized for three cases as shown in Table 3 according to the loading cases and constraint conditions. Optimum results for this problem are given in Table 4. Conventional algorithms are applied for two options with analytic sensitivities and sensitivities from the finite difference method. In three cases, SAM obtain overdesigned values compared to the values from the conventional nonlinear programming algorithms. The numbers of function calculations from the conventional algorithms are smaller than those from SAM when the conventional algorithms utilize the analytic sensitivities. This result does not matter since SAM does not use the analytic sensitivity information which is extremely expensive. However, the numbers are reduced considerably compared to those from the conventional algorithms with sensitivities made by the finite difference method.

Table 3
Loading Cases and Constrains Condition for the Ten-Bar Truss

Case	No.of Loads	Constraints Considered
1	4	Stress
2	2	Stress, Displacement
3	4	Stress, Displacement & Frequency

Table 4
Results of Twenty-Five-Bar Problem

		Case 1				Case 2				Case 3			
		FOV	ITR	FEV	GEV	FOV	ITR	FEV	GEV	FOV	ITR	FEV	GEV
RQP	Analy.	1664.5	10	10	10	5060.8	25	28	28	4783.2	19	22	22
	F.D.M.	1664.5	10	150	0	5060.8	18	594	0	4783.2	19	252	0
CFB	Analy.	1664.5	14	14	14	5060.8	39	46	46	4783.2	32	35	35
	F.D.M.	1664.5	14	194	0	5060.8	50	594	0	4783.2	34	426	0
MFD	Analy.	1664.5	14	109	12	5060.8	19	236	19	4764.3	18	483	16
	F.D.M.	1664.5	11	214	0	5060.8	14	300	0	7278.5	12	103	0
SAM		1910.1	63	89	0	5435.9	74	77	0	4974.5	71	74	0

4.2 Twenty-Five-Bar Truss Problem

The formulation is given in reference [17]. The loading cases and constraint constraints conditions for three cases are illustrated in Table 5. Optimum results for this problem are given in Table 6. The trend of the result is almost same that of Ten-Bar Truss problem. However, it is noted that the numbers of function calculations from SAM are almost same as those from the conventional algorithms with analytic sensitivities. Thus, SAM can be more useful for the bigger problems.

Table 5
Loading Cases and Constrains Condition for the Twenty-Five-Bar Truss

Case	No. of Loads	Constraints Considered
1	4	Stress
2	4	Stress, Displacement
3	4	Stress, Displacement & Frequency

Table 6
Results of Twenty-Five-Bar Problem

		Case 1				Case 2				Case 3			
		FOV	ITR	FEV	GEV	FOV	ITR	FEV	GEV	FOV	ITR	FEV	GEV
RQP	Analy.	91.23	30	47	46	545.0	19	86	16	590.6	26	87	16
	F.D.M.	91.23	25	239	0	545.0	18	174	0	590.6	15	162	0
CFB	Analy.	91.23	44	53	44	545.0	50	61	61	590.6	23	28	28
	F.D.M.	91.23	15	148	0	545.0	15	473	0	590.6	23	219	0
MFD	Analy.	91.23	13	71	13	544.9	13	123	13	589.5	12	60	12
	F.D.M.	91.23	8	110	0	544.9	13	216	0	589.9	10	114	0
SAM		91.78	63	66	0	615.6	42	45	0	744.5	47	50	0

5. CONCLUSIONS

From this research, following conclusions are obtained :

(1) Reasonably good designed values are obtained by SAM for the problems where conventional algorithms cannot make convergence due to the noises or complexities. It is noted that the phenomena are observed from a very small mathematical problems with one design variable and a few constraints. The conventional algorithms are expected to have more difficulties for larger problems. It is extremely difficult to apply optimizations to dynamic or fluid mechanics problems. The sensitivity information is hard to get and the sensitivity information is not useful due to the noises. Therefore, it can be applied to design problems which are known to be impossible to apply the optimization so far.

(2) When standard engineering problems are tested, SAM shows the considerable reduction of the function calculations with a little bigger cost values.

(3) In structural optimization, the results are similar to those from standard engineering problems. SAM can get better performances for the larger scale problems.

(4) SAM has some disadvantages in that it cannot get precise optimum values. In engineering application field, a precise optimum values is not very important since

the mathematical optimum can be more dangerous. Therefore, SAM can be applied easily in practical engineering problems.

(5) The application of SAM is very easy since it does not need the sensitivity calculations. If an analysis program is prepared, it can be easily applied with some effort for the interface. When the selection of the initial values is important, SAM can be applied to find it with less expensiveness.

REFERENCES

1. L.A.Schmit,Structural Synthesis-Its Genesis and Development,AAIA,19 No.10 1249 1263 (1981)
2. G.J.Park and J.S.Arora,Role of Database Management in Design Optimization Systems,J.Aircraft,24 No.11 745 750 (1987)
3. L.A.Schmit and B.Farshi,Some Approximation Concepts for Structural Synthesis, AIAA,12 No.5 692 699 (1974)
4. L.A.Schmit and H.Miura,An Advanced Structural Analysis/Synthesis Capability-ACCESS 2, Int.J.Num.Met.Eng, 12 353 377 (1978)
5. G.N.Vanderplaats,Efficient Algorithm for Numerical Airfoil Optimization, J.Aircraft, 16 No.12 842 847 (1979)
6. V.Sarihan and J.O.Song,Optimization of the Wrist Pin End of an Automobile Engine Connecting Rod with an Interface Fit, J. Mechanical Design, ASME Transaction, 112 406 412 (1990)
7. J.A.Bennet and G.J.Park,Automotive Occupant Dynamics Optimization,Proceedings of ASME Design Technical Conferences,Miami, Florida, U.S.A. 406 412 (1991)
8. M.L.James,G.M.Smith and J.C.Wolford,Applied Numerical Methods for Digital Computation,299 366 (1985)
9. Y.S.Park,G.J.Park and W.I.Lee,An Evaluation of the Second-Order Approximation Method for Engineering Optimization,J.KSME 16 No.2 236-247 (1992) (In Korean)
10. G.N.Vanderplaats,COPES/ADS-A Fortran Control Program for Engineering Synthesis Using the ADS Optimization Program, June,1985
11. J.S.Arora and C.H.Tseng, IDESIGN User's Manual Version 3.3, Optimal Design Laboratory, U.of Iowa, Iowa City, Iowa, U.S.A.
12. P.B.Thanedar,J.S.Arora,C.H.Tseng,O.K.Lim and G.J.Park,Performance of some SQP Algorithms on Structural Design Problems, Int.J.Num.Meth.Eng.23 2187 2203 (1986)
13. O.K.Lim, B.W.Lee and S.I.Cho,Improvement of Recursive Quadratic Programming Using PC, J.KSME,14 No.4 850 856,(1986)(In Korean)
14. G.N.Vanderplaats, Numerical Optimization Techniques for Engineering Design, McGraw-Hill Book Company, New York, 1984
15. J.S.Arora and P.B.Thanedar,User's Manual for Truss Design Problems with IDESIGN 3.3, U.Iowa,Iowa City,Iowa,U.S.A.,1985
16. J.S.Arora,Intoduction to Optimum Design,McGraw-Hill Book Company,New York,1989
17. E.J.Haug and J.S.Arora,Applied Optimal Design, John Wiley & Sons,New York,1979

Reliability and Optimization of Structural Systems, V (B-12)
P. Thoft-Christensen and H. Ishikawa (Editors)
Elsevier Science Publishers B.V. (North-Holland)

Simultaneous Loading Effect of Plural Vehicles on Fatigue Damage of Highway Bridges

M. Sakano*, I. Mikami*, K. Miyagawa**

* Department of Civil Engineering,
Kansai University, Osaka 564, Japan

** Division of Bridge Design,
Hitachi Zosen Corporation, Osaka 554, Japan

Abstract

This study proposes a definition of fatigue design load for highway bridges that uses the root-mean-cube value of truck weights and the simultaneous loading factor obtained through computer simulation analysis.

Keyword Codes: I.6.3; I.6.4; I.6.6
Keywords: Simulation and Modeling, Applications; Model Validation and Analysis; Simulation Output Analysis

1. INTRODUCTION

We have no effective method of fatigue assessment for highway bridge members except orthotropic steel decks in Japanese Specifications for Highway Bridges [1]. A number of highway bridges, however, have experienced fatigue cracking under heavy traffic loading in various structural members besides orthotropic steel decks [2]. Therefore, it is of immediate interest to establish the fatigue design method with the reasonable design load for highway bridges.

Since fatigue cracks in highway bridges are developed by their daily traffic loading, the design load for fatigue assessment should be based not, as in the case of static strength assessment, on the worst loading conditions expected to occur once over the life of bridges, but on typical traffic conditions. In contrast, however, to the case of railway loading, which consists of limited types of trains with the narrow range of weight distributions, in the case of highway bridges, it is difficult to define the fatigue design load which will inflict fatigue damage equivalent to that caused by actual traffic loading, owing to the random-like sequence of various types of vehicles with a wide range of weight distribution.

Miki et al. [3,4] have already proposed a fatigue design load that uses the static design load of T-20 for convenience, and the equivalent number of T-20 loading cycles through computer simulation analyses. Furthermore, by applying the concepts of point load with equal weight and the point process with probability law, Fujino et al. [5] have investigated the effect of the simultaneous presence of plural vehicles on the fatigue damage.

In this study, we consider the fatigue design load to be a large truck which gives the fatigue damage equivalent to that caused by the actual traffic loading; and propose a definition of fatigue design load, using the root-mean-cube value of large truck weight and the simultaneous loading factor of plural vehicles to represent the weight of such a large truck, namely, the equivalent truck weight. The fatigue damage caused by the traffic flow is then obtained by performing computer simulation analyses of highway traffic loading. Traffic conditions and vehicle properties in the simulation analysis are supposed on the basis of field data measured on urban and interurban expressways in Japan.

2. EQUIVALENT TRUCK WEIGHT

2.1. Simulation of traffic loading

The equivalent truck weight can be obtained by performing the computer simulation analysis of highway traffic loading [3,4]. In the simulation, the traffic flow is generated by applying the Monte Carlo method, supposing the type, weight, headway and velocity of vehicles to be random variables with specific probability distributions. Traffic conditions are supposed as follows.

Fig.1 shows five types of vehicle models used in the simulation and their weight distributions. They are the car (C), the small truck (ST), the double-axle large truck (2-LT), the triple-axle large truck (3-LT), and the quadruple-axle trailer truck (TT). The spacing and weight ratio of the axles of each vehicle model are supposed on the basis of actual data measured in Osaka city [6]. The weight distribution of each vehicle model is supposed working from measurements taken at the Ashiya toll-gate on the Hanshin Expressway [7]. In Fig.1 the bar graph indicates the measured distribution and the line graph indicates the assumed distribution. The weight distribution for the car and the small truck is assumed to be the log-normal distribution, and that of the double-axle large truck, the triple-axle large truck, and the quadruple-axle trailer truck is assumed to be the combination of normal and log-normal distributions with two peaks, representing loaded trucks and unloaded ones.

Table 1 gives three types of traffic constitution. Case A is intended to be a traffic flow with a high percentage of large trucks, such as occurs on interurban expressways at night. Case B is intended to be a

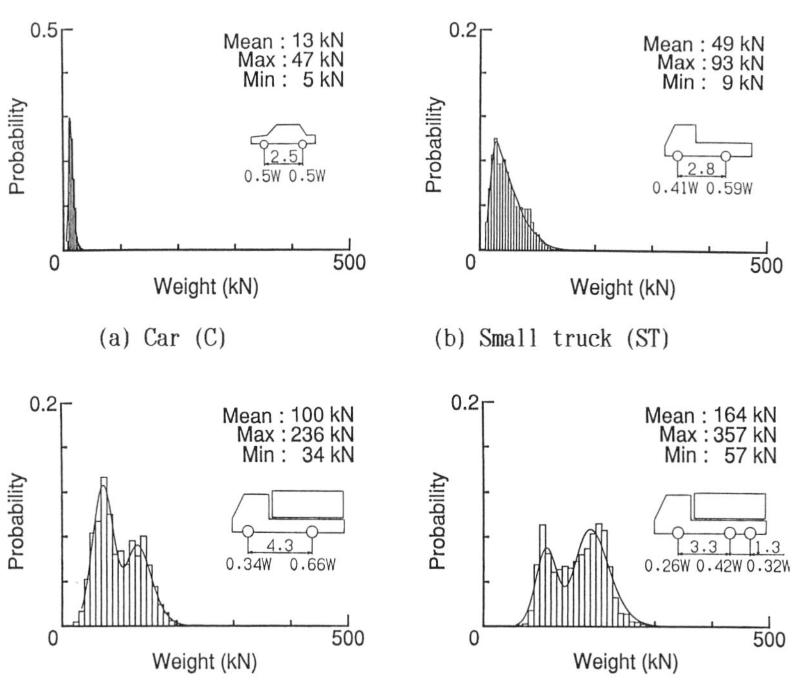

(a) Car (C)

(b) Small truck (ST)

(c) Double-axle large truck (2-LT)

(d) Triple-axle large truck (3-LT)

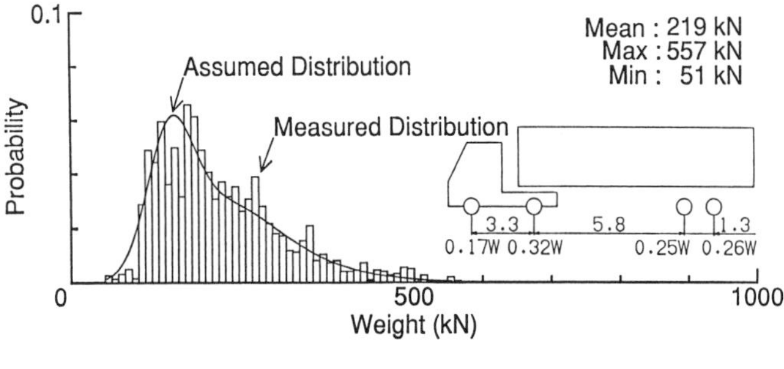

(e) Quadruple-axle trailer truck (TT)

Figure 1. Vehicle models and their weight distributions.

Table 1. Traffic constitution (%)

Traffic Constitution	C	ST	2-LT	3-LT	TT	2-LT + 3-LT + TT
A	10	5	25	50	10	85
B	50	5	20	20	5	45
C	75	12	10	2	1	13

traffic flow with a moderate percentage of large trucks, such as occurs on interurban expressways in the daytime. Case C is intended to be a traffic flow with a high percentage of cars, such as occurs on urban expressways.

The traffic volume is set to be 1500 vehicles/h for the traffic flow which includes a high percentage of trucks. The time headway of vehicles is assumed to obey the Erlang distribution [8]. The degree of freedom is supposed as 3 on a single lane bridge and in the traveling lane of a double lane bridge, and as 2 in the passing lane of a double lane bridge. Considering the traffic flow with a high percentage of large trucks on an expressway, the velocity of each vehicle is assumed to obey normal distribution with a mean value of 70 km/h and a standard deviation of 10 km/h.

2.2. Equivalent truck weight

Equivalent truck weight can be obtained on the basis of fatigue damage caused by stress fluctuations which are produced in bridge members when a simulated traffic flow as passes over the bridge. For stress fluctuations can be substituted the bending moment fluctuation induced in the center of the span by the traffic loading. Only the cases of simply supported girders are investigated in this study, because the influence line dependent on the type of the bridge has little influence on the fatigue damage [4]. The effect of the span length is investigated through comparison between two cases of span lengths of, respectively, 30m and 50m, and the effect of the number of traffic lanes is also investigated through comparison between a further two cases, that of a single lane bridge, and that of a double lane one. Fig. 2 shows an example of the bending moment fluctuation (single lane bridge, span length of 30m, Case A) and Fig.3 shows the moment range histogram obtained by applying the rain flow counting method.

The fatigue damage D can be obtained by applying the linear damage rule and expressed by Eq.(1)

$$D=\Sigma(M_{ri}{}^{m}\cdot n_i) \quad (1)$$

where M_{ri} is the bending moment range and n_i is the number of cycles for M_{ri}. The value of m is assumed to be 3 because the slope of the S-N lines of welded structural members is about -1/3. The equivalent truck weight can be obtained as that of a double-axle large truck, which inflicts a degree of fatigue damage equal to D divided by the total number of large trucks in the highway loading simulation.

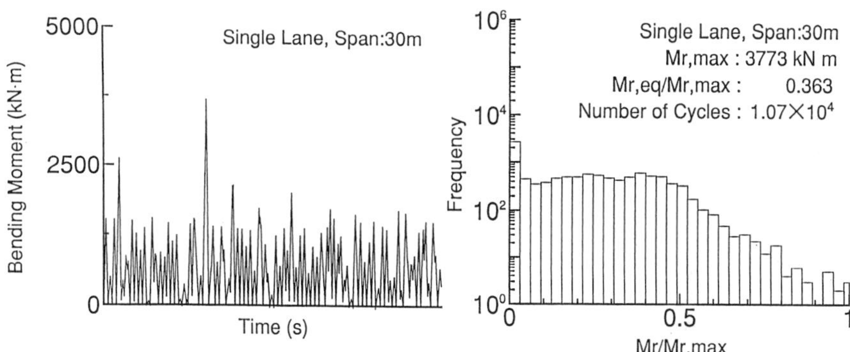

Figure 2. Bending moment fluctuation. Figure 3. Moment range histogram.

Table 2 gives the fatigue damage, the number of large trucks and the equivalent truck weight obtained in each combination of traffic and structural conditions.

Table 2. Fatigue damage and equivalent truck weight.

Number of Lanes	Span Length L (m)	Traffic Constitution	Fatigue Damage D $(kN \cdot m)^3$	Number of Trucks N_{LT}	Equivalent Truck Weight W_{eq} (kN)
1	30	A	1.33×10^{13}	8540	163
		B	6.18×10^{12}	4545	156
		C	1.38×10^{12}	1288	144
	50	A	8.50×10^{13}	8540	178
		B	3.70×10^{13}	4545	166
		C	8.03×10^{12}	1288	152
2	30	A	4.00×10^{13}	17082	187
		B	1.58×10^{12}	9018	170
		C	3.23×10^{12}	2560	152

The effect of simultaneous loading of plural vehicles can be clearly reflected by the equivalent truck weight, since the equivalent truck weight becomes larger when the percentage of large trucks, the span length and the number of traffic lanes increase.

3. ROOT-MEAN-CUBE VALUE OF LARGE TRUCK WEIGHT

The simultaneous loading factor should essentially be defined as the ratio of the equivalent truck weight with simultaneous loading of plural vehicles to that without such loading. The equivalent truck weight without simultaneous multiple loading can be obtained through the simulation analysis with a headway distance longer than the span length so that all vehicles may pass over the bridge independently.

The axle spacing can be ignored and a vehicle can be regarded as a point load when the span length is much larger than the axle spacing. The equivalent truck weight becomes equal to the root-mean-cube value of large truck weight when dimensionless vehicles pass over the bridge separately. The root-mean-cube value of large truck weight (W_{RMC}) can be easily obtained by Eq.(2)

$$W_{RMC} = \{\Sigma(W_{RMCi}^{3} \cdot x_i)\}^{1/3} \tag{2}$$

where W_{RMCi} is the root-mean-cube value of large truck weight obtained from the weight distribution and x_i is the percentage of each type of the large truck.

Table 3. Root-mean-cube value of large truck weight (W_{RMC}) and equivalent truck weight without simultaneous loading (W_{eq}').

Traffic Constitution	W_{RMC} (kN)	W_{eq}' (kN)	
		L=30 (m)	L=50 (m)
A	176	165	168
B	168	154	162
C	145	140	145

Table 3 gives the root-mean-cube value of large truck weight in the case of each type of traffic constitution and the equivalent truck weight without simultaneous loading obtained when the span length is 30m and 50m. The root-mean-cube value of large truck weight and the equivalent truck weight without simultaneous multiple loading become smaller as the percentages of triple-axle large trucks and trailer trucks become smaller (case A → B → C). The root-mean-cube value of large truck weight agrees well with the equivalent truck weight without simultaneous loading. We can therefore use the root-mean-cube value of large truck weight obtained easily from the weight distribution and the percentage of each type of large truck as a basis for defining the simultaneous loading factor instead of the equivalent truck weight without simultaneous loading obtained by performing the simulation analysis.

4. SIMULTANEOUS LOADING FACTOR

We can represent structural properties of bridges such as the span length and the number of traffic lanes by using the simultaneous loading factor γ expressed by Eq.(3), which is defined to be the ratio of the equivalent truck weight (W_{eq}) to the root-mean-cube value of large truck weight (W_{RMC}),

$$\gamma = W_{eq}/W_{RMC} \tag{3}$$

4.1. Effect of the span length

Table 4 gives the value of the simultaneous loading factor γ for the single lane bridge with the span length of 30m and 50m. When the span length is 30m, there is little effect of simultaneous loading since the value of γ is about 1.0 in cases A and B (with the higher percentage of large trucks). When the span length is 50m, the effect of simultaneous loading cannot be disregarded since the value of γ is more than 1.05 in case A.

Table 4. Effect of the span length (single lane).

Span Length L (m)	Traffic Constitution		
	A	B	C
30	0.99	1.01	1.03
50	1.06	1.02	1.05

4.2. Effect of the number of traffic lanes

Table 5 gives the value of the simultaneous loading factor γ for the single lane bridge and the double lane bridge with the span length of 30m. In the double lane bridge, the value of γ is as large as 1.1 in case B and 1.15 in case A. The number of traffic lanes has therefore a great influence on the simultaneous loading of plural vehicles, even when the span length is too short to have any effect of simultaneous loading in the case of a single lane bridge.

Table 5. Effect of the number of traffic lanes (span length : 30m).

Number of Lanes	Traffic Constitution		
	A	B	C
1	0.99	1.01	1.03
2	1.15	1.09	1.06

5. CONCLUDING REMARKS

It has been demonstrated that we can define the reasonable fatigue design load for highway bridges using the simultaneous loading factor and the root-mean-cube value of large truck weight. Further investigations may in the future be required, into various traffic conditions and various types of bridges.

REFERENCES

1. Japanese Association for Road, Specifications for Highway Bridges, (1990) (in Japanese).
2. C.Miki, M.Sakano, K.Tateishi and Y.Fukuoka, Proceedings of JSCE, 392/I-9 (1988) (in Japanese).
3. C.Miki, Y.Goto, J.Murakoshi and K.Tateishi, Journal of Structural Engineering, 32A (1986) (in Japanese).
4. C.Miki, I.Sugimoto, S.Miyazaki and T.Mori, Journal of Structural Engineering, 36A (1990) (in Japanese).
5. Y.Fujino, B.K.Bhartia, C.Miki and M.Ito, Journal of Structural Engineering, 33A (1987) (in Japanese).
6. Kansai Society for Research on Road • Committee for Survey and Research on Highway Bridges, Report of Subcommittee for Fatigue (1989) (in Japanese).
7. Hanshin Expressway Public Corporation, Report of Committee for Highway Design Load (1984) (in Japanese).
8. M.Kubo and M.Shinozuka, Proceedings of the 45th Annual Conference of the Japan Society of Civil Engineers, I-200, (1983) (in Japanese).

Reliability and Optimization of Structural Systems, V (B-12)
P. Thoft-Christensen and H. Ishikawa (Editors)
Elsevier Science Publishers B.V. (North-Holland)

Reliability Analysis of Stochastic Structural Systems using Theory of Random Renewal Pulse Processes

W. Shiraki*, M. Tsunekuni**, S. Matsuho*
* Department of Civil Engineering,
Tottori University, Tottori-shi, Tottori, 680, Japan
** Research Development Department,
Tokyo Electric Power Services Co., Ltd., Tokyo, 100, Japan

ABSTRACT

In this study, a practical reliability evaluation method of structural systems with spatial uncertaintues modeled by two-variate random fields is proposed. Using the results of stochastic finite element method, the response in each finite element of a structural system is modeled as a two-dimensional random pulse space-process. The level crossing theory of two-dimensional pulse space-processes is applied to evaluate the mean upcrossing rate of the response by extending the level crossing theory of one-dimensional pulse processes.

As a numerical example, a bending problem of a steel plate resting on the elastic foundation with statistical non-uniformity such as the bearing coefficient is treated, and the reliability analysis of the plate is performed for the lateral deflection and the bending moment.

Keyword Codes: I.2.1; I.5.4; J.6
Keywords: Artificial Intelligence, Applications and Expert Systems; Pattern Recognition, Applications; Computer-Aided Engineering

1. INTRODUCTION

In structural reliability problems, recently, the uncertainties for loads or material properties of structures have been treated as random fields or random processes of time and space coordinates[1-3]. Such processes that develop simultaneously in space and time are called spatiotemporal processes. In many enginerring problems, two special cases are usually considered such as random function of only time (process) and random function of only space coordinates (random field).

In this paper, a practical reliability evaluation method of structural systems with spatial uncertainties modeled by two-variate random fields is proposed.

First, the probabilistic characteristics such as the expectation functions and the auto-covariance functions of responses of a stochastic structural system are evaluated by using the stochastic finite element method. The set of response in each finite element of the system is modeled as a two-dimensional random pulse space-process.

Next, extending the level crossing theory of one-dimensional random pulse processes, the mean upcrossing rate of the two-dimensional random pulse space-process is evaluated. The evaluation of the mean upcrossing rate of two-variate random fields is very difficult in general. However, the modeling by pulse space-processes makes it easy.

Finally, this proposed method is applied to the reliability analysis of a steel plate resting on an elastic foundation with statistical non-uniformity such as the bearing coefficient which is assumed to be a two-variate random field. In numerical examples, the probability of exceedance for the deflection and bending moment of the plate are respectively evaluated, and the effectiveness of this proposed methods is demonstrated.

2. Discretization of Random Fields Using Local Averages

For the finite element anlysis, it is necessary to discrete the random field of interest into random variables in each finite element. In this study, the random field is discretized by means of the local average which is evaluated by taking spatial average of the field over each finite element, as suggested by vanmarcke [4].

A two-dimensional, statistically homogeneous, continuous random field of the spatial coordinates, t_1 and t_2, is considered and denoted by $X(t_1, t_2)$ here. The local average of a rectangular finite element with the centroid located at (t_1, t_2) is defined as

$$X_A(t_1,t_2) = \frac{1}{A}\int_{t1-T1/2}^{t1+T1/2}\int_{t2-T22}^{t2+T2/2} X(t_1,t_2)dt_1dt_2 \tag{1}$$

where T_1 and T_2 are the side length along the respective coodinate axis and $A=T_1T_2$ is the area of the finite element. When the original random field $X(t_1, t_2)$ has the expectation m and standard deviation σ, the expectation and variance of the local average $X_A(t_1, t_2)$ are represented by Eqs.(2) and (3), respectively.

$$E[X_A] = m \tag{2}$$

$$Var[X_A] = \sigma^2\gamma(T_1,T_2) \tag{3}$$

In Eq.(3), $\gamma(T_1, T_2)$ is called as the variance function of $X(t_1, t_2)$ and given by

$$\gamma(T_1,T_2) = \frac{1}{T_1T_2}\int_{-T1}^{+T1}\int_{-T2}^{+T2}\left(1-\frac{|\tau_1|}{T_1}\right)\left(1-\frac{|\tau_2|}{T_2}\right)\rho(\tau_1,\tau_2)d\tau_1d\tau_2 \tag{4}$$

where τ_1 and τ_2 are the distances between two points of interest along the respective coodinate axis, and $\rho(\tau_1, \tau_2)$ is the auto-correlation function of $X(t_1, t_2)$. Finally, the covariance of the two averaged random fields, X_A and X_A', which correspond to the two rectangular finite elements located arbitrarily, can be expressed as Eq.(5) using the significations shown in Fig.1 as follows:

$$\mathrm{Cov}[X_{Ai}, X_{Aj}] = \frac{1}{A_i A_j} \cdot \frac{\sigma_c^2}{4} \sum_{k=0}^{3} \sum_{l=0}^{3} (-1)^k (-1)^l T_{1k}^2 T_{2l}^2 \gamma(T_{1k}, T_{2l}) \tag{5}$$

Taking advantage of the finite element characteristics assessed by the local averages of spatial uncertainties, the stochastic finite element analysis using the first-order perturbation technique is performed. And the probabilistic characteristics and correlation properties of the responses of the stochastic structural systems.

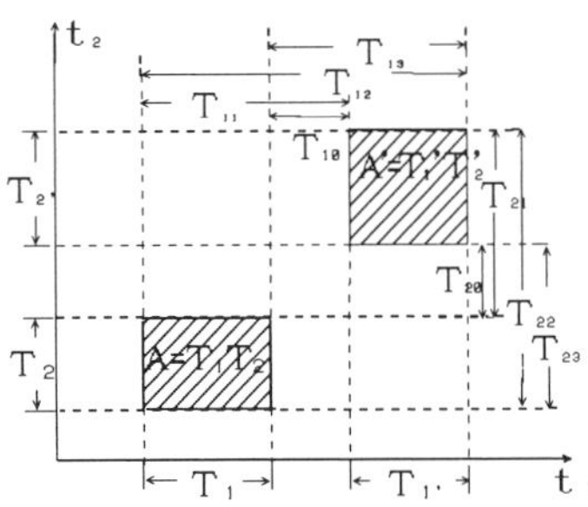

Fig.1 Distances characterizing relative location of rectangular areas A and A'

3. Reliability Analysis Using Theory of Random Pulse Processes

Using the results of the first-order pertubation-based stochastic finite element analysis, in this study, the response in each finite element of a stochastic structural system is modeled as the random pulse space-process. Then, the failure probability of the system is evaluated on the basis of the mean upcrossing rate of the pulse space-process by extending the level crossing theory of one-dimensional pulse processes.

In the followings, the level crossing theory of one-dimensional pulse processes is shown in short at first. Then, the extension of the one-dimensional level crossing problems to the two-dimensional ones is demonstrated. It is very difficult to solve accurately in practice the level crossing problems of random fields or responses of stochastic structual systems. However, it is clarified that the application of the level crossing theory to the stochastic responses modeled as pulse processes makes it easy.

The mean pcrossing rate $\nu_x^+(r)$ of a random rectangular pulse process $X(t)$ of time coordinate, depicted in Fig.2, to a given threshold r is expressed as

$$\nu_X^+(r) = \lim_{\Delta t \to 0} \frac{1}{\Delta t} P[\, X(t) < r \text{ and } X(t+\Delta t) > r \,] \tag{6}$$

For a smooth continuous process, the Rice's formula

$$\nu_X^+(r) = \int_0^\infty \dot{x} f_{X\dot{X}}(r, \dot{x}) d\dot{x} \tag{7}$$

is obtained from Eq. (6). In the above equation $\dot{X}$ is the time derivative process, and $f_{X\dot{X}}$ is the joint probability density function of the value of the original and it's derivative process. For a renewal rectangular pulse process, Eq. (6) is rewritten as follows[5]:

$$\nu_X^+(r) = \lim_{\Delta t \to 0} \frac{1}{\Delta t} \int_{-\infty}^{r} \hat{f}_X(x) \lambda_f \Delta t \hat{G}_S(r) dx \tag{8}$$

In Eq. (8), $\hat{f}_X(x)$ is the arbitary-point-in-time probability density function, and λ_f is the mean arrival rate of X(t). And then, $\lambda_f \Delta t$ means the probability that the new pulse arrives between t and t+Δt. And $\hat{G}_S(r)$ is the probability that the peak of new pulse exceeds r. If the new pulse always arrives at the time interval Δt, the integration in Eq. (8) turns out the product of the probability that the prior pulse remains below r and the probabilty that the new pulse exceeds r.

Analogously to the concepts mentioned above, in this study, the one-dimensional level crossing problems are extended to the level crossing problems of two-dimensional non-homogeneous pulse space-processes. For a two-dimensional space-process, illustrated in Fig. 3, the event that a rectangular space-process with the centroid at coordinated $(t_1+\Delta t_1/2, t_2+\Delta t_2/2)$ exceeds a threshold r along the respective coordinate axis in the area $\Delta t_1 \Delta t_2$ is defined by the condition that $X(t_1-\Delta t_1/2, t_2-\Delta t_2/2)<r$ and $X(t_1+\Delta t_1/2, t_2-\Delta t_2/2)<r$ and $X(t_1-\Delta t_1/2, t_2+\Delta t_2/2)<r$ and $X(t_1+\Delta t_1/2, t_2+\Delta t_2/2)>r$. The mean upcrossing rate in $\Delta t_1 \Delta t_2$ is defined by

$$\nu_X^+(r \mid t_1, t_2) = \frac{1}{\Delta t_1 \Delta t_2} P[\, X_{11}<r \text{ and } X_{12}<r \text{ and } X_{21}<r \text{ and } X_{22}>r \,] \tag{9}$$

where $X_{11}=X(t_1-\Delta t_1/2, t_2-\Delta t_2/2)$, $X_{12}=X(t_1+\Delta t_1/2, t_2-\Delta t_2/2)$, $X_{21}=X(t_1-\Delta t_1/2, t_2+\Delta t_2/2)$ and $X_{22}=X(t_1+\Delta t_1/2, t_2+\Delta t_2/2)$. By introducing the joint probability density function of the four adjoint pulses, denoted by $f_{X_{11}X_{12}X_{21}X_{22}}$, Eq. (9) is rewritten by

$$\nu_X^+(r \mid t_1, t_2) = \frac{1}{\Delta t_1 \Delta t_2} \int_{-\infty}^{r} dx_{11} \int_{-\infty}^{r} dx_{12} \int_{-\infty}^{r} dx_{21} \int_{r}^{\infty} f_{X_{11}X_{12}X_{21}X_{22}}(x_{11}, x_{12}, x_{21}, x_{22})\, dx_{22} \tag{10}$$

If the event of upcrossing of the pulse space-process may be approximated by the Poisson process, the failure probability indicated by $P_f^*(r)$ of the two-dimensional stochastic structural system is evaluated by

$$P_f^*(r) = 1 - \exp\left(-\sum_{i=1}^{n} \nu_{Xi}^+(r \mid t_1, t_2) \cdot \Delta t_{1i} \Delta t_{2i} \right) \tag{11}$$

where n is the number of rectangular pulse, i.e. the number of finite elements in the analysis.

In general, it is very difficult to solve accurately level crossing problems utilizing the classic formula such as the Rice's formula expressed as Eq. (7). This comes from the difficulty of analytical estimations of the probabilistic properties of derivative processes, included in the classic formula. Taking advantage of the concept of pulse processes, however, there are no terms of the derivatives. Consequently, the reliability analysis of stochastic structural systems can be easily performed.

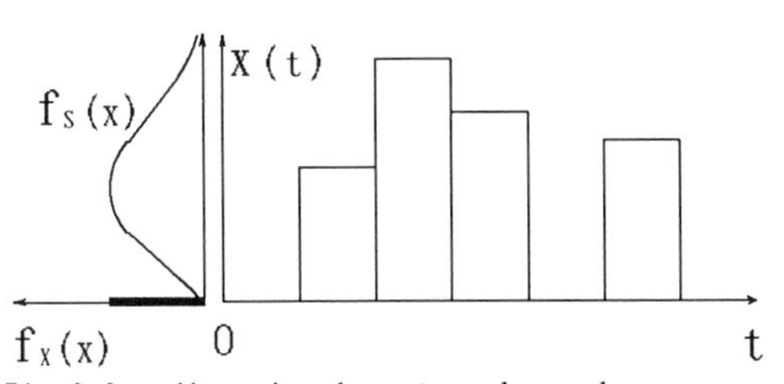

Fig. 2 One-dimensional rectangular pulse process

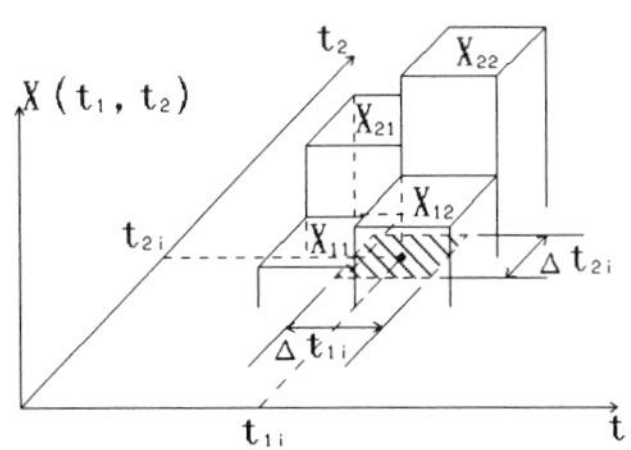

Fig. 3 Level crossing of two-dimensional rectangular pulse-space process X_{22}

4. Numerical Examples

As numerical examples, the bending problems of a steel plate resting on the elastic foundation of the Winkler's modelis considered, as shown in Fig. 4. In the problems, the bearing coefficient of the foundation is modeled as the two-dimensional, statistically homogeneous, continuous Gaussianrandom field, represented by $g(x_1, x_2)$, with the mean m_g=626.5tonf/m^2, the standard deviation σ_g and the auto-correlation function of the form

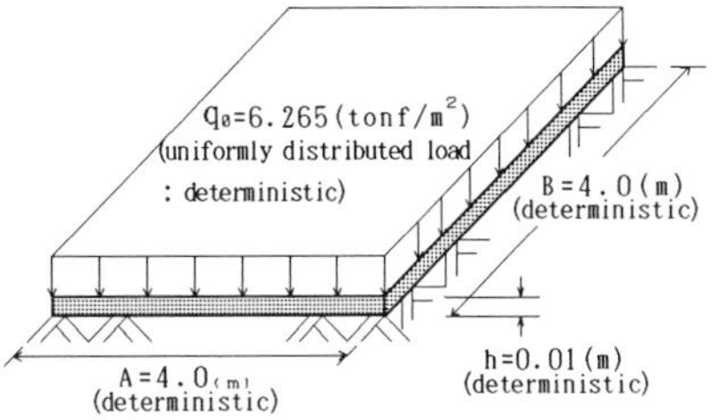

q_0=6.265 (tonf/m^2) (uniformly distributed load : deterministic)

B=4.0 (m) (deterministic)

h=0.01 (m) (deterministic)

A=4.0 (m) (deterministic)

Fig. 4 Analytical model

$$\rho_g(\tau_1, \tau_2) = \exp\left[-\left\{ \left(\frac{\tau_1}{\delta_1}\right)^2 + \left(\frac{\tau_2}{\delta_2}\right)^2 \right\} \right] \tag{12}$$

In Eq. (12), δ_1 and δ_2 are the parameters relating to the correlation distances for x_1 and x_2-direction, respectively, and $\delta_1 = \delta_2$=1.127m is used. And τ_1 and τ_2 are the distances between two points of interest along respective coordinate axis. The steel plate with deterministic flexual rigidity is simply supported and subjected to a deterministic uniformly distributed load whose intensity is q_0=6.265 tonf/m^2. The parameters associated with the steel plate such as Young's modulus, Poisson's ratio and thickness are taken as 2.1×10^7tonf/m^2, 0.3 and 0.01m, respectively, and all these parameters are treated as deterministic quantities.

In the stochastic finite element analysis for the evaluation of the failure probability for the lateral deflection, the system, i.e. the deterministic steel plate and the random elastic foundation, was divided into 64 finite elements with

81 nodal points (8×8 meshes). Then the side length of each square finite element becomes 0.50(m), one-half of the correlation distance of the random elastic foundation.

In the evaluation of the failure probability for deflection, three parameters such as the coefficient of variation (c.o.v.) of the bearing coefficient of the foundation, defined by σ_g/m_g, the type of probability density function (p.d.f.) of the deflection in each finite element and the threshold are considered. Three variations of the c.o.v., i.e. 0.10, 0.15, 0.20, are treated. And the normal distribution type is assumed since the type of p.d.f. of responses can not be obtained by the stochastic finite element method, by which only the first and second order statistical moment of response are obtained. Three thresholds, i.e. level 7, 8 and 9, are determined as follows: the threshold of seven times the value of the maximum standard deviation of all finite elements for the case of c.o.v. of 0.20 is referred to level 7. Levels 8 and 9 are also determined in the same way of level 7. As a result, level 7=0.01354m, level 8= 0.01547m and level 9= 0.01741m are used as the three thresholds in the calculation.

In the reliability analysis for moment, three same parameters as for deflection are taken into consideration, that is the c.o.v. of the bearing coefficient of the foundation, the type of p.d.f. of the moment response and thresholds. The failure probability for the moment perpendicular to x_1-axis is calculated for the case with the threshold 10(=0.1691(tof·m)), the normal p.d.f. and c.o.v. of 0.20. The threshold is determined with respect to the maximum standard deviation of the moment of all finite elements in the same way for deflection. The system is divided into 16×16 meshes.

The integration calculus in Eq.(10) is performed by means of the IFM(Iterative Fast Monte Carlo) procedure[6]. The integration calculus with small error can be carried by the use of the IFM software package. The numerical error is about 2%.

First, the system calculation results of system failure probability for the lateral deflection are listed on Tables through 1 to 3. In these Tables, P_f^0 and P_f^1 are the failure probabilities which are calculated for the assumption that each pluse might be mutually independent and completely correlated, respectively. P_f^0 and P_f^1 are given by Eqs.(13) and (14), respectively.

$$P_f^0 = 1 - \prod_{i=1}^{n} (1 - P_{fi}) \tag{13}$$

$$P_f^1 = \max\{ P_{fi} \} \tag{14}$$

where P_{fi} is the probability that each pulse independently exceeds a given threshold, and n is the number of pulses, i.e. the number of finite elements.

Similarly, Table 4 lists the failure probabilities for the bending moment perpendicular x_1-axis.

From these results, it is clarified that the proposed metod in this study can performe the reliability analysis of the structure in the two-variate random field.

Table 1 Failure probability for lateral deflection for level 7 (Cases 1, 2 and 3)

Pf / C.O.V.	Pf^{1}	Pf^{*}	Pf^{e}
0.10	0.6363E-02	0.1252E-01	0.5885E-01
0.15	0.2750E-05	0.5472E-05	0.1931E-04
0.20	0.2042E-10	0.4151E-10	0.1297E-09

Table 2 Failure probability for lateral deflection for level 8 (Cases 4, 5 and 6)

Pf / C.O.V.	Pf^{1}	Pf^{*}	Pf^{e}
0.10	0.4812E-01	0.1145	0.5364
0.15	0.1209E-02	0.3621E-02	0.1471E-01
0.20	0.5407E-05	0.1708E-04	0.5611E-04

Table 3 Failure probability for lateral deflection for level 9 (Cases 7, 8 and 9)

Pf / C.O.V.	Pf^{1}	Pf^{*}	Pf^{e}
0.10	0.1056	0.2446	0.8949
0.15	0.1147E-01	0.3840E-01	0.1774
0.20	0.4806E-03	0.1906E-02	0.7029E-02

Table 4 Failure probability for bending moment perpendicular to X_1-axis

	$Pf^{(1)}$	Pf^{*}	$Pf^{(e)}$
level 10	0.396178E-04	1.364229E-04	4.953146E-04

5. Summary and Conclusions

In this study a practical reliability evaluation method of structural system with spatial uncertainties modeled by two-variate random fields is proposed.

In numerical examples, the reliability analysis of a steel plate resting on the random elastic foundation of the well-known Winkler's model is performed by the proposed method. In the analysis, the bearing coefficient is treated as two-dimensional random field. And the failure probability of the plate is evaluated for deflection as well as bending moment using the mean upcrossing rate to the given thresholds. The failure probability, obtained by this proposed method was compared as the failure probability obtained on the assumption of the mutually independent deflection (or bending moment) in each finite element and the failure probability obtained on the assumption of the completely correlated deflection (or moment).

From these results, it is clarified that the proposed method in this study can perform the reliability analysis of the structural systems with spatial uncertainties represented by random field, without considering the derivative random process required in the formulation ofthe conventional level crossing problems such as Rice's formula.

REFERENCES

[1] A.D.Kiureghian, and J.-B.Ke, Stochastic Finite Element Method in Structural Reliability, Probability Engineering Mechanics, Vol.3, No.2, 1988, pp.88-91.

[2] F.Yamazaki, and M.Shinozuka, Safety Analysis of Stochastic Finite Element Systems by Monte Carlo Simulation,Structural Eng./Earthquake Eng., Vol.5, No.2, 1988, pp.313-323.

[3] E.Vanmarke, and M.Grigoriu, Stochastic Finite Element Analysis of Simple Beams, Engineering Mechanics, ASCE, vol.109, No.5, 1983, pp.1203-1214.

[4] E.Vanmarke, Random Fields : Analysis and Synthesis, MIT Press,1983

[5] R.D.Larrabee, and A.Cornell, Combination of Various Load Processes, Structural Engineering, ASCE, Vol.107, No.1, 1981, pp.223-239.

[6] W.Shiraki, An Exension of Iterative Fast Monte-Carlo(IFM) Procedure and Its Applications to Time-Variant Structural Reliability Analysis, Proc. of ICOSSAR, 1989, pp.1015-1018.

Reliability and Optimization of Structural Systems, V (B-12)
P. Thoft-Christensen and H. Ishikawa (Editors)
Elsevier Science Publishers B.V. (North-Holland)

Reliability of Structure being Fatigue-Degraded due to Stochastic Excitation

P. Śniady, R. Sieniawska, S. Żukowski

Institute of Civil Engineering, Technical University of Wroclaw, Wybrzeze Wyspiańskiego 27, PL-50-370 Wroclaw, Poland

In the paper the reliability of structure under random excitation treated as the time to the first passage failure with respect to the fatigue change of the threshold is considered. It is assumed that the degradation of the construction is described by the cumulative jump process. In such a case the mean rate of the failure is to be found as a sum of the probabilities of three independent events. The degradation of the structure can also be interpreted as capacity descrease due to another reasons, for example corrosion.

Keyword Codes: G.3; I.6.5; J.2
Keywords: Probability and Statistics, Model development,
Physical Sciences and Engineering

1. INTRODUCTION

The problem of predicting the reliability of systems subjected to random excitation arises in many engineering applications particular in the dynamics of mechanical and structural systems. This problem has received a great deal of attention in the last few years. A solution of reliability problem in dynamics of structure depends on the specification of a suitable failure mechanism. Because of mathematical experience most of the research has been devoted to two essential approaches which usually are investigated separately: barrier crossing (first passage) problem which describes failure occurring instantenteneously and fatigue failure which is formulated as cumulative damage failure [1,2,3,4,5]. The barrier crossing (first passage) problem has a more simple physical interpretation and leads to a more precise mathematical formulation and has been considered for stationary and nonstationary processes with wide and narrow band spectra [6,7]. Fatigue failure (cumulative damage) estimations are usually made using the well-known Palmgren-Miner rule [1]. In many realistic situations both kinds of structure damage mechanisms are mutually related and in many reliability problems they have to be considered together.

In this paper the reliability of structure under random excitation treated as the time to the first passage failure with respect to the fatigue change of the threshold is considered. It is assumed that the degradation of the construction is described by the cumulative jump process. In such a case the mean rate of the failure is to be found as a sum of the probabilities of three independent events. The degradation of the structure can also be interpreted as capacity descrease due to another reasons, for example corrosion.

This research was supported by the Scientific Research Committee, Warsaw, grant number 772309203

2. FORMULATION OF THE PROBLEM

Let the fatigue degradation of the structure be described by the cumulative jump process (Fig.1). The level of structure capacity $R(t)$ is a function decreasing in time. It is assumed that the decreasing of the structure capacity (barier crossing) take place at random times $t_1, t_2, \ldots, t_n$ which constitute a Poisson stochastic process, and is of jumping character.

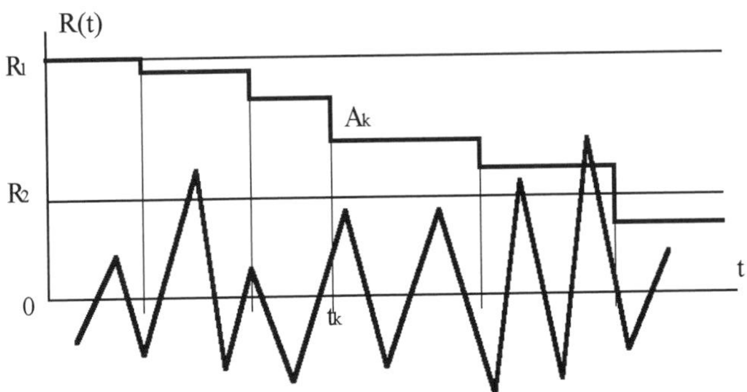

Figure 1. Pattern of the problem

The capacity $R(t)$ of the structure is described by expression

$$R(t) = R_1 - \sum_{k=1}^{N(t)} A_k, \qquad A_k \geq 0 \tag{1}$$

where R_1 is the capacity at time $t = 0$, $N(t)$ is Poisson process with parameter $\lambda(t)$, A_k are the positive, mutually independent random (dropping) variables which describe fatigue reduction of the capacity at time t .

Let the stochastic process $Y(t)$ be the response of a structure under stochastic exitation. If the probability density function (p.d.f.) of the random variables A_k or its cumulative distribution function (c.d.f.) is known the probability density function of crossing the level $R(t)$ or its cumulative distribution function can be derived.

Let us introduce the notation

$$F_R(x,t) = P\{R(t) \leq x\}, \quad f_R(x,t) = \frac{dF(x,t)}{dx} \tag{2}$$

and

$$S_n = \sum_{k=1}^{n} A_k, \quad F_n(x) = P\{S_n \leq x\}, \quad f_n(x) = \frac{df_n(x)}{dx} \qquad \text{for } n = 1,2,\ldots, \tag{3}$$

Since in the equation (1) contains the sum of independent variables the c.d.f. and p.d.f. have forms

$$F_R(x,t) = P\{R(t) \le x\} = p_0(t) \cdot H(R_1 - x) + \sum_{n=1}^{\infty} F_n(R_1 - x) \cdot p_n(t) \tag{4}$$

and

$$f_R(x,t) = p_0(t) \cdot \delta(R_1 - x) + \sum_{n=1}^{\infty} f_n(R_1 - x) \cdot p_n(t) \tag{5}$$

where $H(\cdot)$ and $\delta(\cdot)$ are the Heaviside and the Dirac delta functions, respectively. Because it was assumed that $N(t)$ is the Poisson process then

$$p_n(t) = P\{N(t) = n\} = \frac{\Lambda^n(t)}{n!} e^{-\Lambda(t)} \qquad \text{for} \quad n = 0,1,2,...,, \tag{6}$$

where $\Lambda(t) = \int_0^t \lambda(\tau) d\tau$.

From independency of random variables A_k follows that the functions $F_n(x)$ and $f_n(x)$ can be obtained from relationships [2,8]

$$F_n(x) = \int_0^x F_{n-1}(x-\xi) f_A(\xi) d\xi \tag{7}$$

and

$$f_n(x) = \int_0^x f_{n-1}(x-\xi) f_A(\xi) d\xi \tag{8}$$

For example, let the p.d.f. $f_A(a)$ is of the gamma type

$$f_A(a) = \frac{\gamma^k a^{k-1}}{\Gamma(k)} a^{-\gamma a}, \qquad \text{for} \quad a \ge 0 \tag{9}$$

where constants k and γ are positive parameters and $\Gamma(k)$ is the gamma function. For the gamma distribution function (9) it can be found that

$$f_n(x) = \frac{\gamma(\gamma x)^{nk-1}}{\Gamma(nk)} e^{-\gamma x} \tag{10}$$

and p.d.f. of the crossing level $R(t)$ in this case is equal to

$$f_R(x,t) = e^{-\Lambda(t)} \{\delta(R_1 - x) + \gamma e^{-\gamma(R_1 - x)} \cdot \sum_{n=1}^{\infty} \frac{[\gamma(R_1 - x)]^{nk-1} \Lambda^n(t)}{\Gamma(nk)\Gamma(n+1)}\}, \qquad \text{for } x \le R_1 \tag{11}$$

For other types of p.d.f. $f_A(a)$ one can apply the results presented in the paper [9].

There are three situations when the construction can fail. The first one is when the structure's response $Y(t)$ passes over the the threshold level $R(t)$. The second one is when the capacity $R(t)$ is reduced to the value less than Y(t). The last one is when $R(t)$ decreases to the value less than R_2 and $Y(t)$ is less than R_2, simultaneously. The mean failure rate of the structure in situation described is to be found as a sum of mean failure rates of three independent stochastic events.

Case I.
Let the capacity $R(t)$ is in the range $R_2 \leq R(t) = r \leq R_1$. Consider the problem of crossing the threshold $R(t)$ by stochastic process $Y(t)$ (see Fig.2).

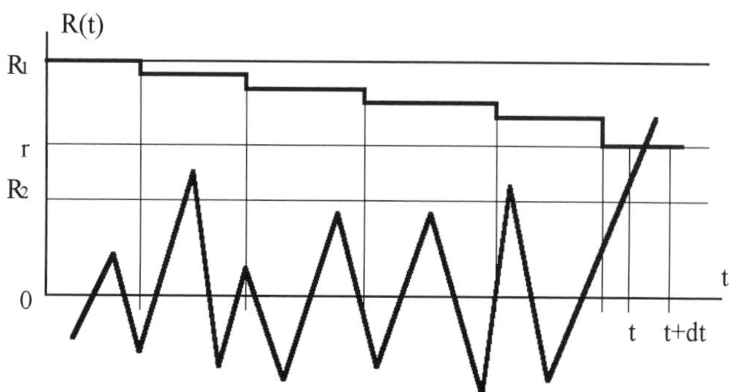

Figure.2. Failure of the structure according case I.

The crossing occurs if in time interval $(t, t+dt)$ the stochastic process $Y(t)$ is satisfying $Y(t) = r$ and $\dot{Y}(t) > 0$. It is well known that the mean crossing rate can be obtained from the Rice's rule

$$\upsilon_{y1}(r,t) = \int_0^\infty \dot{y} f_{y\dot{y}}(r,\dot{y}) d\dot{y} \tag{12}$$

where $f_{y\dot{y}}(y,\dot{y})$ is the joint probability density function of processes $Y(t)$ and $\dot{Y}(t)$.
Because the crossing level r is random, $\upsilon_{y1}(r,t)$ is also random and expected value of the mean crossing rate is equal to

$$E[\upsilon_{y1}(t)] = \int_{R_2}^{R_1} \int_0^\infty \dot{y} f_{y\dot{y}}(r,\dot{y}) f_R(r,t) d\dot{y} dr \tag{13}$$

Case II.
The failure of the structure can take place also in the case when the capacity is in the range $R_2 \leq R(t) = r \leq R_1$, the response process $Y(t)$ satisfies the inequalities $R_2 \leq Y(t) < r$ and in the time interval $(t, t+dt)$ it happens the capacity degradation jump equal to A_i such that $Y(t+dt) > R(t+dt) = r - A_i$ (Fig. 3).

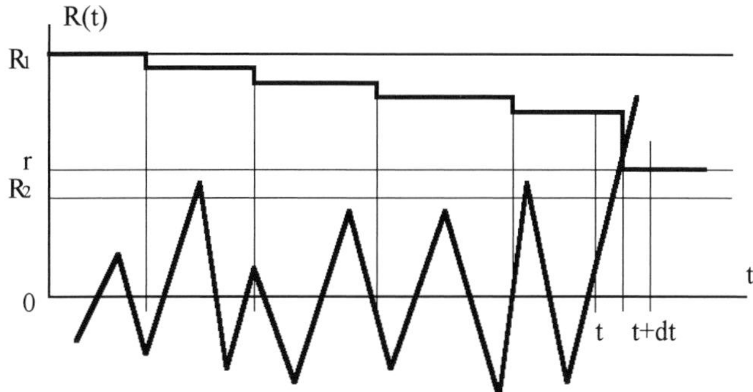

Figure.3. Failure of the structure according case II.

For given crossing level r the mean rate of crossing is equal to

$$\upsilon_{y2}(r,t)=\frac{1}{dt}P\{dN(t)=1\}\cdot P\{R_2\le Y(t)<r,Y(t+dt)>r-A_i|R(t)=r\}=$$
$$=\lambda(t)\int_{R_2}^{r}[1-F_A(r-y)]f_y(y)dy \tag{14}$$

where p.d.f. $f_y(y)=\int_{-\infty}^{\infty}f_{y\dot{y}}(y,\dot{y})d\dot{y}$ and $F_A(x)$ is c.d.f. of random variables A_i.

The mean value of crossing rate is given by

$$E[\upsilon_{y_2}(t)]=\lambda\int_{R_2}^{R_1}\int_{R_2}^{r}[1-F_A(r-y)\cdot f_y(y)\cdot f_R(r,t)dydr \tag{15}$$

Case III.
Let the response process $Y(t)$ is beneath the level R_2 $(Y(t)<R_2))$ and in the time interval $(t,t+dt)$ a capacity degradation jump of the crossing level takes place so that we have $R(t)>R_2>R(t+dt)$ i.e. the structure also fails (Fig. 4).
For given level r we have

$$\upsilon_{y3}(r,t)=\frac{1}{dt}P\{dN(t)=1\}\cdot P\{A_i>r-R_2\}\cdot P\{Y(t)<R_2\} \tag{16}$$

and

$$E[\upsilon_{y_3}(t)]=\lambda(t)\cdot F_y(R_2)\int_{R_2}^{R_1}[1-F_a(r-R_2)]f_R(r,t)dr \tag{17}$$

Taking into account that the structure's failure occurs rarely, it is to be assumed that the failure process is the Poisson process with the intensity $\upsilon_y(t)$, where

$$\upsilon_y(t) = E[\upsilon_{y1}(t)] + E[\upsilon_{y2}(t)] + E[\upsilon_{y3}(t)]. \tag{18}$$

Hence, the reliability of the structure, when $\upsilon_y(t) << 1$, is approximatelly

$$\rho(t) = 1 - e^{-\int_0^t \upsilon_y(\tau)d\tau} \tag{19}$$

The more exact formula is

$$\rho(t) = 1 - \int_{R_2}^{R_1} e^{-\int_0^t (\upsilon_{y1}(r,\tau)+\upsilon_{y2}(r,\tau)+\upsilon_{y3}(r,\tau))d\tau} f_R(r,t)\,dt \tag{20}$$

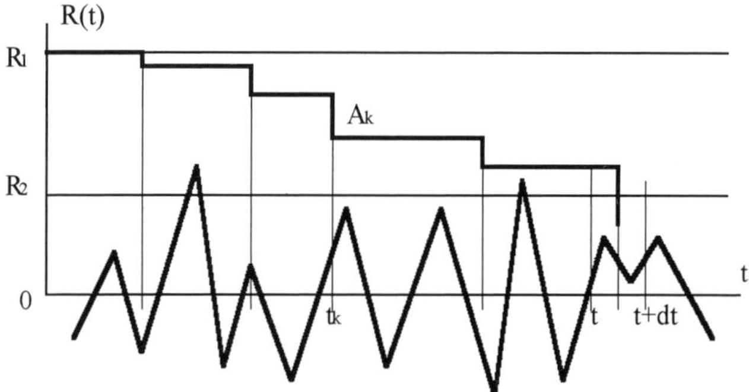

Figure.4. Failure of the structure according case III..

3. NUMERICAL RESULTS AND CONCLUSIONS

Consider the stochastic vibrations of a structure modeled by the lightly damped single degree of freedom dynamical system. Following assumptions have been made: the probability density function of structure response is known and is of Gaussian type

$$f_y(y) = \frac{1}{\sqrt{2\pi\sigma_y}} \exp[-\frac{1}{2}(\frac{y-\mu_y}{\sigma_y})^2] \tag{21}$$

with the mean value equal to $\mu_y = 0$. The probability density function of the fatigue capacity reduction variables (see (1)) is exponential

$$f_A(a) = \gamma e^{-\gamma a}, \qquad a \geq 0 \tag{22}$$

In this case, the probability density function of the structure's capacity is described by expression (11)

By the assumptions made above the expressions (13), (15) and (17) have the forms

$$E[\upsilon_{y1}(t)] = \frac{\sigma_{\dot{y}}}{2\pi\sigma_y} e^{-\lambda t} [e^{-\frac{R_1^2}{2\sigma_y^2}} + e^{-\gamma R_1} \sum_{n=1}^{\infty} \frac{(\gamma\lambda t)^n n}{(n!)^2} \int_{R_2}^{R_1} e^{\gamma r - \frac{r^2}{2\sigma_y}} (R_1 - r)^{n-1} dr] \tag{23}$$

$$E[\upsilon_{y2}(t)] = \frac{\lambda}{\sqrt{2\pi\sigma}_y} e^{-\lambda t - \gamma R_1} [\int_{R_2}^{R_1} e^{\gamma y - \frac{y^2}{2\sigma_y^2}} dy + \sum_{n=1}^{\infty} \frac{(\gamma\lambda t)^n n}{(n!)^2} \int_{R_2}^{R_1} (R_1 - r)^{n-1} \int_{R_2}^{r} e^{\gamma y - \frac{y^2}{2\sigma_y^2}} dr] \tag{24}$$

$$E[\upsilon_{y3}(t)] = \lambda F_y(R_2) e^{-\lambda t + \gamma(R_2 - R_1)} [1 + \sum_{n=1}^{\infty} \frac{(\gamma\lambda t)^n n}{(n!)^2} \int_{R_2}^{R_1} (R_1 - r)^{n-1} dr] \tag{25}$$

respectively.

The expected value of the mean rate crossing (18) was calculated by using different dimensionless parameters and shown in Figures 5 and 6.

All curves are obtained for such values $\lambda = 0.1$, $\gamma = 0.1$, $R_2 = 0.6$, $t = 10$.

In the Figure 5 the results are shown as a function of R_1 calculated for different values of σ_y and $\sigma_{\dot{y}}$. The curve signed by "1" was computed for $\sigma_y = 1$, $\sigma_{\dot{y}} = 10$, the curve "2" for $\sigma_y = 1$, $\sigma_{\dot{y}} = 5$, the curve "3" for $\sigma_y = 0.5$, $\sigma_{\dot{y}} = 5$ and the curve "4" for $\sigma_y = 1$, $\sigma_{\dot{y}} = 1$.

In the Figure 6 the expected value of the mean rate crossing is plotted as a function of σ_y for $R_1 = 1$ and different relation betwen σ_y and $\sigma_{\dot{y}}$. The curve signed by "1" was computed for $\sigma_y / \sigma_{\dot{y}} = 1$, the curve "2" for $\sigma_y / \sigma_{\dot{y}} = 0.5$, the curve "3" for $\sigma_y / \sigma_{\dot{y}} = 0.2$ and the curve "4" for $\sigma_y / \sigma_{\dot{y}} = 0.1$..

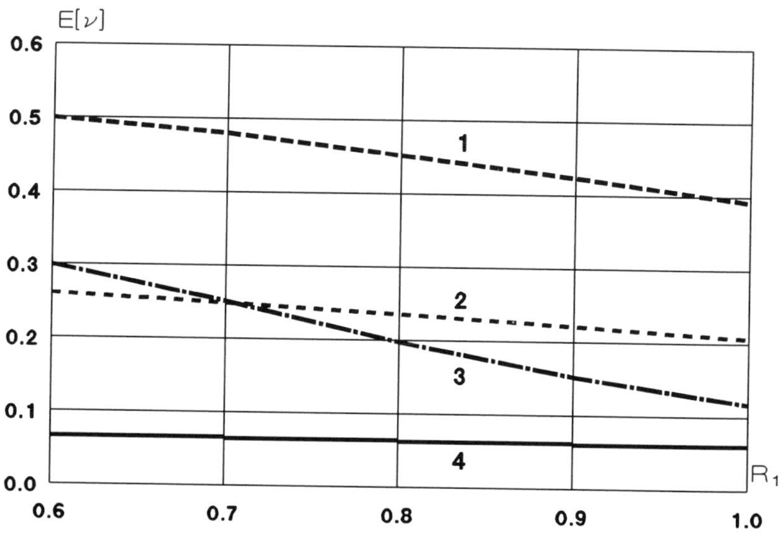

Figure 5. The expected value of the mean rate of crossing v.s. R_1

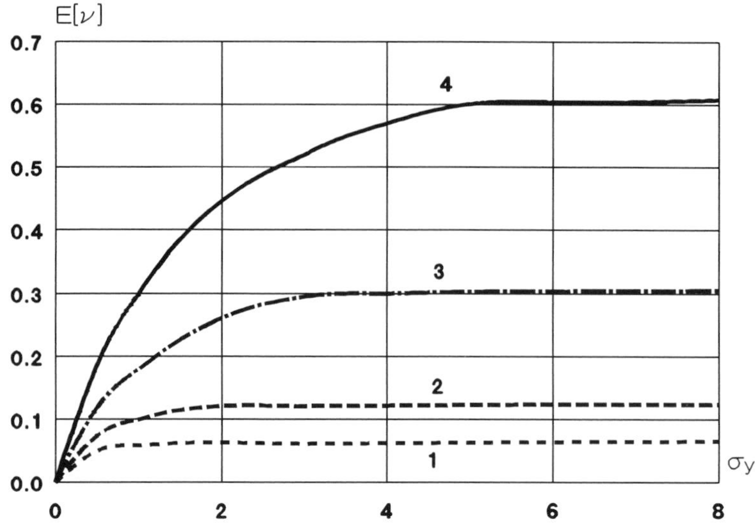

Figure 6. The expected value of the mean rate of crossing v.s. σ_y

REFERENCES

1. Lin Y.K., Probabilistic Theory of Structural Dynamics, McGraw Hill, New York, 1967.
2. Sobczyk K. and Spencer B.T.Jr., Random Fatigue. From Data to Theory, Academic Press Inc., Boston 1992.
3. Sieniawska R. and Sniady P., Life Expectancy of Highway Bridges due to Traffic Flow, J. of Sound and Vibr., 140(1), 31-38, 1990.
4. Yang J.-N., Nonstationary Envelope Process and First Excursion Probability, J. of Struc. Mech., 1, 1972, 231-248.
5. Yang J.-N. and Heer E., Reliability of Randomly Excited Stuctures, AIAA Journal, vol. 9, no 7,1262-1968, 1972.
6. Vanmarcke E.H., On the Distribution of the First Passage Time for Normal Stationary Random Process, J. of Applied Mech., ASME, 42, 1975, 215-220.
7. Roberts J.B., Probability of First Passage Failure for Nonstationary Random Vibration, J. of Applied Mech., ASME, 42, 1975, 716.
8. Sobczyk K. and Trebicki J., Modelling of Random Fatigue by Cumulative Jump Processes, Eng. Fracture Mech., Vol.34,2, 477-193.
9. Ditlevsen O., Asypmtotic First-Passage Time Distribution in Compound Poisson Process, in "Nonlinear Structural Systems under Random Conditions", ed. F. Casciati, I. Elishakoff, J.B. Roberts, Elsevier, 1990, 327-336.

Reliability and Optimization of Structural Systems, V (B-12)
P. Thoft-Christensen and H. Ishikawa (Editors)
Elsevier Science Publishers B.V. (North-Holland)

Structural Design Sensitivity in SFEM Formulation

P. Śniady, M. El-Meligy, S. Żukowski

Institute of Civil Engineering, Technical University of Wroclaw,
Wybrzeze Wyspiańskiego 27, PL-50-370 Wroclaw, Poland

Design sensitivity analysis provides design derivatives which represent trend for the structural systems. A new treatment for the sensitivity analysis within the framework of SFEM is presented. The sensitivities of the response statistics up to the second-order with respect to the distribution parameters of the basic random variables are obtained. The first-order perturbation method is used to incorporate randomness of the structure parameters and loads into the formulation. The sensitivity of the response statistics may be of practical importance for reliability analysis and structural optimization problems of which the objective functions or constraints include response statistics. Numerical example is presented to illustrate the proposed formulation.

Keyword Codes: G.3; I.6.5; J.2
Keywords: Probability And Stastistics, Model Development,
Phycical Sciences And Engineering

1. INTRODUCTION

Design sensitivity analysis provides design derivatives which play an important role in modern structural analysis and design problems. Rather than being important in their own right as they represent trend for the structural performance functions, design derivatives are crucial request of most structural optimization algorithms which are gradient-based. Structural design sensitivity within the context of deterministic analysis is now well defined problem [2]. The fact, that the SFE-based reliability methods [1] make use of the response gradients in order to get the reliability index in a minimization process, facilitates the computation of the so-called reliability sensitivity measures with respect to any desired set of parameters, especially the distribution parameters of the random variables under consideration [4,6]. Consequently, the use of these sensitivity measures gives indication about the relative importance of each variable as well as the relative importance of the uncertainty of each variable [5]. Within the context of SFE-based second moment analysis, the similarities between the design sensitivity analysis and the pertubation methods [6,9] facilitates the evaluation of the expectation and covariances of the sensitivity gradients in a combined analysis [3].

Within the context of SFE-based second moment analysis, many situations exist in which the gradients of the stochastic response functions, with respect the distribution parameters of the basic random variables, ought to be evaluated. These gradients may be of importance in their own right as they indicate the trend of the stochastic response functions or they are needed for quality control through determining the relative importance of the uncertainty of a specific random variable. Furthermore, the evaluation of these gradients may be a request of an optimization problem in which the objective function and/or the constraints include response statistics. Similar to those of the SFE based reliability methods, the evaluation of these gradients, with respect to the expected values of the basic random variables, determine the relative importance of these random variables. Consequently, the unnecessary random variables may be eliminated in any sequential analysis.

The primary goal of this work is to evaluate the sensitivities of the response statistics w.r.t. some distribution parameters, e. g., the expected values and the standard deviations of the basic random variables. A numerical example is introduced to show the application of this treatment to framed structures.

2. STOCHASTIC FINITE ELEMENT FORMULATION

In this section the first-order perturbation method is introduced following the notation used in [8]. Consider a vector of N random variables $\mathbf{a} = [a_1, a_2, \ldots, a_N]^T$ with zero mean. The state equations under static loads are expressed as

$$\mathbf{KU} = \mathbf{F} \tag{1}$$

where $\mathbf{K}$ is the stiffness matrix, $\mathbf{U}$ is the displacement vector, and $\mathbf{F}$ is the load vector. Expanding $\mathbf{K}$ and $\mathbf{F}$ about their deterministic values using matrix Taylor series expansion and truncating after the second term, one can obtain

$$\mathbf{K} = \mathbf{K}^o + \sum_{i=1}^{N} \mathbf{K}_i' \, a_i \tag{2}$$

$$\mathbf{F} = \mathbf{F}^o + \sum_{i=1}^{N} \mathbf{F}_i' \, a_i \tag{3}$$

where $\mathbf{K}^o$ and $\mathbf{F}^o$ are the deterministic stiffness matrix and load vector ,respectively, $\mathbf{K}_i'$ and $\mathbf{F}_i'$ are the first partial derivatives of $\mathbf{K}$ and $\mathbf{F}$ w. r. t. a_i evaluated at $\mathbf{a} = 0$; i. e.,

$$\mathbf{K}_i' = \left.\frac{\partial \mathbf{K}}{\partial a_i}\right|_{\mathbf{a}=0} \tag{4}$$

$$\mathbf{F}_i' = \left.\frac{\partial \mathbf{F}}{\partial a_i}\right|_{\mathbf{a}=0} \tag{5}$$

Similarly, the unknown displacement vector is expressed as

$$\mathbf{U} = \mathbf{U}^o + \sum_{i=1}^{N} \mathbf{U}_i' \, a_i \tag{6}$$

in which the coefficient vectors $\mathbf{U}^o$ and $\mathbf{U}_i'$ can be represented by the following set of recursive equations:

$$\mathbf{U}^{o} = (\mathbf{K}^{o})^{-1}\mathbf{F}^{o} \tag{7}$$

$$\mathbf{U}_i' = (\mathbf{K}^{o})^{-1}(\mathbf{F}_i' - \mathbf{K}_i'\mathbf{U}^{o}) \tag{8}$$

One way of obtaining Eq. (8) is by differentiating Eq. (1) as

$$\frac{\partial \mathbf{K}}{\partial a_i}\mathbf{U} + \mathbf{K}\frac{\partial \mathbf{U}}{\partial a_i} = \frac{\partial \mathbf{F}}{\partial a_i} \tag{9}$$

and setting $\mathbf{a} = 0$.

Applying the expectation operator $\mathbf{E}[.]$ on Eq. (6), the expected value of the displacement vector $\mathbf{E}_U$ is

$$\mathbf{E}_U = \mathbf{E}[\mathbf{U}] = \mathbf{U}^{o} \tag{10}$$

The covariance matrix of the displacement vector $\mathbf{C}_{UU}$ is

$$\mathbf{C}_{UU} = \sum_{i=1}^{N}\sum_{j=1}^{N}\mathbf{U}_i'(\mathbf{U}_j')^T\,\mathbf{E}[a_i a_j] \tag{11}$$

The variance vector of displacements $\mathbf{V}_U$ which is the diagonal components of $\mathbf{C}_{UU}$ is

$$\mathbf{V}_U = \sum_{i=1}^{N}\sum_{j=1}^{N}\mathbf{Diag}\,[\mathbf{U}_i'(\mathbf{U}_j')^T]\,\mathbf{E}[a_i a_j] \tag{12}$$

3. DESIGN SENSITIVITY OF STOCHASTIC STRUCTURAL RESPONSE

Consider a vector of uncorrelated random variables $\mathbf{X} = [x_1, x_2, \ldots, x_N]^T$, with a mean vector μ_x and variance vector $\mathbf{V}_x$. The random vector $\mathbf{X}$ can be expressed as

$$\mathbf{X} = \mu_x + \mathbf{a} \tag{13}$$

where $\mathbf{a}$ is a vector of random variables with zero mean and the same variance vector $\mathbf{V}_x$. Let a stochastic response function be expressed as

$$\Psi = \Psi[\mu_x, \mathbf{V}_x, \mathbf{E}_u(\mu_x), \mathbf{V}_u(\mu_x, \mathbf{V}_x)] \tag{14}$$

The total derivatives of Ψ w. r. t. μ_x and $\mathbf{V}_x$ are

$$\frac{\mathbf{d}\Psi}{\mathbf{d}\mu_x} = \frac{\partial\Psi}{\partial\mu_x} + \frac{\partial\Psi}{\partial\mathbf{E}_u}\frac{\mathbf{dE}_u}{\mathbf{d}\mu_x} + \frac{\partial\Psi}{\partial\mathbf{V}_u}\frac{\mathbf{dV}_u}{\mathbf{d}\mu_x} \tag{15}$$

and

$$\frac{\mathbf{d\Psi}}{\mathbf{dV_x}} = \frac{\partial\Psi}{\partial\mathbf{V_x}} + \frac{\partial\Psi}{\partial\mathbf{V_u}}\frac{\mathbf{dV_u}}{\mathbf{dV_x}} \tag{16}$$

The partial derivatives $\frac{\partial\Psi}{\partial\mu_\mathbf{x}}$, $\frac{\partial\Psi}{\partial\mathbf{E_u}}$, $\frac{\partial\Psi}{\partial\mathbf{V_u}}$, and $\frac{\partial\Psi}{\partial\mathbf{V_x}}$ can be easily evaluated because of the explicit dependence, whereas the derivatives $\frac{\mathbf{dE_u}}{\mathbf{d}\mu_\mathbf{x}}$, $\frac{\mathbf{dV_u}}{\mathbf{d}\mu_\mathbf{x}}$, and $\frac{\mathbf{dV_u}}{\mathbf{dV_x}}$ should be calculated within the context of SFEM. Recalling Eqs. (7)-(10), one can write

$$\frac{\mathbf{dE_\mu}}{\mathbf{d}\mu_{\mathrm{x_i}}} = \mathbf{U}'_\mathbf{i} \tag{17}$$

which is already evaluated before, hence no additional computational effort is required. Differentiating Eq. (12) w. r. t. $\mu_{\mathrm{x_i}}$, hence

$$\frac{\mathbf{dV_u}}{\mathbf{d}\mu_{\mathrm{x_i}}} = \sum_{\mathbf{l=1}}^{\mathrm{N}}\sum_{\mathbf{k=1}}^{\mathrm{N}} \mathbf{Diag}[\mathbf{U}'_{\mathbf{l},\mu_{\mathrm{x_i}}}(\mathbf{U}'_\mathbf{k})^\mathrm{T} + \mathbf{U}'_\mathbf{l}(\mathbf{U}'_{\mathbf{k},\mu_{\mathrm{x_i}}})^\mathrm{T}\,\mathbf{E}[\mathrm{a_l a_k}]] \tag{18}$$

$\mathbf{U}'_{\mathbf{l},\mu_{\mathrm{x_i}}}$ can be obtained by differentiating Eq. (9) as

$$\frac{\partial\mathbf{K}}{\partial\mathrm{a_l}}\frac{\partial\mathbf{U}}{\partial\mu_{\mathrm{x_i}}} + \frac{\partial^2\mathbf{K}}{\partial\mathrm{a_l}\partial\mu_{\mathrm{x_i}}}\mathbf{U} + \frac{\partial\mathbf{K}}{\partial\mu_{\mathrm{x_i}}}\frac{\partial\mathbf{U}}{\partial\mathrm{a_l}} + \mathbf{K}\frac{\partial^2\mathbf{U}}{\partial\mathrm{a_l}\partial\mu_{\mathrm{x_i}}} = \frac{\partial^2\mathbf{F}}{\partial\mathrm{a_l}\partial\mu_{\mathrm{x_i}}} \tag{19}$$

and noting that $\mathbf{U}'_{\mathbf{l},\mu_{\mathrm{x_i}}} = \frac{\partial^2\mathbf{U}}{\partial\mathrm{a_l}\partial\mu_{\mathrm{x_i}}}$, hence

$$\mathbf{U}'_{\mathbf{l},\mu_{\mathrm{x_i}}} = \mathbf{K}^{-1}[\frac{\partial^2\mathbf{F}}{\partial\mathrm{a_l}\partial\mu_{\mathrm{x_i}}} - \mathbf{2}\frac{\partial\mathbf{K}}{\partial\mathrm{a_l}}\frac{\partial\mathbf{U}}{\partial\mu_{\mathrm{x_i}}} - \frac{\partial^2\mathbf{K}}{\partial\mathrm{a_l}\partial\mu_{\mathrm{x_i}}}\mathbf{U}] \tag{20}$$

The evaluation of $\frac{\mathbf{dV_u}}{\mathbf{dV_{x_i}}}$ requires the differentiation of Eq. (12) w.r.t. $\mathrm{V_{x_i}}$ as

$$\frac{\mathbf{dV_u}}{\mathbf{dV_{x_i}}} = \sum_{\mathbf{l=1}}^{\mathrm{N}}\sum_{\mathbf{k=1}}^{\mathrm{N}} \mathbf{Diag}[\mathbf{U}'_\mathbf{l}(\mathbf{U}'_\mathbf{k})^\mathrm{T}]\mathbf{E}[\mathrm{a_l a_k}]_{,\mathrm{V_{x_i}}} \tag{21}$$

in which

$$\mathbf{E}[\mathrm{a}_l\mathrm{a}_k]_{,V_{x_i}} = \begin{cases} 1 & \text{for } \mathrm{l = k = i}, \\ 0 & \text{elsewhere.} \end{cases} \tag{22}$$

The evaluation of the derivatives of the statistics of strains and stresses w. r. t. the distribution parameters of the basic random variables is straightforward following the steps as before and utilizing the element strain-displacement and stress-strain relationships.

4. NUMERICAL EXAMPLE

A computer program was developed by which we could estimate the sensitivities of the response statistics w. r. t. the expected values and the standard deviations of the basic random variables. This code was used to analyse the frame shown in Figure 1.

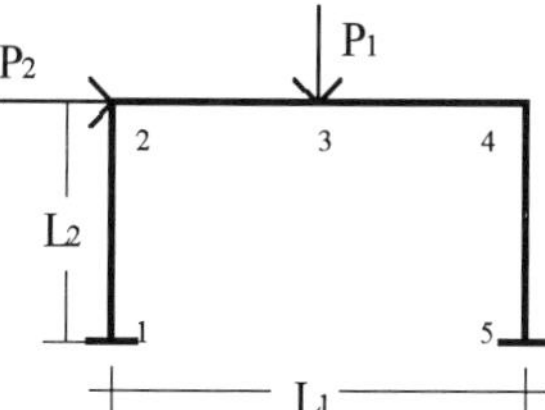

Figure 1. The portal frame

The basic random variables are P_1, P_2, L_1, and L_2. The expected values of the load parameters are determined by $0 \le E[P_1] \le P_1^\circ$, $0.8P_2^\circ \le E[P_2] \le P_2^\circ$, $P_1^\circ = P_2^\circ = P^\circ$. In Figs 2, 3, 4 shown are the sensitivities of the variance of the maximum bending moment in point 3 (V_{M_3}) as a function of the expected value of the load parameter P_1, as a function of the expected value of the length parameter L_1 and as a function of the variation coefficient (V_{P_1}) of P_1, respectively. In Figures 2, 3 and 4 the curves 1, 2, 3 and 4 give the sensitivities w.r.t. the expected values of the parameters P_1, P_2, L_1, and L_2, respectively. As shown in Fig. 2 it is found that the change of the expected value of P_1 has no effect on $\frac{dV_{M_3}}{dE[P_1]}$ whereas the effects is so pronounced on $\frac{dV_{M_3}}{dE[L_1]}$ and moderate on $\frac{dV_{M_3}}{dE[P_2]}$ and $\frac{dV_{M_3}}{dE[L_2]}$. On the other hand, referring to Fig. 3, the change of the expected value of L_1 has a great influence on $\frac{dV_{M_3}}{dE[L_2]}$ and has a very slight influence on $\frac{dV_{M_3}}{dE[P_1]}$, $\frac{dV_{M_3}}{dE[P_2]}$ and $\frac{dV_{M_3}}{dE[L_1]}$. Form safety requirements point of view, investigating the effect of changing the cefficient of variation is more importat. As shown in Fig.4 the changing of V_{P_1} has a crucial effect on $\frac{dV_{M_3}}{dE[P_2]}$, $\frac{dV_{M_3}}{dE[L_1]}$ and $\frac{dV_{M_3}}{dE[L_2]}$ but the effect on $\frac{dV_{M_3}}{dE[P_1]}$ is slight.

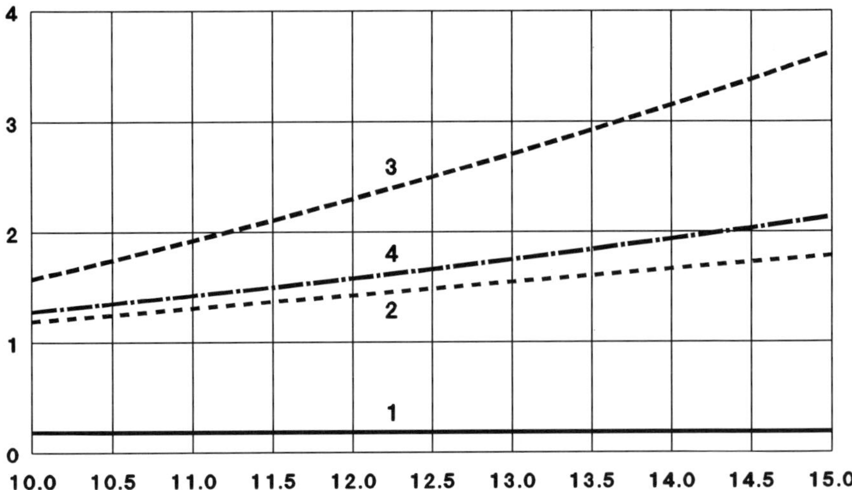

Fig. 2. Sensitivity of the variance of maximum bending moment as a function of P_1

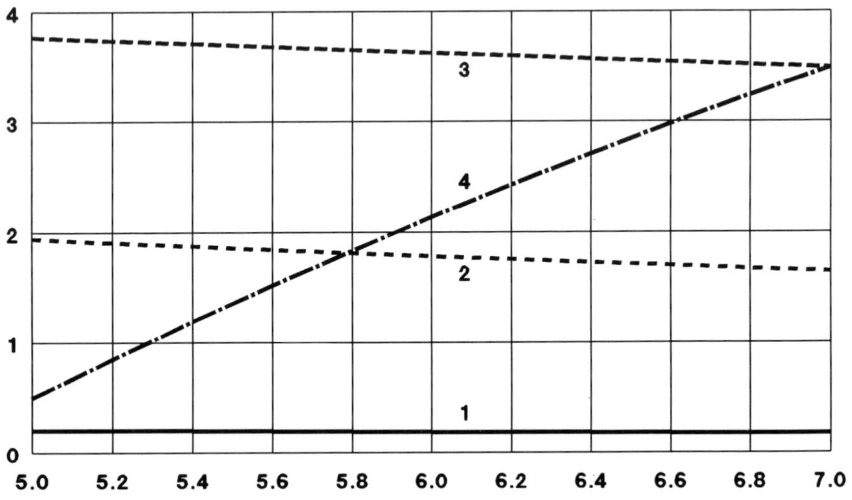

Fig. 3. Sensitivity of the variance of maximum bending moment as a function of L_1.

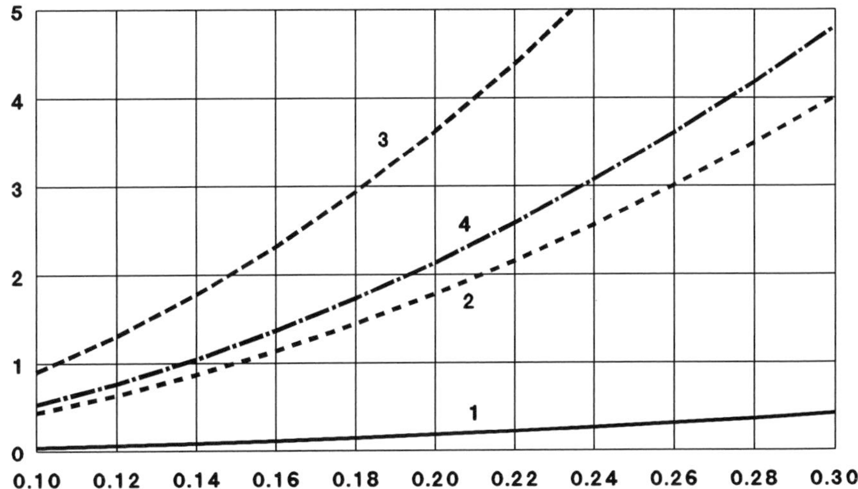

Fig. 4. Sensitivity of the variance of maximum bending moment as a function of VP_1

REFERENCES

1. A. Kiureghian and B.-J. Ke, The Stochastic Finite Element Method in Structural Reliability, Probabilist. Engrg. Mech., 3(2), 1988, pp. 83-91.
2. E.J. Haug, K.K. Choi and V. Komkov, Design Sensitivity Analysis of Structural Systems, Academic Press, New York, 1986.
3. T.D. Hien and M. Kleiber, Stochastic Structural Design Sensitivity of Static Response, Comput. Struct., 38(5/6), 1991, pp. 659-667 .
4. M. Hohenbichler and R. Rackwitz, Sensitivity and Importance Measures in Structural Reliability, Civ. Engrng. Syst., 3, December, 1986, pp. 203-209.
5. P.-L. Liu and A. Kiureghian, Finite Element Reliability of Geometrically Nonlinear Structures, J. Engrg. Mech., ASCE, 117(8), 1991, pp.1806-1825
6. Liu, W. K., Belytschko, T. and Mani, A., "Random Field Finite Elements," Int. J. Numer. Methods Eng., 3, 1986, pp. 1831-1845.
7. H.O. Madsen, S. Krenk and N.C. Lind, Methods of Structural Safety, Prentice-Hall, Englewood Cliffs, N. J., 1986.
8. M. Shinozuka and F. Yamazaki, Stochastic Finite Element Analysis and Introduction, (in) Stochastic Structural Dynamics: Progress in theory and Application, (Eds.) S. T. Ariaratnam, G. I. Schueller and I. Elishakoff, Elsevier, 1988.
9. E. Vanmarke et. al., Random Fields and Stochastic Finite Elements, Structural Safety, 3, 1986, pp. 143-166.

Reliability and Optimization of Structural Systems, V (B-12)
P. Thoft-Christensen and H. Ishikawa (Editors)
Elsevier Science Publishers B.V. (North-Holland)

Reliability-Based Optimal Design of Experiment Plans for Offshore Structures

J. D. Sørensen, M. H. Faber, I. B. Kroon

University of Aalborg, Sohngaardsholmsvej 57, DK-9000 Aalborg, Denmark

ABSTRACT
Design of cost optimal experiment plans on the basis of a preposterior analysis is discussed. In particular the planning of on-site response measurements on offshore structures in order to update probabilistic models for fatigue life estimation is addressed. Special emphasis is given to modelling of uncertainties in the transfer function. An example is presented in which the optimal number of response spectra measurements is sought.

KEYWORDS: Experiment planning, reliability, optimization, fatigue, measurements.

1. INTRODUCTION
Optimal design of experiments for dynamicly sensitive linear structural systems with uncertain transfer functions is considered in this paper. Ideally the objective of performing an experiment is to obtain information such that the total expected costs due to experiments, maintenance and failure in the lifetime of the structure are minimized. In practice, however, the planning of experiments is often performed in a much simpler way, namely with the objective to minimize the standard deviations of the estimates of the measured quantities and with some upper limit on the cost of the experiment, see e.g. Viertl [1] and Ljung [2]. In this way the relation to the application of the structural system is neglected and important information which could be used in the experiment planning is disregarded. In the present paper the preposterior analysis is applied to formulate a consistent framework for experiment planning where the actual application of the structure being observed is taken into account. This is done indirectly by minimizing the total expected costs due to the experiments, future maintenance and failure.

Fatigue failure in offshore steel platforms is a very important failure mode which is considered in this paper. The uncertainties influencing the fatigue crack growth of steel structures include uncertainties in the loading, uncertainties in the structural behaviour and uncertainties in the material characteristics.

For offshore structures the fatigue inducing load is dominated by the wave load and the structural behaviour which is governed by the structural transfer function. For probabilistic fatigue assessment of offshore structures it is important to make a careful modelling of the uncertainties. This aspect has been treated in a number of papers, see e.g. Madsen et al. [3] and Shetty & Baker [4]. However, for the uncertainties associated with the transfer function a certain amount of subjectiveness is involved in the modelling as the real nature of this source of uncertainty in general is unknown. One way to reduce this subjectiveness is by onsite measurements of the structural response. An important question is therefore how to plan such measurements for the updating of probabilistic fatigue models in a cost optimal way. This problem has been discussed in e.g. Sørensen et al. [5] and Faber et al. [6]. For dynamicly sensitive structures the experimental design variables can be the duration of the experiment/measurement and the number, type and position of the sensors. The costs of the experiment will be highly dependent on these parameters.

In section 2 cost optimal planning of experiments is described in general. Based on the preposterior analysis from the classical decision theory an optimization problem is formulated where the optimization variables are the parameters defining the experiment. In section 3 the experiment planning is addressed. Here it is explained how the modelling and updating of the subjectively assessed uncertainty of the transfer function is performed using Bayesian statistics. Finally, in section 4 an example is presented illustrating the application of the proposed methodology.

2. COST OPTIMAL PLANNING OF EXPERIMENTS

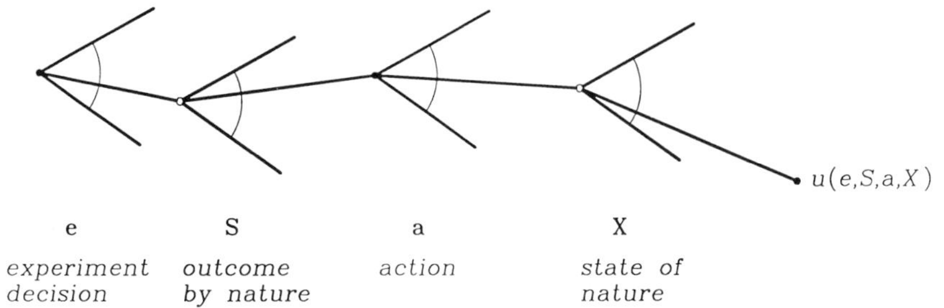

Figure 1. Decision tree.

Experiment planning plays an important role in several places in the decision theoretical approach to structural design and maintenance. The experimental design problem is to choose the experimental conditions such that the information provided by the experiment is maximized. However, as experiments are often connected with considerable expenses it seems reasonable that the additional cost of this new information is considered in the experiment planning. If on behalf of the outcome of the experiments it is possible to take an action reducing the probability of failure or increasing the lifetime of the component, the experimental costs might be justified. The decision variables of the experiment planning problem are denoted **e** and **a**. The corresponding decision tree is illustrated in figure 1. The stochastic variables **S** model the (unknown) outcomes of the experiment and the state of nature **X** model the stochastic variables in the system considered. According to the classical decision theory the optimal experiment plan can be determined by, see e.g. Raiffa & Schlaifer [7] and Benjamin & Cornell [8]

$$\max_{\mathbf{e}} \max_{\mathbf{a}} \quad E_{\mathbf{S}|\mathbf{e}}[E''_{\mathbf{X}|\mathbf{S}}[u(\mathbf{e},\mathbf{S},\mathbf{a},\mathbf{X})]] \tag{1}$$

where $E_{\mathbf{S}|\mathbf{e}}[\cdot]$ and $E''_{\mathbf{X}|\mathbf{S}}[\cdot]$ are the expectation operations with respect to the stochastic variables **S** and with respect to the stochastic variables **X** given $\mathbf{S} = \mathbf{s}$.

A formulation equivalent to (1) minimizing the total expected cost C_{TOT} could be

$$\min_{\mathbf{e}} \min_{\mathbf{a}} \quad C_{TOT}(\mathbf{e},\mathbf{a}) = C_E(\mathbf{e}) + C_A(\mathbf{e},\mathbf{a}) + C_F(\mathbf{e},\mathbf{a}) \tag{2}$$

where C_E is the expected experimental costs, C_A the expected costs due to the action taken and C_F the expected costs due to failure. The problem is a preposterior decision problem as the whole analysis has to be carried out on behalf of prior expectations of the outcome of the experiment before the experiment is actually conducted.

For engineering structures experiment planning enters at a number of different stages, for example in the planning of materials testing (such as fatigue life testing), in selection of response surfaces (approximations of complicated/time consuming functional relationships in terms of simple polynomial representations), in systems identification (identification of models and corresponding parameters for dynamic systems), in maintenance and inspection planning (e.g. times and locations for inspections) and in structural reassessment (such as proof testing). From the literature it appears that the development of these related subjects has taken place relatively independently of each other and that the basic common features of the associated problems have not yet been utilized.

Generally the uncertain quantities related to the modelling of the complete system described above can be divided into the following groups:

a) Quantities used to estimate the probability of failure, P_F.

b) Quantities used to estimate the costs, e.g. the real rate of interest r and the coefficients C_E, C_F.

The uncertainty related to these quantities can be taken into account by:

1) Modelling the quantities by stochastic variables (or stochastic processes). This is typically done for the quantities in group a) and the uncertainty is then taken into account in the reliability calculations using e.g. FORM/SORM.

2) Performing sensitivity analyses. This is typically done for quantities which are not known very well and which are not modelled by stochastic variables. Examples are the cost coefficients in group b), the statistical parameters used to model the stochastic variables and the decision variables.

In the example shown in section 4 both stochastic variables and sensitivity analyses are used in order to take into account the uncertainties in the system considered.

3. MODELLING OF UNCERTAINTIES IN TRANSFER FUNCTIONS

It is assumed that the short-term wave surface elevation spectra $S_{\eta\eta}(\omega, \theta)$ are known for given H_s, T_z where η indicates the water surface elevation, θ is the main wave direction and ω is the frequency. The spectrum is assumed to be one-sided and is discretized in N intervals of equal size $\Delta\omega$, see figure 2.

The bars in the response spectrum $S_{\sigma\sigma}(\omega, \theta)$ (stress spectrum), see figure 4, can be related to the bars in the load spectrum, see figure 2, by

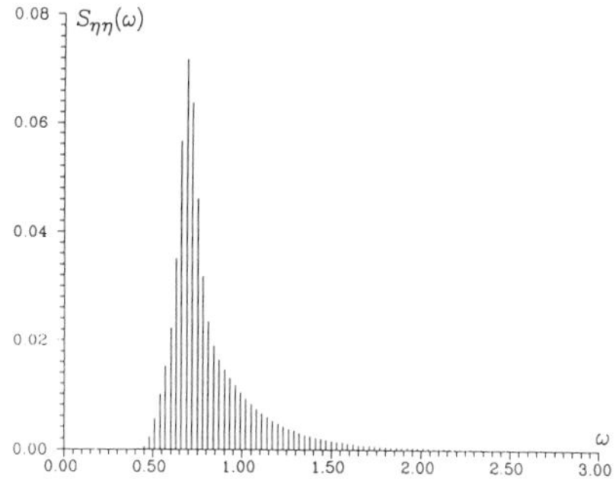

Figure 2. Discretized wave spectrum (JONSWAP for $H_s = 3m$ and $T_z = 7sec$, see Barltrop & Adams [9]).

$$S_{\sigma\sigma}(\omega_k, H_s, T_z, \theta) = |H(\omega_k, \theta)|^2 S_{\eta\eta}(\omega_k, H_s, T_z, \theta) \tag{4}$$

where $H(\omega_k, \theta)$ is the transfer function. Generally the transfer function $H(\omega_k, \theta)$ is only known to a certain degree, i.e. the transfer function is subjected to uncertainty. This uncertainty is modelled by, see e.g. Winterstein et al. [10]

$$H(\omega_k,\theta) = (1+Z_k)\hat{H}(\omega_k,\theta) \tag{5}$$

where $Z_k, k = 1, ..., N$ are stochastic variables and $\hat{H}(\omega_k)$ is the best unbiased estimate of the transfer function, e.g., estimated from general structural dynamics analysis. In (5) it is assumed that the transfer function uncertainty is independent of the wave direction θ. $Z_k, k = 1, ..., N$ are modelled as zero mean normally distributed variables with standard deviation V_k. $V_k, k = 1, ..., N$ are modelled as subjective stochastic variables with prior density function $f'_{V_k}(v)$.

The best unbiased estimate of the transfer function and the response spectrum are illustrated in figure 3.

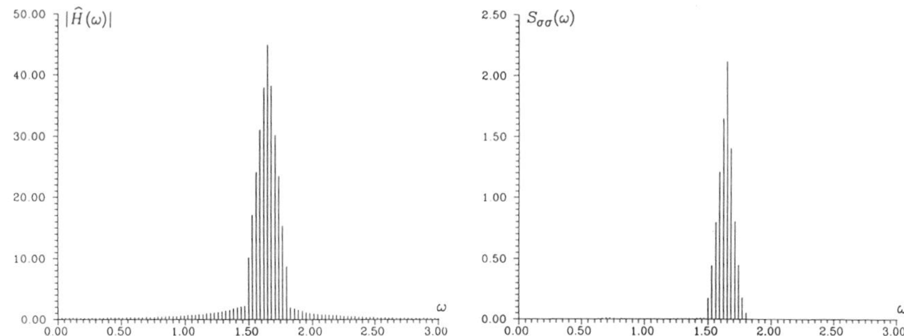

Figure 3. Discretized transfer function and discretized response spectrum.

The subjective uncertainty in the transfer function V_k is assumed to be updated on basis of observations $\hat{s}_k$ of the stress spectrum $S_{\sigma\sigma}(\omega_k)$. The estimates S_k are assumed to be obtained by averaging over n estimates of the stress spectrum. When spectral estimates are based on averaging the statistical uncertainty of the spectral estimates are modelled by

$$S_k = \hat{s}_k \frac{Y_k}{2n} \tag{6}$$

where $\hat{s}_k$ is the actual observation (assumed in this planning phase to be the same in all n estimates) and Y_k is χ^2 distributed with $2n$ degrees of freedom, i.e. the expected value is $2n$ and the coefficient of variation is $1/\sqrt{n}$, see e.g. Bendat & Piersol [11]. Due to limited knowledge of the second derivatives of the stress spectrum the bias of the spectral estimates is neglected. If the response measured is not directly the stresses then the associated transfer function has to be established.

Assuming that the spectral density estimates of the stress spectrum are obtained for known sea state (H_S and T_Z) and known main wave direction θ then the subjective uncertainty associated with $H(\omega_k,\theta)$ can be updated by updating the distribution for the distribution parameter V_k of the random variable Z_k.

The updated (posterior) density function of V_k is obtained from

$$f''_{V_k}(v_k|\text{observation}) = \frac{1}{c_k} P(\text{observation}|V_k = v_k) f'_{V_k}(v_k) \tag{7}$$

where the likelihood function becomes

$$\begin{aligned} P(\text{observation}|V_k = v_k) &= (P(S_{\sigma\sigma}(\omega_k, H_S, T_Z) = S_k|V_k = v_k))^n \\ &= (P((1+Z_k)^2|\hat{H}(\omega_k)|^2 S_{\eta\eta}(\omega_k, H_S, T_Z) = \frac{Y_k}{2n}\hat{s}_k|V_k = v_k))^n \end{aligned} \tag{8}$$

c_k is estimated from

$$c_k = \int_0^\infty P(\text{observation}|V_k = v_k) f'_{V_k}(v) dv \tag{9}$$

It should be noted that the density function of V_k is updated for given H_S, T_Z and θ. The response characteristics governing fatigue crack growth are functional relations involving the spectral moments of the stress spectrum. These spectral moments can be estimated by

$$\begin{aligned}\lambda_M(H_s, T_z, \theta) &= \sum_{k=1}^{N} \omega_k^M S_{\sigma\sigma}(\omega_k, H_s, T_z, \theta) \Delta\omega \\ &= \sum_{k=1}^{N} \omega_k^M (1 + Z_k)^2 |\hat{H}(\omega_k, \theta)|^2 S_{\eta\eta}(\omega_k, H_s, T_z, \theta) \Delta\omega\end{aligned} \tag{10}$$

The density function used to model V_k in Z_k is the posterior density f''_{V_k}. However, as the spectrum for the water surface elevation $S_{\eta\eta}(\omega_k, H_S, T_Z)$ depends on the sea-state the functional relations must be averaged over the possible outcomes of the sea-states. If also several main wave directions are considered the expectation operation must be extended to include also θ.

4. EXAMPLE

In the following example a fatigue sensitive detail from an offshore structure is considered. If the response spectrum is narrow banded then the standard deviation σ_σ of the stress process and expected rate of stress peaks can be estimated from

$$\sigma_\sigma(H_s, T_z, \theta) = \sqrt{\lambda_0(H_s, T_z, \theta)} \tag{a}$$

$$\nu_0(H_s, T_z, \theta) = \frac{1}{2\pi} \sqrt{\frac{\lambda_2(H_s, T_z, \theta)}{\lambda_0(H_s, T_z, \theta)}} \tag{b}$$

The limit state function corresponding to the SN approach is formulated by

$$g = \Delta - \int_0^{2\pi} \frac{T\nu_0(H_s, T_z, \theta)}{K} (2\sqrt{2})^m \sigma_\sigma(H_s, T_z, \theta)^m \Gamma(1 + m/2) f_\theta(\theta) d\theta \tag{c}$$

The stochastic parameters in this limit state function are

- The significant wave height H_s, here assumed Weibull distributed with mean value $3.0m$ and standard deviation $0.6m$.
- The mean wave period T_z. For given $H_s = h_s$ the conditional distribution of T_z is assumed to be normally distributed with mean value $\sqrt{h_s + 20.8}$ and standard deviation $0.3888 \cdot (\sqrt{(h_s + 4.2)/0.12} - \sqrt{(h_s + 1.6)/0.13})$, see Faber et al. [12].
- The material parameter K is assumed to be log-normally distributed with mean value $1.05 \cdot 10^{15} MPa^m$ and a C.O.V. of 50%
- The threshold Δ is normally distributed with mean value 1.0 and a C.O.V. of 10%.
- The uncertainty of the transfer function $Z_k, k = 1, ..., N$ is zero mean normally distributed with standard deviation $V_k, k = 1, ..., N$.
- $V_k, k = 1, ..., N$ is modelled by the posterior density function f''_{V_k}. V_k is strongly dependent on the number of samples n.

The expected lifetime is $T = 10$ years and the material constant m is assumed to be deterministic with $m = 4.1$. In this example the wave direction θ is not taken into account. The transfer function $|H(\omega_k)| = (1+Z_k)|\hat{H}(\omega_k)|, k = 1,2,..,N$ and the load and response spectra, see figures 2-3, are discretized in $N = 12$ frequencies around the peak of the transfer function. The water surface elevation for given sea states are described by the JONSWAP spectrum, see Barltrop & Adams [9].
Numerical calculations show that the probability of failure

$$P_f = P(g \leq 0) \tag{d}$$

is strongly dependent on the choice of the prior density function for V_k, $f'_{V_k}(v_k)$, and of the number of measured spectra to be averaged over, n. In table 1 the reliability index $\beta = -\Phi^{-1}(P_f)$ is shown for various n and a normal prior density function with mean value 0.25 or 0.50 and a C.O.V. of 20%. $n = 0$ corresponds to no measurements. In the calculations the prior density function $f'_{V_k}(v_k)$ is used for V_k when $n = 0$.

n	$\mu_{V'_k} = 0.25$	$\mu_{V'_k} = 0.50$
0	2.73 (2.82)	2.20 (2.77)
1	2.74 (2.82)	2.24 (2.78)
4	2.76 (2.83)	2.43 (2.80)
10	2.80 (2.84)	2.98 (2.84)

Table 1. Reliability indices.

The reliability indices are calculated using both fully dependent and independent uncertainties Z_k modelling the transfer function $|H(\omega_k)|$. Numbers in () are for independent uncertainties $Z_k, k = 1,2,..,12$. It is seen that generally the reliability increases when the uncertainties Z_k of the transfer function are assumed independent. In table 1 the actual observation of the stress spectrum is assumed to take the fixed value $\hat{s}_k = 0.8 \cdot |\hat{H}(\omega_k)|^2 \cdot S_{\eta\eta}(\omega_k, H_s = 3m, T_z = 7s)$.

The change in posterior density function $f''_{V_k}(v_k)$ for various n is shown in figure 4 for the case when $\mu_{V'_k} = 0.50$, the uncertainties $Z_k, k = 1,2,..,12$ on the transfer function are assumed dependent and the observations $\hat{s}_k$ are defined as above. The posterior density functions are calculated at the respective design points. It is seen that there is a transition towards smaller posterior mean and standard deviation for increasing n, i.e., less uncertainty on the transfer function when more samples are available. The updated reliability index taking into account the observations $\hat{s}_k$ can be expected to be rather sensitive with respect to the actual observations $\hat{s}_k$. If $n = 4$, $\mu_{V'_k} = 0.50$ and fully dependent uncertainties of the transfer function are used a change in the scale factor S_{scale} and in the mean wave period T_z in the actual observation of the stress spectrum $\hat{s}_k = S_{scale} \cdot |\hat{H}(\omega_k)|^2 \cdot S_{\eta\eta}(\omega_k, H_s, T_z)$ leads to relatively small changes in β, see table 2.
Another possible way to model the observed stress spectrum is to model the measured stress spectrum by $\hat{s}_k = S_{\sigma\sigma}(\omega_k, H'_s, T'_z)$, where H'_s and T'_z are stochastic variables modelled similarly to, but independently of H_s and T_z. If expectation with respect to H'_s and T'_z is performed a reliability index $\beta = 2.29$ for fully dependent uncertainties of the transfer function, $n = 4$ and $\mu_{V'_k} = 0.50$ is obtained.

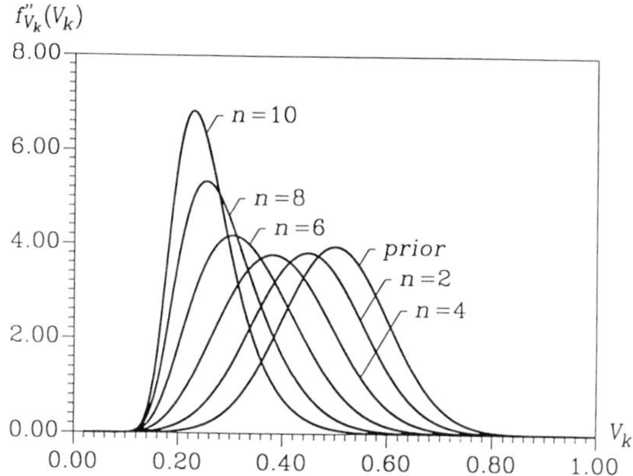

Figure 4. Prior density function $f'_{V_k}(v_k)$ and posterior density functions $f''_{V_k}(v_k)$ for $n = 2, 4, 6, 8$ and 10.

S_{scale}	$(H_s, T_z) = (3m, 7s)$	$(H_s, T_z) = (3m, 5s)$
0.8	2.43	2.31
1.0	2.42	2.28
2.0	2.35	2.18

Table 2. Reliability indices β for $n = 4$, $\mu_{V'_k} = 0.50$ and observation $\hat{s}_k = S_{scale} \cdot |\hat{H}(\omega_k)|^2 \cdot S_{\eta\eta}(\omega_k, H_s, T_z)$.

The experiment planning is seen to be governed by the number n of spectra to be averaged over. The optimal number of n is determined from (2). The following assumptions are made: A linear model for the experimental costs is used, the action costs are not dependent of the basic uncertain quantities $\mathbf{X}$ and the failure costs are proportional to the updated probability of failure. If the minimization with respect to the action is already performed then the optimal value of n can be determined from the optimization problem

$$\min_n C_T = C_{EXP}(n) + C_F P_f(n) + C_A(n)$$
$$= (C_{E,0} + C_{E,1}\ n) + C_F P_f(n) + C_A(n) \qquad \text{(e)}$$

where C_T is the total expected costs, C_F are the cost of failure, C_{EXP} are the experiment costs, C_A are the action costs, $C_{E,0}$ are the fixed experiment costs and $C_{E,1}$ is a factor used to model the marginal experiment costs.

The optimal number n is thus determined from

$$\frac{dP_f}{dn} = -\frac{C_{E,1} + \frac{dC_A}{dn}}{C_F} \qquad \text{(f)}$$

n	dP_f/dn
2	$-1.672 \cdot 10^{-3}$
4	$-1.529 \cdot 10^{-3}$
6	$-1.141 \cdot 10^{-3}$
8	$-0.668 \cdot 10^{-3}$
10	$-0.332 \cdot 10^{-3}$

Table 3. Derivatives of P_f with respect to n.

In table 3 dP_f/dn is shown for some values of n for $\mu_{V'_k} = 0.5$ and for a given result of the measurement $\hat{s}_k = 0.8 \cdot |\hat{H}(\omega_k)|^2 \cdot S_{\eta\eta}(\omega_k, H_s = 3m, T_z = 7s)$. Equation (f) and the numbers in table 3 can be used to determine the optimal number of measured spectra. The above results are for given outcome of the experiment. According to (1) an expectation operation with respect to the experimental outcome should be performed. However, if the optimal number n is not very sensitive to the experimental outcome the procedure described above can be used. Further, if the action costs are dependent of the basic variables $\mathbf{X}$ then more comprehensive computations are needed to determine the optimal action and experiment.

5. CONCLUSION

Cost optimal experiment planning of dynamicly sensitive linear structural systems has been addressed. In order to model the expected costs the uncertainties related to the load and the structural behaviour and material characteristics have to be taken into account. In this paper special effort has been put into the modelling of the uncertainties in the structural transfer function. Bayesian statistics have been used for this purpose making it possible to formulate a model including prior knowledge of the uncertainties. In the formulated model the uncertainties of the transfer function can be updated with respect to actual measurements of the structural response. An example of the fatigue application has been presented showing the usefulness in experiment planning.

ACKNOWLEDGEMENTS

Part of this paper is supported by the research project "Risk Analysis and Economic Decision Theory for Structural Systems" granted by the Danish Technical Research Council.

REFERENCES

[1] Viertl, R.: *Statistical Methods in Accelerated Life Testing.* Applied Statistics and Econometrics, H. 32, Vandenhoeck & Ruprecht, Gottingen, 1988.

[2] Ljung, L.: *System Identification: Theory for the User.* Prentice Hall, Englewood Cliffs, 1987.

[3] Madsen, H.O., Krenk, S. and Lind, N.C.: *Methods of Structural Safety.* Prentice-Hall, 1986.

[4] Shetty, N.K. and Baker, M.J.: *Fatigue Reliability of Tubular Joints in Offshore Structures.* (3 papers), Proc. 9th OMAE Conference, Houston, 1989.

[5] Sørensen, J.D., Faber, M.H. and Kroon, I.B.:*Risk Based Optimal Fatigue Testing.* Probabilistic Mechanics and Structural and Geotechnical Reliability, Proceedings of the Sixth Speciality Conference, Denver, Colorado, 1992.

[6] Faber, M.H., Sørensen, J.D. and Kroon, I.B.:*Optimal Fatigue Testing - a Reassessment Tool.* Presented at the IABSE Conference, Copenhagen, Denmark, 1993.

[7] Raiffa, H. and Schlaifer, R.: *Applied Statistical Decision Theory.* Harward University Press, Cambridge, Mass., 1961.

[8] Benjamin, J.R. and Cornell, C.A.: *Probability, Statistics and Decision for Civil Engineers.* McGraw-Hill, 1970.

[9] Barltrop, N.D.P. and Adams, A.J.: *Dynamics of Fixed Marine Structures.* Third edition, Butterworth-Heinemann Ltd, Oxford, 1991.

[10] Winterstein, S.R., Kroon, I.B. and Ude, T.C.: *Fatigue of Floating Offshore Structures: Modelling Uncertainty in Hydrodynamics and Fatigue Properties.* Proceedings of ASCE Structures Congress, Irvine, California, April 1993.

[11] Bendat, J.S. and Piersol, A.G.: *Random Data: Analysis and Measurement Procedures.* John Wiley & Sons, Inc., 1971.

[12] Faber, M.H., Sørensen, J.D., Rackwitz, R., Thoft-Christensen, P. and Bryla, P.: *Reliability Analysis of an Offshore Structure: A Case Study - 1.* Proceedings of the 11th OMAE conf., Calgary, Canada, 1992.

Reliability and Optimization of Structural Systems, V (B-12)
P. Thoft-Christensen and H. Ishikawa (Editors)
Elsevier Science Publishers B.V. (North-Holland)

Structural Reliability Estimation Based on Simulation Outside the β-Sphere

M. Yonezawa*, S. Okuda**

* Department of Industrial Engineering, Kinki University,
341, Kowakae, Higashi-Osaka, Osaka 577, Japan

** Kumano Technical College, Kinki University,
2800, Arima-cho, Kumano, Mie 519-42, Japan

Abstract

This paper is concerned with an assessment of structural failure probability based on the Monte Carlo simulation with partition of the region technique. The basic random variables are assumed to be stochastically independent and normally distributed and the reliability index ß to be known. The defining domain of basic variables of the structural system is divided into two disjoint regions: one is a region inside a k-dimensional sphere of radius ß, refered to the ß-sphere and the other one is the region outside of it. The ß-sphere is excluded from the sampling of simulations and samples are generated just outside the ß-sphere. Each sample is produced by combining a random radius value with a truncated chi-square p.d.f. defined for outside the ß-sphere and a random direction vector value generated from standard normal variates. The proposed method is illustrated with some numerical examples.

Key word Codes: J.2; J.6; J.m
Keywords: Computer Applications, Physical Science and Engineering; Computer-Aided Engineering; Miscellaneous

1. INTRODUCTION

The structural failure probability is defined, for time-independent case, by the following multi-dimensional integral[1]:

$$P_f = \int_{g(x)\leqq 0} f(x)dx \qquad (1)$$

where $X = (X_1,X_2,...,X_k)$ is the vector of basic random variables in the k-dimensional space, $f (x)$ is the joint probability density function (p.d.f.) of basic random variables and $g(x) = g(x_1,x_2,...,x_k)$ is the limit state function. For any realization x of basic random variables X, the set $\{ x \mid g(x)\leqq 0 \}$ implies the failure region.

The integral in Eq. (1) is generally difficult to calculate because the domain of integration has often irregular shape and $f(x)$ is generally complex. Therefore, simulation methods are widely adopted to estimate the integral.

Using an indicator function $I[(\cdot)]$ defined by

$$I[(\cdot)] = 1 \text{ if } (\cdot) \leqq 0$$
$$I[(\cdot)] = 0 \text{ if } (\cdot) > 0 \quad (2)$$

Eq. (1) can be rewritten as follows:

$$P_f = \int_{\text{all } x} I[g(x)] f(x)\, dx \quad (3)$$

then it is expressed as an expected value of $I[g(x)]$ with respect to $f(x)$.

The estimate of failure probability P_{fe} and its variance $Var(P_{fe})$ is evaluated through simulations with generating samples $x^{(i)}$ (i=1,2,...,N) with the p.d.f. $f(x)$ as follows:

$$P_{fe} = (1/N) \sum_{i=1}^{N} P_f^{(i)} \quad (4)$$

$$Var(P_{fe}) \doteqdot \sum_{i=1}^{N} [P_f^{(i)} - P_{fe}]^2 / \{N(N-1)\} \quad (5)$$

$$P_f^{(i)} = I[g(x^{(i)})] \quad (6)$$

However, a large number of simulations are generally required to estimate the integral for the case with low structural failure probability. Then variance reduction techniques such as the importance sampling[2]-[5], [9] and partition of the region technique[6]-[8], have been inevitably introduced to estimate the structural failure probability. In this paper, a sampling method is proposed which is based on partition of the region technique to exclude the ß-sphere from the sampling of simulations.

2. PARTITION OF THE REGION

Consider that the defining domain of k basic random variables is divided into two disjoint parts, that is, one is a region inside the k-dimensional ß-sphere and the other one is outside of it as shown in Figure 1, where ß is the reliability index[1] and adopted as a radius of the k-dimensional hypersphere, refered to ß-sphere, which has its center at the origin of standardized coordinates and is tangent to the limit state function in the safety region.

The failure probability P_f is expressed by regarding these disjoint domains as follows[3],[6]:

$$P_f = \int_{|x| \leqq \beta} I[g(x)]f(x)\,dx + \int_{|x| > \beta} I[g(x)]f(x)\,dx \tag{7}$$

As the first term of Eq. (7) is equal to zero, P_f can be evaluated only by the second term of Eq. (7). Then introducing a truncated p.d.f. $f_t(x)$ defined for outside the ß-sphere, Eq. (7) is rewritten as follows:

$$P_f = (1-P)\int_{|x| > \beta} I[g(x)]f_t(x)\,dx \tag{8}$$

where P is the probability of the ß-sphere and the truncated p.d.f. $f_t(x)$ is defined for outside the ß-sphere as follows:

$$f_t(x) = f(x)/(1-P) \quad : |x| > \beta$$
$$= 0 \quad : |x| \leqq \beta \tag{9}$$

and $f_t(x)$ must satisfy the following of integral,

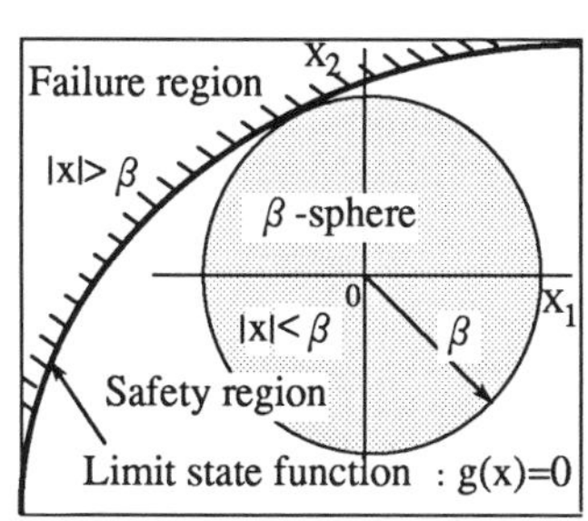

Figure 1. Partition of the region basic variables space[3],[6].

$$\int_{|x| > \beta} f_t(x)\,dx = 1 \tag{10}$$

Using random variables $x_t \in \{x \mid |x| > \beta\}$ with the truncated p.d.f. $f_t(x_t)$, Eq. (3) can be rewritten as follows:

$$P_f = \int_{\text{all } x_t} (1-P)I[g(x_t)]f_t(x_t)\,dx_t \tag{11}$$

then P_f is expressed as an expected value of $(1-P)I[g(x_t)]$ with respect to $f_t(x_t)$.

Generating samples $x_t^{(i)}$ ($i = 1,2,\ldots N$) with the truncated p.d.f. $f_t(x_t)$, the estimate of failure probability P_{fe} and its variance Var (P_{fe}) is evaluated by Eqs. (4) and (5) with Eq. (12) in place of Eq. (6).

$$P_f^{(i)} = (1-P)I[g(x_t^{(i)})] \tag{12}$$

3. SAMPLES GENERATION FROM A TRUNCATED CHI-SQUARE P. D. F.

Since basic variables are assumed to be independent normal variates, a square of the magnitude of the vector of them,

$$z = r^2 = \sum_{i=1}^{k} (x_i)^2 \tag{13}$$

follows a chi-square distribution with k degrees of freedom, that is,

$$f_{\chi^2}(z) = (z)^{(k/2)-1} \exp(-z/2) / \{2^{k/2} \Gamma(k/2)\} \tag{14}$$

and r is a magnitude of the vector of basic random variables.

The probability P of the ß-sphere can be obtained by evaluating the chi-square distribution,

$$P = \text{prob}[z \leqq ß^2] = F_{\chi^2}(ß^2) \tag{15}$$

where $F_{\chi^2}(z)$ is the cumulative distribution function of the chi-square distribution.

Therefore, the truncated chi-square p.d.f. is defined for outside the ß-sphere as follows:

$$\begin{aligned} f_{\chi^2_t}(z) &= f_{\chi^2}(z)/(1-P) & &: z \geqq ß^2 \\ &= 0 & &: z < ß^2 \end{aligned} \tag{16}$$

and $f_{\chi^2_t}(z)$ satisfies the following integral.

$$\int_{z \geqq ß^2} f_{\chi^2_t}(z)\,dz = 1 \tag{17}$$

Noting that a magnitude of the vector of basic random variables corresponds to a radius r in polar coordinates, random values of the k-dimensional basic variables x_t with the truncated p.d.f. $f_t(x_t)$ can be easily obtained by sampling a random radius value with the truncated p.d.f. of Eq. (16) and a k-dimensional random direction vector.

First, the sampling region outside the ß-sphere is classified with respect to the frequency of z, where the interval of class is q. Then the truncated p.d.f is composed as a form of a frequency histogram of the chi-square p.d.f. which is

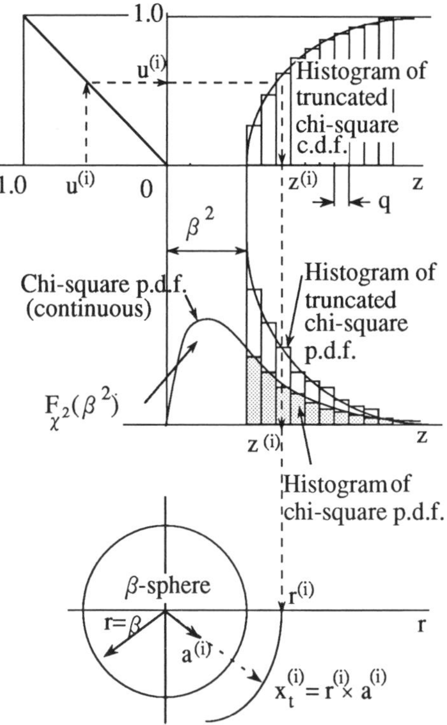

Figure 2. Generation of a random radius from the truncated chi-square distribution.

numerically integrated over each interval of class boundary and multiplied by 1 / (1−P) for the region outside the ß-sphere.

Summing the frequency histogram, then a cumulative frequency function is obtained as shown in Figure 2. For an i-th uniform random variate $u^{(i)}$ of a i-th simulation cycle, the inverse method gives an i-th square of a random magnitude of basic variable vector $z^{(i)}$ is obtained, the square root of which yields an i-th random radius $r^{(i)}$ in polar coordinates. Therefore, an i-th sample of basic variables, $x_t^{(i)} = (x_{1t}^{(i)}, x_{2t}^{(i)}, \ldots, x_{nt}^{(i)})$ for outside the ß-sphere is generated by the product of an i-th random radius $r^{(i)}$ with the truncated chi-square p.d.f. and an i-th unit random direction vector $\mathbf{a}^{(i)}$, which is generated by using k random standard normal variates

$V^{(i)} = (V_1^{(i)}, V_2^{(i)}, \ldots, V_k^{(i)})$, that is

$$x_t^{(i)} = r^{(i)} \mathbf{a}^{(i)} \tag{18}$$

$$\mathbf{a}^{(i)} = (V_1^{(i)}, V_2^{(i)}, \ldots, V_k^{(i)}) / |V|^{(i)} \tag{19}$$

$$|V|^{(i)} = \{ \sum_{j=1}^{k} (V_j^{(i)})^2 \}^{1/2} \tag{20}$$

For the case of structural systems with multi mode failure, a minimum ß should be selected for the radius of ß-sphere as shown in Figure 3. Even if the ß point can't be obtained, an appropriate radius of sphere may be adopted as an excluded region in the common safety region.

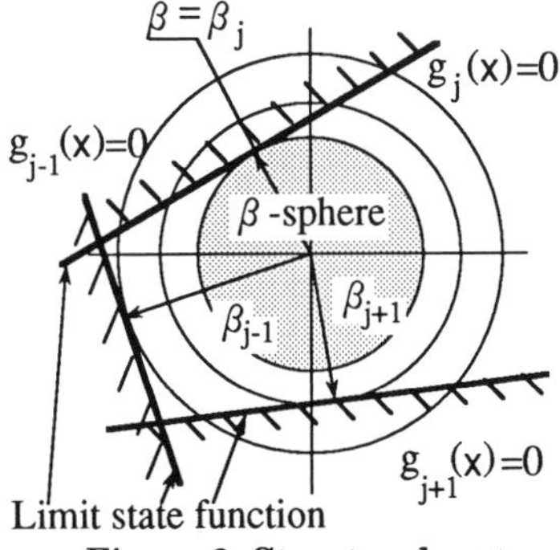

Figure 3. Structural system with multi mode failure.

4. NUMERICAL EXAMPLES

The structural failure probabilities for structures with limit state functions given by Eqs. (21) and (22) are estimated by the proposed sampling method, the crude Monte Carlo and the importance sampling(ISPUD) for multi mode failure[4].

$$\text{No. 1}: g(x) = (1 - x_2 - x_1^2) \cup (1 + x_2 - x_1^2) \quad \text{...[4]} \tag{21}$$

$$\text{No. 2}: g(x) = -0.125(x_1^2 + x_2^2 + x_3^2) - x_4 + 4 \quad \text{...[3],[6]} \tag{22}$$

The estimates of the failure probability and their coefficient of variations (c.o.v.) vs. sample size N for each limit state function are shown in Figures 4 and 5, and also given in Tables 1 and 2. Following legends are used in the

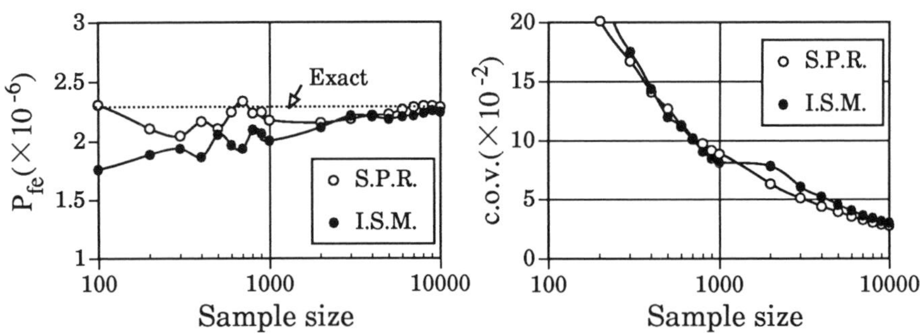

Figure 4. Estimates of P_f and c.o.v for the limit state function of Eq. (21).

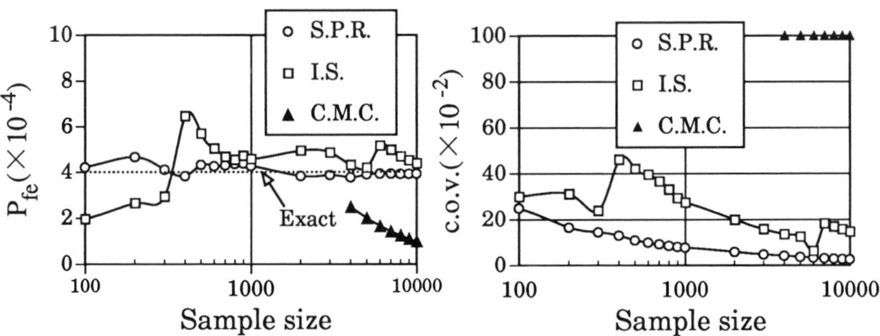

Figure 5. Estimates of P_f and c.o.v for the limit state function of Eq. (22).

Figures and Tables: C.M.C. stands for the Crude Monte Carlo, S.P.R. for the Simulation with Partition of the Region technique (proposed method), I.S for the Importance Sampling and I.S.M. for the Importance Sampling for Multi Mode failure. The basic random variables x_i's are treated as independent normal variates, that is, non-normal variates are transformed to normal variates by Rosenblatt transformation[11].

The probabilities of the ß-sphere for limit state functions of Eqs. (21) and (22) reach to 0.99998 and 0.99698 respectively. It can be said that a large amount of samples which might be lost are applied effectively by the proposed method to the region outside the ß-sphere. Results are not obtained with the crude Monte Carlo method as seen in Figure 4, because all of samples fall in the safety region and no sample in the failure region for this case.

The proposed method gives better estimates for relatively less sample size than any other methods.

5. CONCLUSION

A sampling method is presented which is based on the exclusion from the sampling of the ß-sphere in the standard normal space of basic variables. It is easy to apply the proposed method for simulation approachs of the reliability analysis for structures with multi mode failure.

Table 1 Estimates of P_f for the the limit state function of Eq. (21)

Random variable	mean value	standard deviation	p.d.f.
X_1	0.1	0.1	normal
X_2	0.05	0.2	normal

Methods	C.M.C.		S.P.R. (proposed method)		I.S.M.	
Sample size:N	P_{fe} $\times 10^{-6}$	c.o.v. $\times 10^{-2}$	P_{fe} $\times 10^{-6}$	c.o.v. $\times 10^{-2}$	P_{fe} $\times 10^{-6}$	c.o.v. $\times 10^{-2}$
100	----	----	2.31	27.2	1.76	44.2
500	----	----	2.11	12.7	2.06	12.0
1000	----	----	2.18	8.8	2.01	8.1
5000	----	----	2.23	3.9	2.19	4.5
10000	----	----	2.30	2.7	2.25	3.0

(P_{fe}= 2.3 $\times 10^{-6}$, Schuëller and Stix[4])

Table 2 Estimates of P_f for the the limit state function of Eq. (22)

Random variable	mean value	standard deviation	p.d.f.
X_i, (i=1,2,3,4)	0	1	normal

Methods	C.M.C.		S.P.R. (proposed method)		I.S.	
Sample size:N	P_{fe} $\times 10^{-4}$	c.o.v. $\times 10^{-2}$	P_{fe} $\times 10^{-4}$	c.o.v. $\times 10^{-2}$	P_{fe} $\times 10^{-4}$	c.o.v. $\times 10^{-2}$
100	----	----	4.22	24.9	1.96	30.0
500	----	----	4.33	10.9	5.69	42.1
1000	----	----	4.28	7.7	4.58	27.4
5000	2.0	100	3.92	3.6	4.20	12.5
10000	1.0	100	3.94	2.5	4.40	14.6

($P_{fe} = 4.0 \times 10^{-4}$, Harbitz[6])

REFERENCES

[1] Thoft-Christensen, P. and Baker, M. J. : Structural Reliabilty Theory and Its Applications, Springer-Verlag, Berlin, Heidelberg, New York ,1982.

[2] Rubinstein R. Y. : Simulation and The Monte Carlo Method, John Wiley & Sons, New York, 1981.

[3] Bourgund U., Ouypornprasert W., and Prenninger, P. H. W. : Advanced simulation methods for the estimation of system reliability, Internal Working Report No.19, University of Innsbruck, Austria, 1986.

[4] Schuëller, G. H., Stix, R. : A critical appraisal of methods to determine failure probabilitis, Structural safety, 4 (1987) 293-309.

[5] Bourgund, U. and Bucher, C. G. : Importance sampling procedure using design point(ISPUD), Institute of Engineering Mechanics, University of Innsbruck, Austria, Report 9-86, 1986.

[6] Harbitz, A. : An efficent sampling method for probability of failure calculation, Structural Safety 3 (1986) 104-115.

[7] Csenki, A. : A new Monte Carlo technique in structural reliability with a plastic frame example, Reliability and Optimization of Structural Systems'88, Proc. of the 2nd IFIP WG 7.5 Conference, London (1988) 47-64.

[8] Csenki, A. : An Improved Monte Carlo Method in Structural Reliability, Structural Safety 24 (1989) 275-292.

[9] Bucher, C. G. : Adaptive sampling, an iterative fast Monte Carlo procedure, Structural Safty, 5 (1988) 119-126.

[10] kendall, M. G. : A Course in the Geometry of N dimensions, Griffin's Statistical Monographs & Courses, Charles Griffin, London, 1961.

[11] Ang, A. H-S and Tang, W.H. : Probability Concepts in Engineering Planning and Design, Vol. II, Decision, Risk, and Reliabilty, John Wiley & Sons, New York, 1984.

Reliability and Optimization of Structural Systems, V (B-12)
P. Thoft-Christensen and H. Ishikawa (Editors)
Elsevier Science Publishers B.V. (North-Holland)

Reliability Approach to the Tensile Strength of Unidirectional CFRP Composites by Monte-Carlo Simulation

K. Goda

Department of Mechanical Engineering,
Hiroshima University, Higashi-Hiroshima, 724, Japan

ABSTRACT

A numerical technique based on the Monte-Carlo method in a shear-lag model is developed to simulate the tensile strength and fracture process for a unidirectional carbon fiber reinforced plastic composite. The technique improves on the conventional approach by using an r_{min} method that can determine reasonably the stresses working on the fiber and matrix elements as the damage progresses. The r_{min} method is based on tracking the incremental ratio of the strength of an element to its stress. The present model includes the effect of the sliding frictional forces around fiber breaks caused by debonding between the fiber and matrix. Statistical properties of the tensile strength were obtained through simulation runs involving one hundred samples for each value of the frictional force parameter. Also studied was the size effect in composite strength with increasing numbers of fibers.

1. INTRODUCTION

The fracture of fiber reinforced composite materials involves a complicated damage accumulation process resulting from random fiber break and debonding between the fiber and matrix. Several theoretical models for investigating the probability distribution for strength, based on the load sharing rules called LLS and TLS*1, have been created [1]-[3]. The models were evaluated as a function of the probability distribution function for fiber strength, so it was enough to comprehend the relation between the random fiber strength and composite strength. However, the models cannot be applied for the composite associated with the debonding because of the idealized load sharing rules. The Monte-Carlo simulation technique coupled with a stress analysis method is one of the most effective tools for understanding the fracture process including the effect of the debonding. Several researchers have attempted the Monte-Carlo simulation technique from a micromechanical viewpoint. For example, Oh [4] developed the technique by applying the finite difference method to the shear-lag model [5], and Kimpara, et al. [6] improved the technique by taking the effect of matrix fracture (debonding) into account. Goda and Fukunaga [7] also simulated the tensile strength of metal matrix composites using the similar technique. In these papers, a Weibull distribution was typically used for generating the random fiber breaks, but the deformation behavior of the matrix was not realistic in that the frictional force at the interface after debonding was neglected. According to the pull-out test involving a single filament embedded in a matrix block [8], the relation between the load and the displacement indicates that the interface region first deforms linearly, and then the load drops in an instance to a lower constant value as soon as the debonding occurs. The load values suggest an appreciable value for the sliding frictional stress working at the interface. Thus, it is necessary to incorporate the effects of both random fiber breakage and the complicated debonding behavior into the Monte-Carlo simulation technique.

In the present study, the fracture process for a unidirectional carbon fiber reinforced plastic composite (CFRP) is studied. The shear-lag model for simulating the mechanics of the tensile fracture process in the composite takes into account the sliding friction of the fiber at the interface after debonding. The effect of the value of the frictional force parameter on the statistical properties of the tensile strength was evaluated through many simulation runs, and compared to the case of a composite without debonding.

---

*1. LLS and TLS are the abbreviations for " Local Load Sharing " and " Tapered Load Sharing ", respectively. The former means that two fibers adjacent to the broken fiber(s) sustain the load lost by the break(s), and the latter means that the nearest four fibers sustain the load.

Furthermore, the size effect on the statistical properties of the strength with increasing numbers of fibers was also studied.

2. ANALYTICAL METHOD

2.1 Shear-lag model In the shear-lag model [5], it is assumed that the fibers sustain the axial force, and the matrix transfers through shear the force lost at broken fibers to the neighboring fibers. Since the model does not take into account the normal stress working in the matrix, the stress working on the composite is assumed to be the stress that is applied directly to the fibers. In polymer composites, the difference is typically small. Furthermore, the problem can be solved easily in comparison with the other stress analysis methods, because there is just one force equilibrium condition along the fiber axis. Additionally, it has been shown that the stress distributions determined by the shear-lag model were in good agreement with those given by 3-D finite element methods, when the matrix deforms elastically [9]. Therefore, the shear-lag model has often been used as the stress analysis method for unidirectional composite materials.

The shear-lag equations of force equilibrium for a circular composite mono-layer with N reinforcing fibers parallel to the x-axis are given as follows:

$$\begin{aligned} &A\frac{d\sigma_1}{dx}+h\tau_1-h\tau_N=0\\ &\quad\vdots \\ &A\frac{d\sigma_N}{dx}+h\tau_N-h\tau_{N-1}=0 \end{aligned} \qquad (1)$$

where, σ_i (i=1,..., N) is the normal stress working along the fiber axis, A is the cross-sectional area of the fiber, h is the thickness of the layer, and τ_i (i=1,...,N) is the shear stress in the matrix between the i-th and (i+1)-th fiber. The reason why a circular mono-layer was chosen is that the edge effects due to overstress concentrations caused by fiber breakage on a planar edge are eliminated from the statistics. Figure 1 shows the relation between the stress and the strain working in the constituents used in this study. As shown in Fig. 1.(a), a fiber element is assumed to be a linear elastic body with a tensile strength X which varies statistically. Also, the dynamic effects of fiber breakage are assumed to be negligible. The relation between the stress σ_i and the displacement u_i is given as follows:

$$\sigma_i=E\frac{du_i}{dx} \qquad (2)$$

where, E is the fiber elastic modulus. On the other hand, the shear deformation of the matrix is also assumed to behave elastically, as shown in Fig.1.(b). However, if the shear stress reaches the interfacial shear strength τ_m, debonding between the fiber and matrix will occur. After debonding occurs, the fiber slides with a certain interfacial frictional force on the matrix [8]. Therefore, the matrix may be approximated as if it behaves elastic-perfectly-plastic, as shown in Fig. 1.(c). The shear stress in the matrix is given as follows:

1) If the matrix deforms elastically,

$$\tau_i=\frac{G}{d}(u_{i+1}-u_i)+\Delta\tau_a \qquad (3a)$$

2) If the matrix deforms with the fiber sliding,

$$\tau_i=\tau_s \qquad (3b)$$

where, d is the distance between fibers, G is the shear elasticity of the matrix and τ_s is the constant shear stress on the matrix generated by the interfacial friction (where a minus should be put in the front of t_s when negative sliding occurs). Also $\Delta\tau_a$ is the apparent stress increment given by $\Delta\tau_a = \tau_s - G\gamma^*$ where γ^* is the shear strain on the matrix element at the beginning of the unloading; this term is needed once sliding has occurred and unloading is possible so that the matrix then behaves elastically upon unloading. By substituting eqns (2) and (3a) or (3b) into eqn (1), second order, ordinary simultaneous differential equations are obtained.

2.2 Approach to finite difference method It is possible to solve the simultaneous differential equations mentioned above. But it would take a prohibitively long time to solve them because it is necessary to take increasingly complex boundary conditions into consideration as fiber breakage and debonding increase. So, the finite difference method developed by Oh [4] and Kimpara et al. [6] was applied to the present study. Figure 2 shows the model in which a fiber element F(i,j) is located between the nodal point (i,j-1) and (i,j), and a matrix element M(i,j) is located between the nodal point (i,j) and (i+1,j). By denoting the displacement of the nodal point (i,j) by $u_{i,j}$, the second order differential form in eqn (1) is approximated as follows:

$$\frac{d^2u_i}{dx^2}=\frac{u_{i,j-1}-2u_{i,j}+u_{i,j+1}}{(\Delta x)^2} \qquad (j=1,\cdots,K) \qquad (4)$$

where Δx is the distance between nodal points, namely the length of a fiber element, and K is the number of fiber elements along a fiber. The finite differential form given by eqn (4) can be applied only for an unbroken fiber element. If fiber element F(i,j) breaks, where we assume this break to be at the center point (i,j-1/2) of the fiber element, then the relation between the displacements $u_{i,j}$ and $u_{i,j-1/2}$ is given by the following equation because of zero strain:

$$u_{i,j}=u_{i,j-1/2} \qquad (5)$$

Therefore, we can rewrite the finite differential form on the nodal point (i,j) as follows:

$$\frac{d^2u_i}{dx^2}=\frac{4}{3(\Delta x)^2}(u_{i,j+1}-u_{i,j}) \qquad (6)$$

Thus, by accounting for the possibility of breakage of the fiber element F(i,j+1), we express eqns. (4) and (6) as one equation as follows:

$$\frac{d^2u_i}{dx^2}=\frac{4\{o_{i,j+1}(u_{i,j+1}-u_{i,j})-o_{i,j}(u_{i,j}-u_{i,j-1})\}}{(2+o_{i,j}+o_{i,j+1})(\Delta x)^2} \qquad (7)$$

where, $o_{i,j}$ is a constant which is determined by whether the fiber element breaks or not. That is,

$$o_{i,j}=H\{X_{i,j},\sigma_{i,j}\} \qquad (8)$$

where, $H\{\cdot\}$ is the Heaviside step function, $X_{i,j}$ is the tensile strength of the fiber element F(i,j), and $\sigma_{i,j}$ is the normal stress working on the fiber element F(i,j) given by the following equation:

$$\sigma_{i,j}=E\frac{u_{i,j}-u_{i,j-1}}{\Delta x} \qquad (9)$$

We reformulate the shear stress working on the matrix element M(i,j), in order to deal with eqns (3a) and (3b) as one equation, as follows:

$$\tau_{i,j}=p_{i,j}\left(G\frac{u_{i+1,j}-u_{i,j}}{d}+\Delta\tau_{a\ i,j}\right)+(1-p_{i,j})\tau_s\cdot \mathrm{sgn}(u_{i+1,j}-u_{i,j}) \qquad (10)$$

where, $p_{i,j}$ is equal to 1 when the matrix element deforms elastically, even if the matrix element has debonded from the fiber, and, $p_{i,j}$ is equal to 0 when the matrix element shows a constant shear stress governed by the sliding friction. Also we define sgn(·) as the transform function, which is 1 if the value in parentheses is positive, which is -1 if this value is negative, and which is 0 if this value is zero. By substituting eqns (7) and (10) into eqn (1), we obtain

$$c_{1(i,j)}u_{i,j-1}+c_{2(i,j)}u_{i-1,j}+c_{3(i,j)}u_{i,j}+c_{4(i,j)}u_{i+1,j}+c_{5(i,j)}u_{i,j+1}=C_{(i,j)} \qquad (11)$$

where

$$c_{1(i,j)}=\frac{4o_{i,j}}{2+o_{i,j}+o_{i,j+1}}$$

$$c_{2(i,j)}=\frac{p_{i-1,j}\cdot Gh(\Delta x)^2}{EAd}$$

$$c_{3(i,j)}=\frac{4(o_{i,j}+o_{i,j+1})}{2+o_{i,j}+o_{i,j+1}}-\frac{(p_{i-1,j}+p_{i,j})\cdot Gh(\Delta x)^2}{EAd}$$

$$c_{4(i,j)}=\frac{p_{i,j}\cdot Gh(\Delta x)^2}{EAd}$$

$$c_{5(i,j)}=\frac{4o_{i,j+1}}{2+o_{i,j}+o_{i,j+1}}$$

and

$$C_{(i,j)}=\frac{h(\Delta x)^2}{EA}\{(1-p_{i-1,j})\tau_s\cdot \mathrm{sgn}(u_{i,j}-u_{i-1,j})-(1-p_{i,j})\tau_s\cdot \mathrm{sgn}(u_{i+1,j}-u_{i,j})$$
$$+p_{i-1,j}\Delta\tau_{a\ i-1,j}-p_{i,j}\Delta\tau_{a\ i,j}\}$$

Since the fiber breakage and the debonding are determined by the constants $o_{i,j}$ and $p_{i,j}$, which are constants describing the current states of the fiber and the matrix elements, hereafter we call these constants "constants of state". In the present study, the following boundary conditions are used:

$$\left.\begin{array}{l} u_{i,0}=0 \\ u_{i,K}=U \end{array}\right\} \qquad (i=1,..,N) \qquad (12)$$

where U is the forced displacement. By giving the constants of state according to the state of each element, eqn (11) can be written by a set of linear equations which consists of N x (K-1) unknown displacements. Thus, the displacement on each nodal point can be numerically solved for by the Gaussian reduction method.

2.3 Simulation procedure The present simulation was carried out as follows. First, the tensile strength of each fiber element was generated following a 2-parameter Weibull distribution $F(\sigma)$ as follows:

$$F(\sigma)=1-\exp\{-\frac{L}{L_0}(\frac{\sigma}{\sigma_0})^{m_f}\} \qquad (13)$$

where, m_f and σ_0 are Weibull shape and scale parameters, respectively. Note that L_0 is the original gage length, at which the single filament tension tests were performed and the Weibull parameters estimated, and L is the extrapolated fiber length. In this study the length L is taken as the fiber element length Δx. In the simulation procedure the following reverse function based on eqn (13) was used as the operator assigning the random Weibull number:

$$X_{i,j}=\sigma_0\{\frac{L_0}{\Delta x}\ln(\frac{1}{1-z})\}^{1/m_f} \qquad (14)$$

where z is a random number taken from the uniform distribution on the interval [0,1].

Before the weakest fiber element breaks, the stress is uniform throughout all the fiber elements. Thus, calculations at stresses less than the strength of the weakest fiber element are unnecessary, and the first displacement calculation is for the stress level equivalent to this strength. After the weakest fiber element breaks, however, it is more difficult to determine directly the next key boundary stress being the stress level producing the next fiber break or debonding. Since the relation between the stress and strain in each element is linear with a certain slope, unless changes occur in the constants of state, the following method is reasonable for evaluating the exact stress working on each element. This method, which is called "r_{min} method", determines the exact stresses through evaluating the minimum value of all incremental ratios obtained from the relation between the element strengths and their stress increments. The incremental ratio $r_{f(i,j)}$ for an unbroken fiber element and the incremental ratio $r_{m(i,j)}$ for an elastically deforming matrix element are given at the n-th displacement calculation, respectively, as follows:

$$r_{f(i,j)}=\frac{X_{i,j}-\sigma_{i,j}^{(n-1)}}{\sigma_{i,j}^{(n)*}-\sigma_{i,j}^{(n-1)}}, \quad r_{m(i,j)}=\frac{\tau_m-\tau_{i,j}^{(n-1)}}{\tau_{i,j}^{(n)*}-\tau_{i,j}^{(n-1)}} \quad \text{or} \quad r_{m(i,j)}=\frac{\tau_s-\tau_{i,j}^{(n-1)}}{\tau_{i,j}^{(n)*}-\tau_{i,j}^{(n-1)}} \qquad (15)$$

where, the superscript '*' means that the stress was evaluated virtually. After determining r_{min} from eqn (15), that is, the smallest of the values taken over all unfailed elements, the following two procedures depending on the value of r_{min} are carried out to give a reasonable boundary condition:

1) If r_{min} is less than or equal to one, it means that the fiber element breaks, or that the matrix element debonds from the fiber or that the fiber slides again on the matrix element. Thus, the stresses working on each fiber element and matrix element just before the above change of state can be determined as follows:

$$\left.\begin{array}{l} \sigma_{i,j}^{(n)}=\sigma_{i,j}^{(n-1)}+r_{min}(\sigma_{i,j}^{(n)*}-\sigma_{i,j}^{(n-1)}) \\ \tau_{i,j}^{(n)}=\tau_{i,j}^{(n-1)}+r_{min}(\tau_{i,j}^{(n)*}-\tau_{i,j}^{(n-1)}) \end{array}\right\} \qquad (16)$$

The constant of state for the element producing r_{min} is thus changed at this stage. In the next calculation the boundary condition used in the current calculation is still used, because there is a possibility of additional fiber breakage or debonding in some other elements. Thus this procedure is repeated until the r_{min} value exceeds one.

2) If r_{min} is more than one, no change of state occurs. That means, the calculated displacements and stresses have already been decided. In the subsequent calculation, therefore, it is necessary to give an increased applied stress as the boundary condition to change the state of some element. Then, the procedure 1) is carried out again because the r_{min} value becomes definitely less than one.

The above procedures are the essence of the r_{min} method. These procedures are carried out until the composite loses 30% of the maximum stress, which is the failure criterion used here. We note that when one or more sliding matrix elements are unloaded at the n-th calculation, the elements are regarded as unloading elements. At this stage, it is not necessary to evaluate r_{min} and the exact stresses. Only the constant of state for the element, which has the most decreased strain in the positive sliding region or the most increased strain in the negative sliding region, is changed from zero to one. Then, the calculation is carried out again under the stress distribution existing in the (n-1)-th calculation. This last procedure is repeated until the occurrence of the unloaded element disappears.

2.4 Material constants Table 1 shows the material constants of the unidirectional CFRP used in this simulation. The constants for the carbon fiber were based on the constants in Netravali et al. [10]. The constants for the matrix were chosen from the typical values of the properties of epoxy resin [11]. For simplicity, in this study the shear strength of the epoxy matrix was used instead of the interfacial shear strength τ_m. The composites consisted of N fibers transversely where N was 10 fibers for the debonding cases. Also 20 basic finite difference elements were taken along the fiber direction, where one fiber element length Δx is 0.057mm, or ten times the fiber diameter. In addition, seven extra elements were added to the top and bottom respectively to reduce edge effects caused by fiber breaks there. The extra fiber elements had infinite tensile strength, but the extra matrix elements were allowed to debond except for the elements in the highest and the lowest positions. (Therefore, the actual computation was done on composite size of 10 x 34.) Three kinds of frictional forces, i.e., as the ratio to the interfacial shear strength, τ_s/τ_m=0, 0.2 and 0.5, were selected, in order to understand the effect of the frictional force on the statistical properties of the tensile strength. The volume fraction of fiber in the composite was about 60% when the distance between fibers was 0.003mm.

3. RESULTS AND DISCUSSION

3.1 Stress-strain diagrams Figure 3 shows examples of the stress-strain diagram obtained by the present technique, in which each was carried out under the same Weibull random numbers. In the shear-lag model only the fibers sustain the axial stress, as mentioned above, so the whole composite stress was replaced by the mean fiber stress. Figures 3(a), (b) and (c) show the results of the composites with the ratios of τ_s to τ_m, 0, 0.2 and 0.5, respectively. Figure3(d) shows the result of the composite without the debonding (calculated as $\tau_m = \infty$), in order to simulate the composite with the load sharing similar to the LLS and TLS rules. Hereafter, these are denoted by D0, D2 and D5 composites, and ND composite, respectively. It is shown that the tensile strength of the D0 composite is the lowest of all. Once a fiber breaks in the composite, the debondings occur along the fiber away from the broken element. Therefore, the composite loses a part of the stress (See the serrated part in the figure(a)), and the stresses in the other fibers were enhanced over a wide region. As a result, the first fiber break causes successively the other fiber breaks, when the applied stress is increased to some degree. The similar diagram is also seen in Fig.3(b) because of the low value of τ_s. Since in the case of τ_s/τ_m=0.5 the unloaded region on the broken fiber decreased, the stress does not drop so largely in the serrated parts of Fig.3(c), compared with the Figs. 3(a) and (b). And the tensile strength is also higher than D0 and D2 composites. The ND composite has the largest tensile strength of all, although many fiber breaks occur, as shown in the serrated parts of Fig.3(d). In the ND composite the stress concentrates to some degree only on a few elements around the broken element, but the stress distribution is unperturbed as a whole. Therefore, the composite is able to raise the stress again and again until the other fiber breaks.

3.2 Average tensile strengths and coefficient of variation Table 2 shows the average strength (Ave.) of a hundred runs, the standard deviation (S.D.) and the coefficient of variation (C.V.). Also, in this section the composite strength was replaced by the mean fiber stress at the composite failure. The results show that as the ratio τ_s/τ_m increases, the average strength increases and the coefficient of variation decreases. That is, the composites with a high sliding frictional force at the interface are able to give more reliable strength properties in the range of these small numbers of fibers, as compared to the composites with a low sliding frictional force. In fact, the ND composite gave the highest average strength of all and the lowest coefficient of variation. In Table 2 the average accumulated number, N_B, of fiber breaks before the composite fractures is also shown. The number is closely related to the average strength and the coefficient of variation. That is, the average strength increases and the coefficient of variation decreases with an increase in the number. These results suggest that a more cumulative fracture process actually raises the mean strength and decreases the variability in strength. Intuitively one might anticipate that a more cumulative fracture process gives higher strength as the applied stress can be increased further. The following mention on order statistics might be able to support the change of the variability in strength. Namely, the distribution of the first order statistic (the weakest fiber strength) is the largest of all statistics and the distribution becomes gradually small with an increase in the order.

3.3 Size effect in composite fracture The above descriptions were concerning the composite with the small number of fibers. As the number of fibers increases, the fracture process associated with debonding should become more cumulative. Therefore, the variability in strength might become smaller. To study the size effect in composite strength, changes in the average strength and the coefficient of variation with increasing numbers of fibers were studied through simulation. The results are shown in Fig.4. It is clear that both the average strength and the coefficient of variation of the ND composites decrease as the number of fibers increases. Bundles are considered to be one of the debonding composites in which complete debondings occur. For comparison, the theoretical mean value and coefficient of variation in bundle strength for a classic Daniels bundle[12] (with length equal to the full composite length corresponding to complete debonding) are shown in Fig.4. Also, the strengths of 100 bundles were simulated using the r_{min} method under the same boundary condition. The average and the coefficient of variation are shown in the same figure, where the simulated bundle values approach the Daniels bundle theory values asymptotically around N=200 as the numbers of fibers increases. For both the average and the coefficient of variation, simulation values for the D0, D2 and D5 composites are located between the bundle values, shown by the solid lines and the black triangles, and the values from the simulations of the ND composites, shown by the white circles. Since the debondings never occur in the ND composite and the fibers in the bundle debonds perfectly, the composites with the debonding would continue to be located between these two extreme quantities, even if the number of fibers increased further.

Table 3 shows the ratio of N_B to the number of fibers for the ND composites and the simulated bundles. The value N_B of the ND composite increases with an increase in the number of fibers, but the ratio clearly decreases. Thus, the degree of the accumulation consequently diminishes. It is known that a large cluster (a large group consisting of fiber break points) often causes the fracture of CFRP with a large number of fibers, because of the severe stress concentration [13]. It is considered that, therefore, the occurrence of clusters and the unstable growth causes the decrease of the ratio N_B/N. Since the number of weaker fiber elements increases as the composite size increases, fiber breaks occur at lower stress levels than for composites with small numbers of fibers. That is, a certain large cluster is formed and the unstable growth develops in the ND composite with the large number of fibers, at an even smaller applied stress. In other words, the larger cluster dominates the failure of ND composite with a large number of fibers, which is different from the failure mechanism of ND composite with a small number of fibers. Thus, the average strength would continue to decrease with the number of fibers. Table 3 also showed that for the simulated bundle strengths, this ratio changes minimally and rapidly approaches a constant as the number of fibers increases. Since the composite with debonding become dominated by the failure mechanism similar to the bundle's, the average strength would little change as compared to the ND composite. Hence, the difference of strength between the composites with debonding and non-debonding might become small. By the way, larger clusters would be created at a more stable stress level (larger groups of broken fibers have less variability in strength [*2]) in the ND composite, so the coefficient of variation decreases. However, the rate of the decrease is slower than that for the bundle, as shown in Fig.4. Therefore, the composite with debonding would eventually show a relative smaller variability in strength as compared to the ND composite. Thus, it is an open question as to whether the statistical properties for the debonding composite might eventually be a little more reliable than for the non-debonding composite. Such a change of properties would be desirable for practical composites with debonding. Quantitative simulation studies of the statistical properties of the composite strength under debonding for a large number of fibers in the composite will be a subject for the future.

4. CONCLUSIONS

A Monte-Carlo simulation technique based on the shear-lag model has been developed to simulate the ten-

---

*2. According to the theoretical model [3] based on the LLS rule, the Weibull shape parameter ρ_c for composite strength is given as follows;

$$\rho_c = k\,\rho$$

where, ρ is the Weibull shape parameter for fiber strength, and k is the size of cluster (the number of broken fibers), and increases with an increase in the composite size.

sile fracture process and the tensile strength of unidirectional carbon fiber reinforced plastic composites by taking into account the effect of sliding friction after debonding near a fiber break. The technique improves on the conventional approach by using the r_{min} method, that can reasonably estimate the stress working on the elements in the composite as damage accumulates. The fracture processes and the strengths were evaluated through a hundred simulation runs for each case. In the case of debonding, the fracture process became less cumulative than the process in the non-debonding composite. The average strength increased with an increase in the frictional force, and the non-debonding composite showed the highest average strength. The coefficient of variation decreased as the frictional force increased, and the non-debonding composite also yielded the smallest coefficient of variation. In the non-debonding composite consisting of a large number of fibers, the degree of the accumulation for the fiber breaks decreased, although the bundle failure simulation did not change the degree. So, it was discussed that a cluster (a group of fiber break points) was related to the degree of the accumulation, so the statistical properties of a composite with non-debonding might eventually become less reliable than that of a composite with debonding.

ACKNOWLEDGEMENT

The present study was carried out in Cornell University, when the author stayed there as a visiting scientist. He would like to express his hearty thanks to Professor S. Leigh Phoenix in the Department of Theoretical and Applied Mechanics, who gave him an important comment for this study.

REFERENCES

[1] For example, B. W. Rosen, J. AIAA, 2(1964), 1985., C. Zweben, ibid, 6[1968), 2325. and D. G. Harlow and S. L. Phoenix, 17(1981), 181., 17(1981), 601.
[2] R. E. Pitt and S. L. Phoenix, Int. J. Frac., 22(1983), 243.
[3] S. L. Phoenix, and R. L. Smith, Int. J. Solids Structures, 19(1983), 479.
[4] K. P. Oh, J. Compo. Mater., 13(1979), 311.
[5] J. M. Hedgepeth, NASA Tech. Note, D-882(1961), 1.
[6] I. Kimpara, T. Ozaki and S. Takada, J. Soc. Mater. Sci. JAPAN, 34(1985), 280 (in Japanese).
[7] K. Goda and H. Fukunaga, Compo. Sci. Tech., 35(1989), 181.
[8] D. Hull, An Introduction to Composite Materials, (1981), 142, Cambridge.
[9] E. D. Reedy, Jr., J. Compo. Mater., 18(1984), 595.
[10] A. N. Netravali, R. B. Henstenburg, S. L. Phoenix and P. Schwartz, Polymer Compo., 10(1989), 226.
[11] B. D. Agarwal and L. J. Broutman, Analysis and Performance of Fiber Composites, (1980), 57, Wiley-Interscience.
[12] H. E. Daniels, Proc. Roy. Soc., A183(1945), 405.
[13] P. W. Manders and M. G. Bader, J. Mater. Sci., 16(1981), 2246.

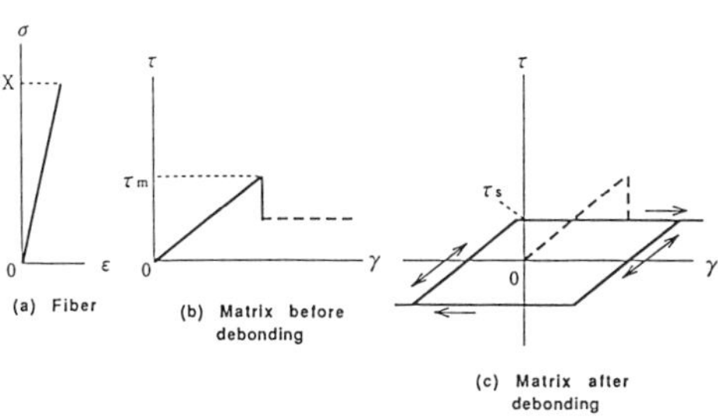

Fig.1 Stress/strain diagrams of fiber and matrix

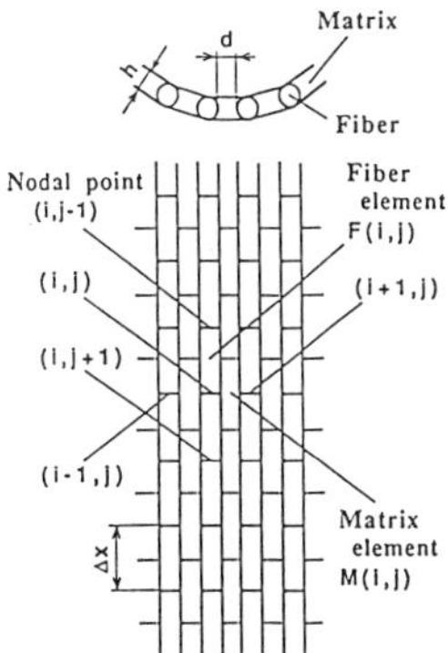

Fig.2 Finite difference model used in the present simulation

Table 1. Material conatants used in the present simulation.

Elastic modulus of fiber: E	278 GPa
Cross-sectional area of fiber: A	2.55×10^{-5} mm^2
Distance between fibers: d	0.003 mm
Weibull shape parameter	4.95
Weibull scale parameter	5762 MPa*
Shear modulus of matrix: G	1327 MPa
Interfacial shear strength: τ_m	39.8 MPa
Thickness of composite: h	0.0057 mm

* This value is a measured one at 10 mm gage length.

Table 2. Average(Ave.), standard deviation(S.D.) and coefficient of variation(C.V.) in simulated tensile strength. (N_B: Average accumulated number of fiber breaks before the composite fractures.)

	Ave. GPa	S.D. GPa	C.V. %	N_B
D0	6.09	0.67	10.92	2.22
D2	6.32	0.68	10.74	2.22
D5	6.53	0.59	9.02	2.82
ND	7.83	0.62	7.86	7.29

Table 3. Ratio of average accumulated number of fiber breaks to total number of fibers.

N	ND composite		Bundle	
	N_B	N_B/N	N_B	N_B/N
5	3.78	0.76	1.43	0.29
10	7.29	0.73	2.28	0.23
20	12.7	0.63	3.83	0.19
30	17.1	0.57	5.29	0.18
50	25.5	0.51	9.03	0.18
100	49.2	0.49	18.1	0.18

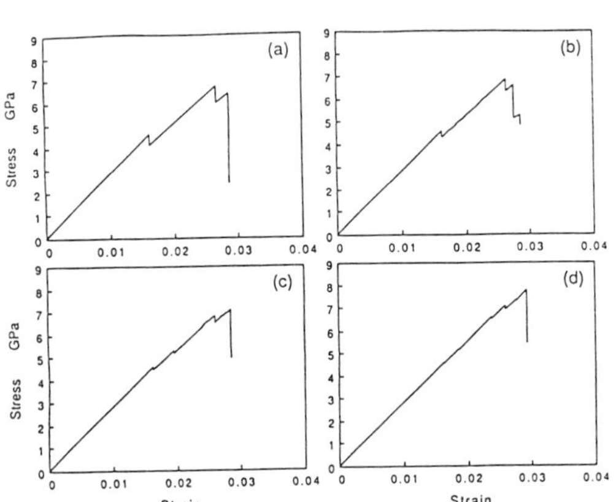

Fig.3 Stress/strain diagram of simulated composites.
(a) D0 (b) D2 (c) D5 (d) ND

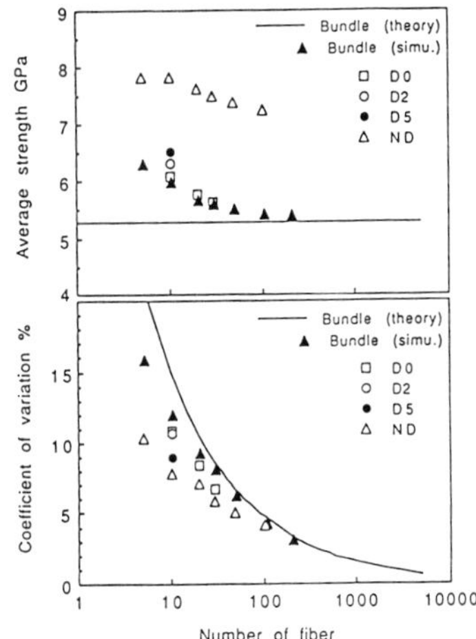

Fig. 4 Change in average strength and co[...] of variation with increases in the number

Reliability and Optimization of Structural Systems, V (B-12)
P. Thoft-Christensen and H. Ishikawa (Editors)
Elsevier Science Publishers B.V. (North-Holland)

Probabilistic Estimation of Member Forces of Steel Girder Bridge under Load Combinations

M. Kawatani*, H. Furuta**, A. Hirose•, M. Hoshika••

* Department of Civil Engineering, Osaka University, Japan

** Department of Civil Engineering, Kyoto University, Japan

• Design Dept. No. 3, Chio Fukken Consultants Co., Ltd., Osaka, Japan

•• Design Dept. No. 3, Civ. Eng.Techn. Div., Obayasi Co., Tokyo, Japan

Abstract

In the design of highway bridges, various load effects and their simultaneous effects on member forces should be considered. A diagonal member of end sway bracing of plate girder bridges is focused here, because of the variety of its dominant load combinations in design. The member forces due to various loads are probabilistically estimated through three-dimensional FEM analysis and traffic simulation, and compared with the design values based on current specifications. Probability of exceeding the allowable stress of the diagonal member under load combinations is also investigated.

Keyword Codes: G.3; I.6.6; J.2
Keywords: Probability and Statistics; Simulation Output Analysis; Physical Sciences and Engineering

1. INTRODUCTION

In the strength design of bridges, various load effects and their simultaneous effects on member forces should be considered. The loads acting on bridges such as wind, earthquake and traffic loads have probabilistic properties, and these load effects must be probabilistically estimated. In this study, a diagonal member of end sway bracing of plate girder bridges is focused because of the variety of its dominant load combinations in design [1], that is of interest from a viewpoint of load combinations. The member force due to various loads is computed by means of three–dimensional FEM analysis, especially simulating traffic loads. Probabilistic values of the member force are compared with the design values through current specifications. Probability of exceeding the allowable stress of the diagonal member under load combinations is also examined.

2. ANALYTICAL PROCEDURE

2.1. Analytical Model

A bridge model is a plate girder bridge with hinged ends as shown in Figure 1. In this bridge a field test of static loadings was already carried out, and experimental results were compared with analytical ones by means of three–dimensional(3–D) FEM using only usual thin shell elements [2]. In the 3–D FEM analysis applied here, solid elements that include thickness effects as well as flexural rigidity are used for a RC (reinforced concrete) slab and RC wall handrails instead of thin shell elements as shown in Table 1. The analytical model of 1742 panel points is shown in Figure 2.

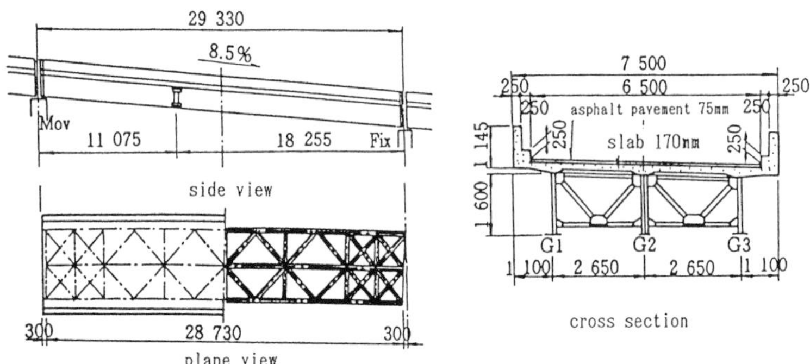

Figure 1. General view of bridge.

Table 1
Employed elements.

member	kinds of element	number of elements
web of main girder	thin wall shell	312
flange of main girder	thin wall shell	312
concrete slab and railing	3-D solid	572
sway bracing and lateral bracing	3-D beam	106
vertical stiffener	3-D beam	300

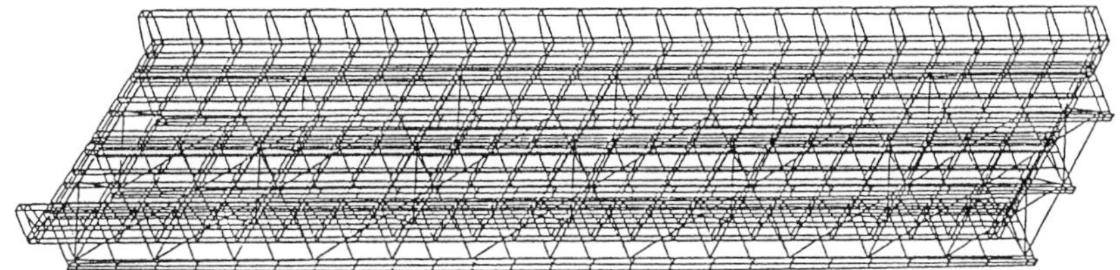

Figure 2. Three-dimensional FEM model.

2.2. Validity of Analytical Method and Model

In order to verify the validity of the analytical method and the model, the analytical results are compared with the field test results. Under the loading condition of a test vehicle as shown in Figure 3, the analytical values of deflection and stress distribution of main girders at the span center much more agree with the experimental values in comparison with another analytical results using the usual FEM model, as shown in Figures 4 and 5. The analytical values of an axial force of the diagonal member focused here also successfully correspond to the measured values as shown in Figure 6. Due to these agreements between the analytical results and the experimental ones, the validity of the analytical method and the model can be examined.

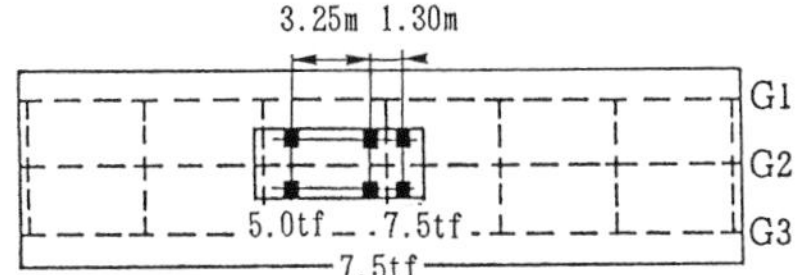

Figure 3. Loading condition of test vehicle.

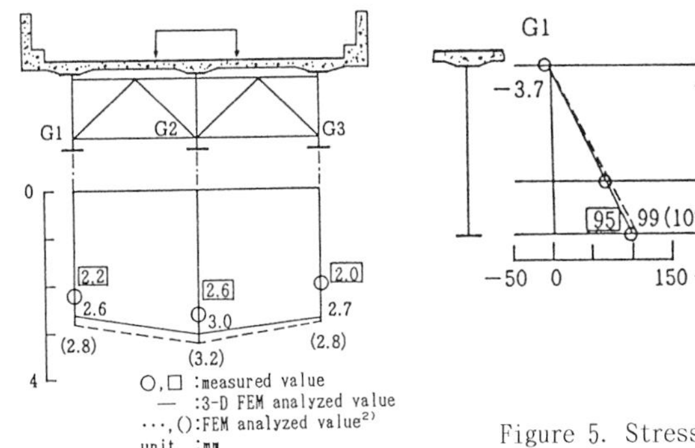

Figure 4. Deflection of main girder.

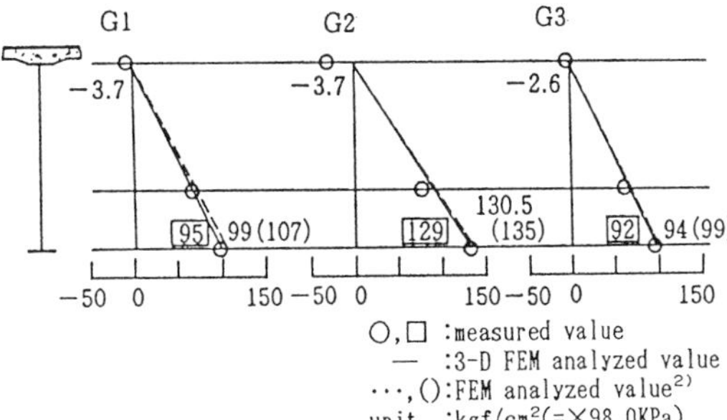

Figure 5. Stress distribution of main girder.

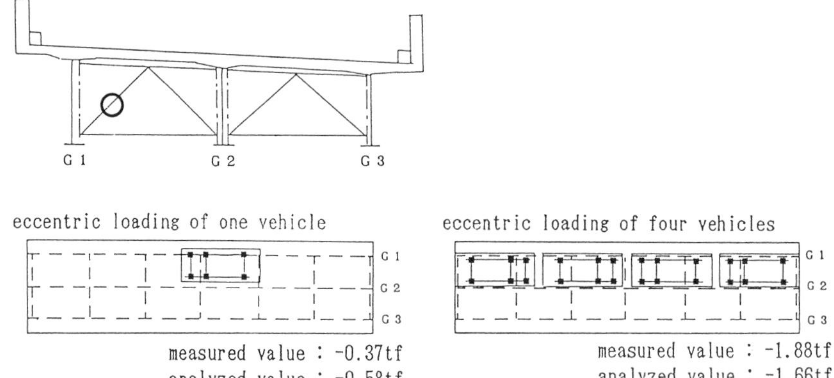

Figure 6. Axial force of diagonal member.

3. LOAD MODELS

As representative loads for the design of bridges, dead(D), live(L), wind(W) and earthquake(E) loads are employed. The large amount of observed data accumulated by Hanshin Expressway Public Corporation on traffic behavior as well as wind and earthquake loading conditions are used for modeling the actual probabilistic loads [3]. Temporal occurrence of these loads are modeled after Borges-Castanheta model [4]. Occurrence of the load is assumed to follow the Poisson's law.

3.1. Live Load Simulation

Live load effects are modeled, especially simulating traffic loads by means of Monte Calro technique based on stochastic data of traffic flows measured on Hanshin Expressway as shown in Table 2 [3]. Axle arrangements and ratios of axle loads of vehicle models are shown in Table 3. Two types of traffic load effects, that are an ordinary congestion and an accidental congestion, are considered as shown in Table 4.

The traffic loads generated by Monte Calro method are acted on the influence surface of an axial force of the end sway bracing as shown in Figure 7, which is analyzed by 3-D FEM. In the case of probability of exceeding an allowable stress, a compression force is calculated acting vehicle loads on the line 25cm away from the end of roadway specified in current codes. These lines are indicated by lines 3' and 4' as shown in Figures 7 and 8. Based on 100 times simulations of the monthly maximum value of the axial force, the extreme value distribution can be modeled as Gumbel distribution functions as shown in Figure 9 and parameter values are listed in Table 5.

Table 2
Vehicle classification and characteristic values.

vehicle classification		proportion of vehicle types (%)			total vehicle weight			vehicle length	
		ashiya	proportion of large size trucks		dist.	μ(t)	σ(t)	μ(m)	σ(m)
			20%	40%					
2 axles large size trucks	unloaded	1.52	1.81	3.62	NOR	7.36	1.90	8.73	1.86
	loaded	1.13	1.34	2.68	LOG	14.05	2.32		
	overloaded	0.01	0.01	0.02	EXP	23.14	3.14		
3 axles large size trucks	unloaded	4.12	4.90	9.80	NOR	11.21	1.89	10.44	1.23
	loaded	7.59	9.02	18.04	LOG	19.77	3.18		
	overloaded	0.05	0.06	0.12	EXP	32.24	2.24		
trailer	unloaded	0.94	1.12	2.24	NOR	14.60	3.50	13.55	1.97
	loaded	1.46	1.74	3.48	LOG	26.98	8.63		
mid size trucks		18.14	17.45	13.09	LOG	5.11	3.01	6.21	1.58
passenger cars		65.04	62.55	46.91	LOG	1.31	0.34	4.00	0.38

*) Can not be seperated because of few data
μ : Mean value, σ : Standard deviation
NOR : Normal distribution, LOG : Log-normal distribution, EXP:Exponential distribution.
Properties of vehicle length is idealized as β-distribution.

Table 3
Vehicle models and ratio of axle loads.

W : total weight

vehicle classification		vehicle model	first axle	second axle	third axle
2-axles large size trucks	unloaded	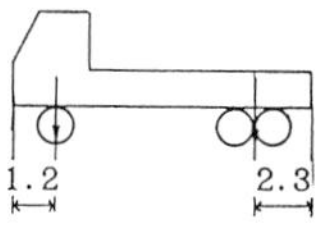 1.2 2.3	0.109W + 3.22	0.981W - 3.22	
	loaded				
	overloaded				
3-axles large size trucks	unloaded				
	loaded				
	overloaded				
trailers	unloaded	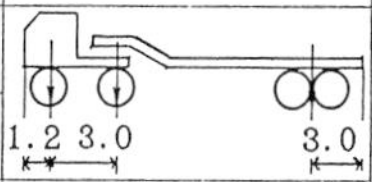 1.2 3.0 3.0	0.024W + 4.19	0.360W + 0.70	0.616W - 4.89
	loaded				
mid size trucks		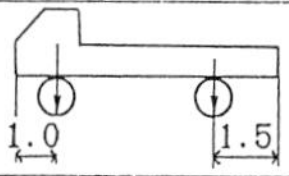 1.0 1.5	0.182W + 1.38	0.818W - 1.38	
passenger cars		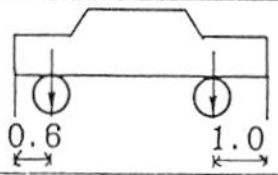 0.6 1.0	0.501W + 0.03	0.499W - 0.03	

Table 4
Models of traffic conditions.

kind of congestion	occurrence rate	duration (hours)	proportion of large size trucks	vehicular gap (m)		
				distribution	mean	st. dev.
accidental	50 $year^{-1}$	1~2	40 %	Log-normal	μ=2.71	σ=1.49
ordinary	1 day^{-1}	6	20 %	Log-normal	μ=8.05	σ=3.93

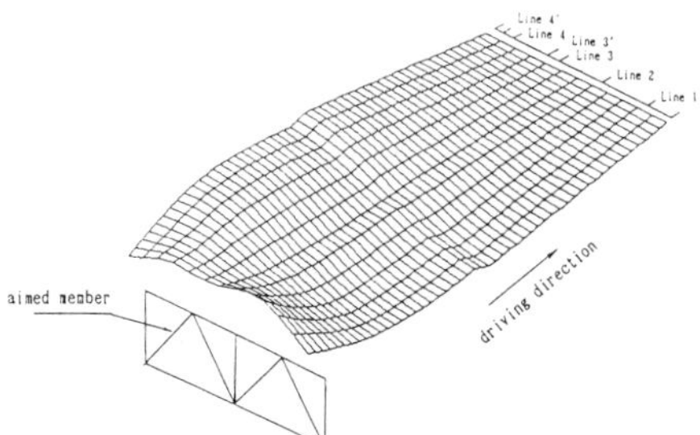

Figure 7. Influence surface of axial force of end sway bracing.

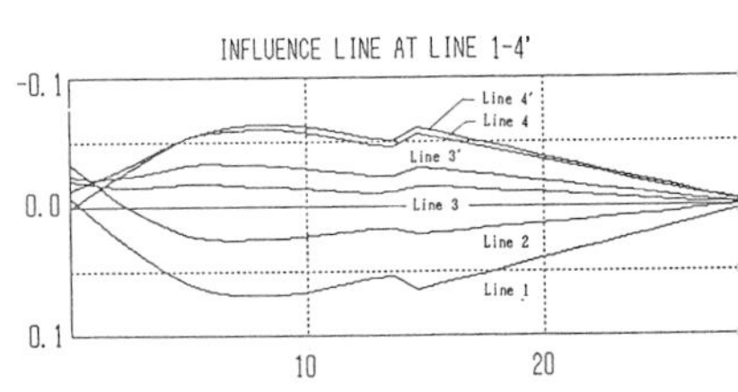

Figure 8. Influnece lines of axial force

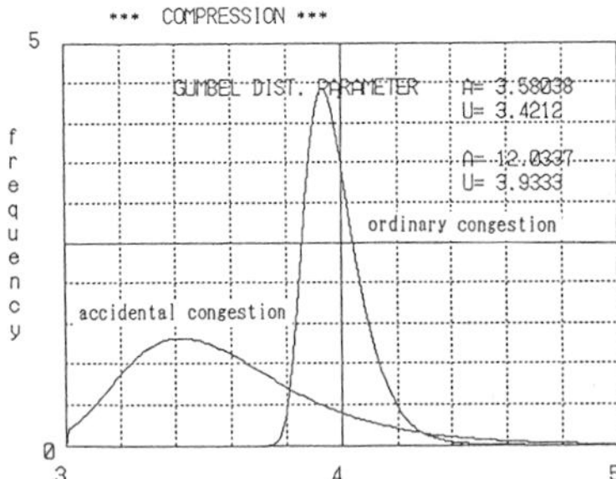

Figure 9. Extreme value distribution of axial force due to traffic load.

Table 5
Parameters of type I extreme value distribution from simulation.

	accidental congestion		ordinary congestion	
	tension	compression	tension	compression
distribution of monthly maximum	α=3.2981	α=3.5804	α=11.2360	α=12.0337
	μ=1.6641	μ=1.6345	μ= 3.2590	μ= 3.4017
distribution of 50 years maximum	α=3.2981	α=3.5804	α=11.2360	α=12.0337
	μ=3.6036	μ=3.4212	μ= 3.9517	μ= 3.9333
0.95 fractile value	4.5039 tf	4.2508 tf	4.2160tf	4.1801 tf

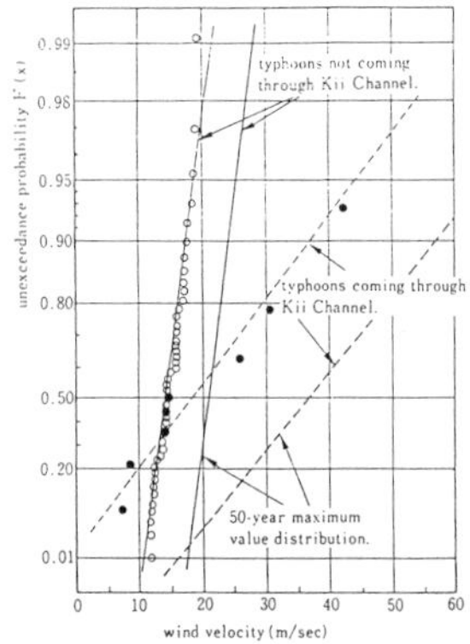

Figure 10. Fitting of wind velocity data to extreme value distribution.

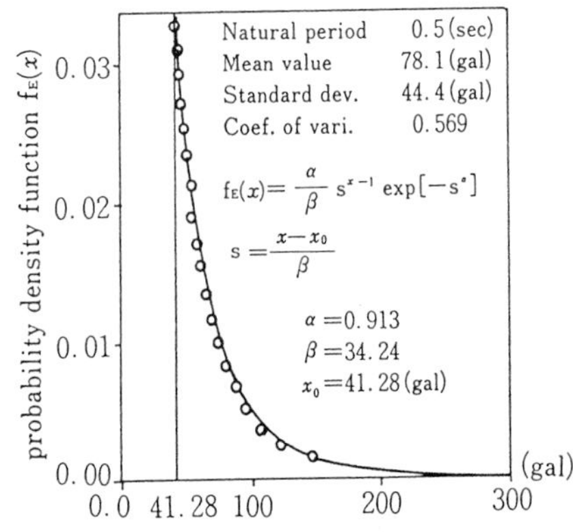

Figure 11. Observed data and extreme value distribution of earthquake load.

3.2. Wind Load and Earthquake Load

Based on the observed data, Gumbel and Weibull distribution functions are chosen to represent wind and earthquake load effects on bridges as shown in Figures 10 and 11, respectively.

4. MEMBER FORCES DUE TO EACH LOAD

Analytical results of probabilistic values in the extreme distribution during 50 years of the axial force of a diagonal member of the end sway bracing due to each load are shown in Table 6, together with the design values through current specifications. The axial force due to the dead load, however, is deterministically estimated from the FEM analytical value. The 0.95 fractile value of the axial force due to the live load is quite smaller than that in the current design, and this result shows that the structural model in design is considerably rough. On the other hand, the 0.95 fractile values due to the wind or the earthquake load are larger than these design values.

Table 6
Axial force of diagonal member due to current code and modeled load. (unit:tf)

load		current code	0.95 fractile value
dead load		-0.725	-0.798*
live load	accidental	-11.129	-4.251
	ordinary		-4.180
wind load		-4.755	-6.416
earthquake load		-7.018	-8.084

*FEM analyzed deterministic value due to dead load

Table 7
Probabilities of exceeding allowable stress.

load combination	probability
D+L	smaller than 10^{-39}
D+W	2.552×10^{-3}
D+E	9.708×10^{-6}
D+L+W	2.715×10^{-6} (accidental) 5.079×10^{-6} (ordinary)
D+L+E	2.706×10^{-8} (accidental) 9.071×10^{-8} (ordinary)

σ_a : allowable stress

5. MEMBER FORCES UNDER LOAD COMBINATIONS

For estimating the axial force under load combinations, probabilities of exceeding the allowable stress of the diagonal member during 50 years are calculated by means of the technique presented by Wen [5]. The allowable stress of the employed member is 537kgf/cm^2 of compression which is smaller than 1400kgf/cm^2 of tension, therefore the probability of exceeding the allowable stress is estimated in compression as shown in Table 7.

In the combination of dead and live loads the probability value of exceeding the allowable stress is very small, because the design value of the axial force is remarkably in safety side as shown in Table 6. In the combinations of D+W and D+E the probability values are the order of 10^{-3} and 10^{-5}, respectively, because of increase in allowable stress and smallness in the probabilities of simultaneous occurrences of two loads, even though the 0.95 fractile values due to the wind or the earthquake load are lager than the design values as shown in Table 6. In the combinations of D+L+W and D+L+E the probability values are further small of the order of 10^{-6} and 10^{-8}, respectively, because of further smallness in the probabilities of simultaneous occurrences of three loads.

6. CONCLUSIONS

i) The design value of the axial force due to the live load is remarkably in safety side compared with 0.95 fractile value analyzed simulating actual traffic loads, and this result shows that the structural model in design is considerably rough.

ii) The 0.95 fractile values due to the wind or the earthquake load are larger than these design values.

iii) Noting the probability of exceeding the allowable stress in load combinations, the probability value in D+L is very small. In the combinations of D+W and D+E, however, the probability values are the order of 10^{-3} and 10^{-5}. This result means that investigations of various load combinations are important.

REFERENCES

1. H.Furuta, M.Kawatani, A.Hirose, Y.Kanbara and J.Sakai, J. Structural Engg., JSCE, 37A (1991) 593 (in Japanese).
2. S.Oshita, I.Suzuki, H.Hayashi and N.Natori, Proc. 45th Annual Conference of JSCE, I-327 (1990) 684 (in Japanese).
3. I.Konishi, H.Kameda, K.Matsuhashi, S.Emi and M.Kitazawa, Structural Safety, 7 (1990) 35.
4. F.Borges and J.Castanheta: "Structural Safety," 2nd ed., pp.151–216, Laboratorio Nacional de Engengaria Civil, Lisbon, 1972.
5. Y.K.Wen, J. Structural Div., Proc. ASCE, 103, No.ST5 (1977) 1079.

INDEX OF AUTHORS

SUBJECT INDEX

IFIP

The INTERNATIONAL FEDERATION FOR INFORMATION PROCESSING is a multinational federation of professional and technical organisations (or national groupings of such organisations) concerned with information processing. From any one country, only one such organisation – which must be representative of the national activities in the field of information processing – can be admitted as a Full Member. In addition a regional group of developing countries can be admitted as a Full Member. On 1 October 1992, 45 organisations were Full Members of the Federation, representing 62 countries.

The aims of IFIP are to promote information science and technology by:

- fostering international co-operation in the field of information processing;
- stimulating research, development and the application of information processing in science and human activity;
- furthering the dissemination and exchange of information about the subject;
- encouraging education in information processing.

IFIP is dedicated to improving worldwide communication and increased understanding among practitioners of all nations about the role information processing can play in all walks of life.

Information technology is a potent instrument in today's world, affecting people in everything from their education and work to their leisure and in their homes. It is a powerful tool in science and engineering, in commerce and industry, in education and adminstration. It is truly international in its scope and offers a significant opportunity for developing countries. IFIP helps to bring together workers at the leading edge of the technology to share their knowledge and experience, and acts as a catalyst to advance the state of the art.

IFIP came into official existence in January, 1960. It was established to meet a need identified at the first International Conference on Information Processing which was held in Paris in June, 1959, under the sponsorship of UNESCO.

Organisational Structure
The Federation is governed by a GENERAL ASSEMBLY, which meets once every year and consists of one representative from each Member organisation. The General Assembly decides on all important matters, such as general policy, the programme of activities, admissions, elections and budget.

IFIP ORGANISATION CHART

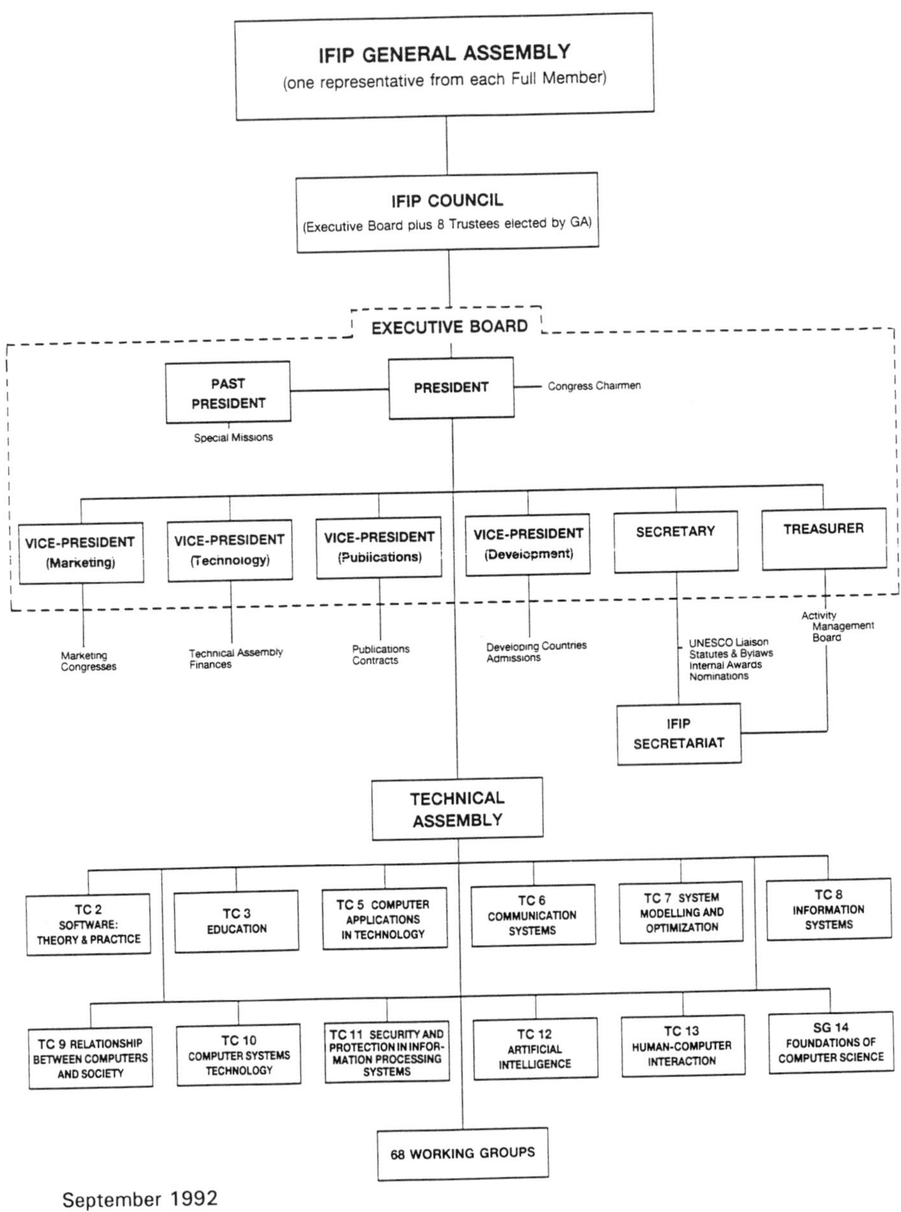

September 1992

The day-to-day work of IFIP is directed by its Officers: the President, Vice-Presidents, Secretary and Treasurer, who are elected by the General Assembly and together constitute the EXECUTIVE BOARD.

The COUNCIL, consisting of the Officers and up to eight Trustees elected from the General Assembly, meets twice a year and takes decisions which become necessary between General Assembly meetings.

The headquarters of the Federation are in Geneva, Switzerland where the IFIP Secretariat administers its affairs.

For further information please contact:

IFIP Secretariat
attn. Mme. GWYNETH ROBERTS
16 Place Longemalle
CH-1204 Geneva, Switzerland
telephone: 41 (22) 28 26 49
facsimile: 41 (22) 781 23 22
Bitnet: ifip@cgeuge51

IFIP's MISSION STATEMENT

IFIP's mission is to be the leading, truly international, apolitical organisation which encourages and assists in the development, exploitation and application of Information Technology for the benefit of all people.

Principal Elements

1. To stimulate, encourage and participate in research, development and application of Information Technology (IT) and to foster international co-operation in these activities.
2. To provide a meeting place where national IT Societies can discuss and plan courses of action on issues in our field which are of international significance and thereby to forge increasingly strong links between them and with IFIP.
3. To promote international co-operation directly and through national IT Societies in a free environment between individuals, national and international governmental bodies and kindred scientific and professional organisations.
4. To pay special attention to the needs of developing countries and to assist them in appropriate ways to secure the optimum benefit from the application of IT.
5. To promote professionalism, incorporating high standards of ethics and conduct, among all IT practitioners.
6. To provide a forum for assessing the social consequences of IT applications; to campaign for the safe and beneficial development and use of IT and the protection of people from abuse through its improper application.

7. To foster and facilitate co-operation between academics, the IT industry and governmental bodies and to seek to represent the interest of users.
8. To provide a vehicle for work on the international aspects of IT development and application including the necessary preparatory work for the generation of international standards.
9. To contribute to the formulation of the education and training needed by IT practitioners, users and the public at large.

Note to Conference Organizers

Organizers of upcoming IFIP Working Conferences are urged to contact the Publisher. Please send full details of the Conference to:

Mrs. STEPHANIE SMIT
Administrative Editor – IFIP Publications
ELSEVIER SCIENCE PUBLISHERS
P.O. Box 103, 1000 AC Amsterdam
The Netherlands
telephone: 31 (20) 5862481
facsimile: 31 (20) 5862616
email: s.smit@elsevier.nl

IFIP TRANSACTIONS

IFIP TRANSACTIONS is a serial consisting of 15,000 pages of valuable scientific information from leading researchers, published in 35 volumes per year. The serial includes contributed volumes, proceedings of the IFIP World Conferences, and conferences at Technical Committee and Working Group level. Mainstream areas in the IFIP TRANSACTIONS can be found in Computer Science and Technology, Computer Applications in Technology, and Communication Systems.

From 1993 onwards the IFIP TRANSACTIONS are only available as a full set.

IFIP TRANSACTIONS A:
Computer Science and Technology
1992: Volumes A1–A19
1993: Volumes A20–A40
ISSN 0926-5473

IFIP Technical Committees that are involved in IFIP TRANSACTIONS A
Software: Theory and Practice (TC2)
Education (TC3)
System Modelling and Optimization (TC7)
Information Systems (TC8)
Relationship Between Computers and Society (TC9)

Computer Systems Technology (TC10)
Security and Protection in Information Processing Systems (TC11)
Artificial Intelligence (TC12)
Human-Computer Interaction (TC13)
Foundations of Computer Science (SG14)

IFIP TRANSACTIONS B:
Applications in Technology
1992: Volumes B1–B8
1993: Volumes B9–B14
ISSN 0926-5481

IFIP Technical Committee that is involved in IFIP TRANSACTIONS B
Computer Applications in Technology (TC5)

IFIP TRANSACTIONS C:
Communication Systems
1992: Volumes C1–C8
1993: Volumes C9–C16
ISSN 0926-549X

IFIP Technical Committee that is involved in IFIP TRANSACTIONS C
Communication Systems (TC6)

IFIP TRANSACTIONS FULL SET: A, B & C
1992: 35 Volumes. US $ 1892.00/Dfl. 3500.00
1993: 35 Volumes. US $ 2340.00/Dfl. 3885.00
The Dutch Guilder prices (Dfl.) are definitive. The US $ prices mentioned above are for your guidance only and are subject to exchange rate fluctuations. Prices include postage and handling charges.

The volumes are also available separately in book form.

Please address all orders and correspondence to:

ELSEVIER SCIENCE PUBLISHERS
attn. PETRA VAN DER MEER
P.O. Box 103, 1000 AC Amsterdam
The Netherlands
telephone: 31 (20) 5862602
facsimile: 31 (20) 5862616
email: m.haccou@elsevier.nl